Residential Property Appraisal

Residential Property Appraisal

Phil Parnham and Chris Rispin

London and New York

First published 2001
by Spon Press
11 New Fetter Lane, London EC4P 4EE

Simultaneously published in the USA and Canada
by Spon Press
29 West 35th Street, New York, NY 10001

Spon Press is an imprint of the Taylor & Francis Group

Typeset in Sabon by
HWA Text and Data Management, Tunbridge Wells
Printed and bound in Great Britain by
TJ International Ltd, Padstow, Cornwall

Hot source material at the
leading edge

British Library Cataloguing in Publication Data
A catalogue record for this book is available from the British Library

Library of Congress Cataloging in Publication Data
Parnham, Phil
Residential property appraisal / Phil Parnham and Chris Rispin.
p. cm.
Includes bibliographical references and index.
1. Residential real estate–Valuation–Great Britain. I. Rispin, Chris. II. Title.

HD1389.5.G7 P37 2001
333.33′82′0941–dc21 00–041988

ISBN 0-419-22570-6

Contents

Acknowledgements

General acknowledgements

The kind permission is gratefully acknowledged for the use of the following:

Consumers' Association for *Which? Special Report: Home Improvements*, October 1997; table using data which assess the impact on value of different home improvements. *Which?* Published by Consumers' Association, 2 Marylebone Road, London NW1 4DF, for further information phone 0800 252100.

NHER for tables outlining insulation levels, heating levels and approximate ratings of a semi-detached house, and table illustrating the changes to SAP ratings that arise from changes to the dwelling; taken from *Program Manual for NHER HomeRater* version 3.10.

Building Research Establishment for material from *Defect Action Sheet*, *Digest 351*, *Housing Design Handbook*, *Defect Action Sheet 93* (1987), *Digest 251* (1990), BRE 251 – Decision Chart produced by BRE and NHBC (1981). Reproduced courtesy of BRE.

Structural Engineers Trading Organisation Ltd for reproduction of figure 6.15 from *Subsidence of Low Rise Buildings* first edition, published by SETO 1994.

English Nature for summary of the comparative sizes of bat, rat and mouse dropping from *Bats in Roofs – A Guide for Surveyors*, published by English Nature.

Her Majesty's Stationery Office for material from The Building Regulations 1991; Part A Diagram 21, Proportions for Masonry Chimneys; Part J Diagram 3, Hearth Sizes; Part K Diagram 6 and Diagram 8 showing typical arrangements on staircases; Part N Diagram 1, critical locations in internal and external walls. Crown copyright is reproduced with the permission of the Controller of Her Majesty's Stationery Office.

Usborne Publishing Ltd for the summary of the 'Leaf Key'. Reproduced and adapted from *Usborne Guide to Trees of Britain and Europe* by permission of Usborne Publishing, 83–85 Saffron Hill, London EC1N 8RT. Copyright © 1981 Usborne Publishing Ltd.

Personal acknowledgements

Special thanks go to all my colleagues at Sheffield Hallam University who bridged the many gaps in my knowledge. In particular I am very grateful to Peter Westland who helped to develop the original book proposal and was so very close to being another co-author.

I would also like to acknowledge the contribution of the two thousand surveyors that have attended the various seminars, workshops and courses I have delivered. Their critical but constructive contributions have hopefully given this book the practical edge it needs.

I need to thank my family for two reasons: first for buying houses that provided me with so many building defect case studies; and second (and more important) to Sue, Laura and Jennie for yet again providing me with the time to complete this work.

Phil Parnham

I would like to thank all those surveyors who have unwittingly contributed to this book, as they will help all of us to appreciate what the customers have had to go through.

For totally different reasons I would like to thank Mike Fisher for his professionalism and the attention to detail, Ewan McCaig for his commitment to quality management and contributions in that area.

However, the professional input was equalled by the support of my family, particularly my wife Vicky, supportive as ever, my elderly teenage sons, Ben and Robin, who will only understand this in years to come; and to my daughter Jenny – we will now have more time to play tennis, promise – Love Dad.

Chris Rispin

Part 1

The appraisal process

Chapter 1

Introduction

1.1 Context

As we move into the new millennium, evidently the pace of change is increasing all the time. The surveying profession has to respond to the challenges if it is to continue to play an influential role in the property industry of the 21st century. Various catalysts are currently influencing the direction of this profession and in respect of residential property they include:

- the government are looking to make home buying easier through the establishment of 'Vendors Surveys' and other initiatives;
- the RICS has an 'Agenda for Change';
- through new technologies the tools are now available to facilitate change in a way never experienced before.

This book looks at the core skills that a surveyor needs to carry out appraisals of residential properties in a changing world. It is not restricted to experienced surveyors alone. The book is written in a way to encourage the student/trainee from whatever background to develop the skills associated with inspecting the residential home.

Looking back over the last ten years it is clear to us why there is a need for this book. We have been involved with organising, producing, delivering and evaluating training events for all kinds of Chartered Surveyors and other professionals involved in the appraisal of residential property. The events have included seminars, workshops and conferences for both in-house staff groups and independent surveyors attending speculative continuing professional development (CPD) sessions. The one enduring impression is that technical topics are always very popular. This is not because participants want to achieve their allocation of CPD. We suggest there are other reasons for this popularity:

- The focus of the knowledge and experience of many professionally qualified surveyors has been closely associated with the valuation of property and not its physical state. On the other hand the building surveyor or other similar qualified professional's focus is the reverse of this. Few courses have catered for the equal combination of skills.
- Many educational courses lack a sound technical grounding. The professional institutions often call for broader, more flexible surveyors armed with business and commercial skills. As a consequence, a number of traditional disciplines have

disappeared from the curricula. Building studies and building defects are two such subjects.

- Technical advice and guidance that are currently published have a broad audience. A lot of literature is specifically written for those who carry out in-depth surveys and investigations of residential property. It is often difficult to identify which part of this advice is best suited to the professional practice of residential valuation and appraisal.
- Many of the recent court cases that have struck fear into the hearts of surveyors have been associated with disputed technical assessments of dwellings. This may have added to the sense of collective professional inadequacy!

To try and meet this demand and plug the skill gaps, the authors have had to rewrite, interpret or specially create training materials that match the professional role of the participant surveyors. Many commentators that write or lecture on technical matters appear to consider that the sole purpose of residential valuations and surveys are to correctly diagnose every hidden defect in every dwelling. Alternatively the less technical standard guidance tends to concentrate on procedure and process, offering little in the way of practical detail.

In addition there are two other sources of change that are appearing on the horizon:

- The structural changes of many lenders have resulted in different business priorities. New technology has revealed opportunities not available previously and exposed different working practices abroad that could be imported to this country. The structure of the residential market in countries like the USA, Australia and New Zealand is much less complex and uncertain. As the global economy has a greater influence on the domestic market and more financial institutions operate internationally, pressure will be applied to simplify the process.
- Customer expectations are changing. The typical 'customer' is becoming far more sophisticated. They are better educated and used to more customer-focused service in all the 'other' products they purchase. Evidence from consumer surveys, media sources and pressure groups suggests that the standard of service that many surveyors provide falls well below current expectations.

This book sets out to meet these challenges. What is needed is a publication that has a clear technical focus directly related to the professional role of its readership. Because this role is closely associated with value, then part of the book must acknowledge the actual process of valuation. Because the commercial world is never static all this must be set against a backcloth of change and increasing expectations from the people who matter – the fee payers!

1.2 Objectives of the book

Based on this contextual review, the objectives of this book are as follows:

- to provide surveyors with sufficient practical and detailed information so the condition of residential dwellings can be appropriately assessed. In this case 'appropriately' would equate to the standard currently expected of the Mortgage Valuation and Homebuyers Survey;

- to help surveyors further develop the skills of communicating the results of these assessments to their customers;
- to provide an overview of the valuation process with a particular emphasis on how the condition of a property affects its value;
- to highlight the changing nature of the residential property appraisal process and initially identify some of the techniques and mechanisms that may help surveyors adapt to this changing environment.

1.3 Definitions

Clearer definitions of the principle terms employed in this book may be useful:

Residential property – any property that is used as, or is suitable for use as, a residence. This book is restricted to domestic dwellings owned by an individual(s) who relies on finance obtained from a commercial lending institution.

Residential appraisal – an act or process of estimating the value, worth or quality of a residential property.

Customer or client – in this book these two terms are used to describe the end user of any survey or inspection report and will usually be a private individual.

Surveyors – this generic term has been used to describe Chartered Surveyors.

1.4 Who this book is for

This book has been written for a broad range of surveyors whose primary interest is with the appraisal of residential property. The book assumes that the reader:

- has already or is close to satisfying the academic requirements of their chosen professional institution. This would have included a course of study that introduced participants into how dwellings are designed and constructed and the principle agents responsible for the deterioration of the building fabric, and;
- has had some professional experience of assessing a range of different properties.

In terms of qualifications and level of experience, this book should be suitable for:

- student general practice surveyors in the later stages of their academic course or are on the sandwich placement or year-out stage;
- building surveying students in earlier stages of their education who are looking for an introduction to the assessment of residential properties;
- students on courses of a more specialised technical nature such as NVQ or HND routes that need an understanding of how to carry out an appraisal of residential property;
- surveyors that are working towards their professional assessment and need to refer to written guidance and technical information on a regular basis;
- those more experienced surveyors who may be changing their professional emphasis away from solely valuation activities to the more challenging pre-purchase surveys;
- qualified and experienced surveyors that need to carry around a source of reference so they can refer to standard guidance when novel situations are encountered.

1.5 The philosophy of the book

The guidance contained in this publication aims to be challenging to the reader in two ways:

- to outline processes and techniques that may potentially take the surveyor beyond the parameters of current standard surveys. This will enable surveyors to more effectively provide those services and better cope with any changes to standard practice in the future;
- to engage with the surveying process and positively advise clients about the suitability of their potential new home. The book is a challenge to 'defensive surveying' and encourages surveyors to focus on the client's needs rather than protecting their own liability.

This book is not a guide to any particular standard form of survey promoted by any particular professional institution. For something more specific then the reader should refer to the publisher of the particular product.

1.6 Technical content – a cautionary note

In an effort to achieve the objectives set out in section 1.2 a massive amount of technical information has been included in this book. This includes such things as indicative joist and rafter sizes, typical ratings for boilers and average ventilation values for airbricks. Every effort has been made to use up-to-date and accurate information but standards and regulations change over time and soon outdate publications like this. Additionally, the authors have purposely 'summarised' some of the technical guidance in a pragmatic effort to give surveyors digestible and broad-brush guidance that can help them make judgements on an everyday basis. Therefore, if more precise guidance on specific topics is required then please refer to the documents identified in the reference section at the end of each chapter.

1.7 Contents of the book

The book is split into three parts:

- **Part I** – this part provides an overview of the valuation process itself. Particular emphasis is placed on the role of condition in determining value. This part also contains advice about how to approach and carry out a survey.
- **Part II** – this is the main focus of the book and so is the largest part. It covers all the defects normally associated with residential property and outlines a strategy to resolve these problems.
- **Part III** – the main practical guidance relates to good practice in report writing. This includes a number of case studies to illustrate both good and bad practices. The part finishes with an assessment of how professional practice in this area may change in the future and suggests a few strategies for turning these threats into opportunities.

Chapter 2

The appraisal process

2.1 Introduction

In the case of Roberts v. J. Hampson & Co (1989) Ian Kennedy described an important aspect of a Building Society Valuation:

> It is a valuation and not a survey, but any valuation is necessarily governed by condition.

Consequently this book will focus on those residential valuations and surveys where condition plays a key role.

This chapter will look at the appraisal process for residential property of which valuation forms a key part. It will do this by reference to basic definitions and then review the Comparable Method of valuation which is one of the common techniques used. The focus of this chapter is also to develop the more scientific approach as this removes some of the subjectivity from the results. It will never be possible to produce a black and white solution, but hopefully this approach will narrow the 'grey' areas.

Anyone undertaking a valuation of property for any credible purpose should have working knowledge of the RICS *Appraisal and Valuation Manual* (RICS 1995), in association with the Incorporated Society of Valuers and Auctioneers and the Institute of Revenues Rating and Valuation, or otherwise known as the 'Red Book'. Unlike this key reference manual, this book attempts to help the inexperienced and to give an insight into the methodology necessary to achieve a competent appraisal of residential property.

An appraisal usually takes place when a property changes ownership but the form of advice may vary. This depends upon whether the parties to the transaction are buyers, sellers, estate agents, conveyancers, financial intermediaries or lenders. An underlying theme is establishing whether it is a good deal for them. For example, the buyers want to know if it is good value for money and whether they will get a mortgage. The seller, on the other hand, wants to know what is the highest price that the property will sell for. The lender is more likely to want to know what is the most realistic price that can be achieved on resale and what is the risk associated with achieving that resale. This publication concentrates on the form of advice that those parties may need by specific reference to the condition of the property. It is for the practitioner to establish the precise needs of the client and to provide the advice in the form requested.

2.2 The appraisal process

2.2.1 An overview

To begin with it will be useful to briefly review the meaning and interpretation of the key words by reference to the Red Book (1995). It refers principally to commercial property and therefore it has been necessary to interpret the concepts for residential property. To assist in this process the dictionary definitions of a number of terms that appear in the Red Book are compared (see figure 2.1).

The linkage between all these terminologies is that the surveyor must go through a recognised process **(the appraisal)** to accumulate information that is sufficient to establish the property's relative quality **(value)** and its specific range of benefits to the client **(worth)**. This is so a monetary figure can be attributed **(price)** which could be considered as compensation should the client be deprived of those benefits.

Market forces ultimately determine price and the surveyor interprets which ones will govern that figure. How this is done depends on the client because different clients will gain different benefits. Numerically, the two most featured clients to a transaction of residential property are the lender and the buyer who both have an interest in its future saleability.

The experienced surveyor may well produce a figure for a property that fairly reflects market value intuitively. In the background is a complex process of comparison that is done almost subconsciously. The basis of that analysis is explained in section 2.3 but valuing intuitively can be accounted for through the process of expertise acquisition (see chapter 3). This section will be of particular relevance to newly qualified surveyors but it is hoped that even those more experienced may find this approach a stimulus to review their existing practices and reflect whether they meet current-day needs!

2.2.2 The lender's view

At the time of writing there is a fundamental change taking place in how lenders view property as a security for lending purposes within the UK. For many years Building Societies legislation governed the procedure that had to be adopted when valuing a property for mortgage purposes. This covered the majority of transactions and required a valuation for all property to ensure that it provided a reasonable security for the loan. Any factors that materially affected the valuation had to be reported. This still applies but for those lenders that fall outside the legislation (i.e. banks, etc.) there is an increased emphasis on the customer's ability to repay the loan. Ensuring the property forms a good security can be of secondary importance. This is especially true for the increasing number of overseas lenders who are more comfortable with this 'customer approach'. As a consequence the emphasis on the valuation report has changed.

This can be illustrated by the changes in inspections associated with the sale and purchase of property. A few years ago there were only main three types:

- the mortgage valuation
- the Homebuyers report (now known as the Homebuyers Survey and Valuation, licensed by the RICS and ISVA)
- the Building Survey.

Term	RICS Red Book (prefaced)	Dictionary definition
Value	Open market value – The best price at which a sale could be achieved subject to various assumptions (PS9 sect 4.4).	Worth, a fair equivalent, recognition of such worth, the degree of this quality.
Worth	The net monetary worth of the benefits and cost of ownership to specific person (D3).	That quality which renders a thing valuable, importance.
Appraisal	The valuation process (in this instance relating to a residual valuation), assimilation of information sufficient to create an audit trail (GN1·7).	The setting of a price, to value with a view to a sale, to estimate the worth of.
Price	The amount of money someone would have to pay in the open market to replace something they have been deprived of (PS 4.3.2).	The amount of money for which a thing is sold or offered.

Figure 2.1 Comparison of key valuation terms.

More recently there has been an increase in alternatives to the 'Homebuyers report'. Additionally there has been an increase in the use of desktop and drive-past assessments. Although these have always been used on a small scale, recent changes may result in an increase in use.

Without an internal inspection it is impossible to provide a valuation within the meaning of the Building Society legislation. However, those lenders that fall outside of this do not need a valuation especially where the loan is relatively low compared with the transaction price (say 60%). In these cases, the lender merely needs to establish that:

- the property falls into the estimated price bracket, and;
- there appears to be acceptable risks.

In some cases the loan may be covered by the value of the land alone. So as long as the land exists the lender need not make any further inquiries. The legal process alone can often establish this fact. The use of a property database may also give sufficient assurance that the value is, in all probability, realistic. Because this book focuses on the condition of the property being appraised, the forms of assessments that do require an internal inspection are not considered any further.

Despite the recent change in emphasis, condition is still seen as the major area in the process of determining whether the value is sustainable. It is still the most commonly reported and often provides a pre-condition to the loan.

2.2.3 The purchaser's view

The purchaser should have a wider view than the lender although this is not always the case. A significant proportion of customers assume that if the lender is satisfied with the property as security for a loan then so should they be. This was confirmed in the case of Smith v. Eric S Bush (1990) where Lord Griffiths stated among other things that:

> If the valuation is negligent and is relied upon damage in the form of economic loss to the purchaser is obviously foreseeable. The necessary proximity arises from the Surveyor's knowledge that the overwhelming probability is that the purchaser will rely upon his valuation. The evidence was that surveyors knew that approximately 90% of purchasers did so and the fact that the surveyor only obtains the work because the purchaser is willing to pay his fee.

This clear ruling has been slightly clouded, as the purchaser will not always pay a fee, as some lenders do not make a charge directly. Because the surveyor will still be paid, the debate will continue. If the cost of the survey is included within the cost of the mortgage, does the purchaser pay it indirectly? The recent move by some lenders to non-disclosure of reports places purchasers in a further dilemma. Can they continue to rely on the action of the lender? There may be a significant difference between the loan and the purchase price. This margin can accommodate a wide range of deficiencies in the property that may in some cases cost many thousands of pounds to rectify. Purchasers may be very disappointed with this gap between the lenders' criteria and their own expectations. Any disgruntled purchasers might be denied remedies in law. If this is the case they will remain disillusioned with the professional who provided that report as well as that person's professional Institution. To counteract this, professional organisations associated with the sale and purchases of property have mounted a campaign to educate the public. They urge people to obtain reports that more fully advise about the factors that will affect their decision to purchase. Such a report requires more detail than that provided by the lender's own surveyor. Appreciating that this is not the only approach the RICS have launched the Agenda for Change to encourage the surveying profession to adapt to the needs of the customers. In addition the Government has launched a campaign to make home buying easier.

2.2.4 Future issues

Do surveyors' reports address the key points that are important to the lender? The lender assesses a loan on the basis of the borrower's ability to repay. However, there is also the fall-back of what is the probability that it will be possible to realise the money on loan should this prove necessary.

The property is only a part of that equation but a significant one hence the reason for the variations in views by some lenders. Take the comparison with a company making a loan on a car. They do not ask for a survey of the car so why should they with a property? It is clearly the purchaser of the car who needs to be sure that the car can perform as sold. The lender only needs to know that the borrower can pay. A property is a major investment to an individual, but should the protection be legally enforced? For this reason it is suggested that there is a clear change of emphasis at the present time. This may of course change.

The problems of negative equity experienced in the late 1980s early 1990s clearly identified a shortcoming within the existing valuation process. Predicting market movements or anticipating a change in price prove to be very difficult. A lot of work has been undertaken on statistical analysis of the housing market both in America and the Pacific Rim but in this country the analysis has generally been retrospective. The ability to forecast accurately is an extremely complex process, especially when global influences on economies have to be taken into account. So it is hardly surprising that the local valuation surveyor struggles with some of the modern-day aspirations.

2.3 The valuation process

The definitions of value for mortgage purposes have changed over time. The current one is known as Open Market Valuation. The other types are noted in the most recent edition of the Red Book and include Estimated Realisation Price (ERP), Estimated Restricted Realisation Price (ERRP) and Market Value (MV). It is important to clarify the type of valuation that is most appropriate to the work being undertaken. Before the various nuances of the valuation definitions can be discussed there are a number of basic principles associated with the valuation process that must be understood.

The appraisal process for residential property consists of a number of stages:

- collection of relevant information
- inspection of the property and its environment
- an interpretation of the extent and nature of the interest
- measurement of development potential or general enhancement of the property to maximise its benefits
- market analysis
- calculation of price/value/worth.

These may appear to be a complex series of questions that would take a significant amount of time to complete. In reality for a mortgage valuation it can take on average 30–45 minutes and for a 'Homebuyers Report' 2–3 hours. The experienced surveyor may find it difficult to recognise the individual stages in the process, as so much of it will be done subconsciously. Also, in the majority of cases the property will meet the normal expectations. In other words fewer investigations will be needed, speeding up the appraisal process even further. Each of these factors will be considered in turn.

2.3.1 Collection of relevant information

It is a requirement of the Royal Institution of Chartered Surveyors (RICS) that corporate members should be able to demonstrate that they have good local knowledge of a particular type of property and the market. Consequently the surveyor should have access to a comparable record of residential property transactions. The size and nature of this record will vary but should have reference to some of the key attributes that will affect the value. These headline attributes are given below (Rees 1992):

- location
- accommodation
- type of property
- the state of repair, appearance and quality of finish
- the quality and quantity of fixtures and fittings
- the potential for improvement
- the supply and demand for property at a point in time
- the exposure of the property in the market place.

The term 'attributes' can be defined as:

- characteristics
- the qualities of a property and its immediate environment
- its positive and negative features relating to the market in which the property is based
- legal aspects that affect the benefits for the owner/occupier.

Additionally, within each of these there are other important factors. For example, under 'type of property':

- size of property
- the dwelling's attachment, e.g. semi-detached, detached, terrace, etc.

The relative importance of each one of these attributes may vary considerably between different geographical areas:

- Character of location can play an overriding role regardless of type of property.
- School catchment areas may influence price regardless of other characteristics.
- Number of bedrooms is probably the key attribute within the accommodation section within the UK.

Usually only limited information is recorded and the analysis hinges on the interpretation of how those attributes combine. This is done by identifying those circumstances that differ from the norm (or the surveyors' expectations of it). In essence the surveyor must be able to record the key attributes of comparable properties so that a full evaluation can take place on-site. Knowing what information to collect is based on a sound understanding of what you are trying to achieve at the outset, i.e. comparison of the subject property with those that are:

- in a similar location
- similarly designed
- in a similar condition
- having similar amenities
- in similar market conditions.

The best comparisons have established values. Therefore the collected information is used to establish the differences between the subject property in its location from the best comparisons known to the surveyor that have been sold in the open market at a point in time. The next step is to determine how important those differences are.

The process does not start at the property but by considering whether anything has changed both nationally, regionally or locally since the last inspection. This may include the density of 'For Sale' boards or on a more macro scale a change in interest rates. So information collection starts by reading local and national newspapers.

Special factors may also impact upon the price. The information given to the surveyor at the outset may not include details of why the property is being bought. A simple example is where someone is moving to be close to a new job. Moving to be close to a family member is another example that may not be as common. This may result in the individual paying more for that particular property in order to be sure of getting the location. This may produce a figure in excess of the general market value and therefore may not be repeatable. Regrettably this level of detail may not be apparent to surveyors as their only personal contact is often limited to the vendor.

2.3.2 How to use the information

Having determined what the objectives are for collecting the data, the next stage is to collect the key attributes of the property, as identified previously. It is important to record any issues that are unique to the property. The reasons for this can be listed as follows:

- An experienced surveyor will have in mind a comparison for the subject property. The similarity will be based upon the expertise of the surveyor and the quality of the existing records. The process will involve an objective and subjective ranking. Referring to established scores for each of the key attributes does this. Generally this is done without consciously thinking. A few comments are made in site-notes usually where there are deviations from the norm.
  The collection of data adds to existing records and the knowledge base of the surveyor.

In addition it is important to establish those features within the property that require enhancement to bring the dwelling up to the expected standard for its type and age. Where the clients are known and the surveyor has had the chance to evaluate their needs then an on-site assessment that measures the real or potential benefits to that client would also be significant.

The process can be summarised as:

- collection of data to use for comparison purposes
- a ranking of the attributes and assessment of them by reference to existing comparables
- identification of defects and enhancement potential
- evaluation of the match between inherent benefits and client needs.

In essence the surveyor has to determine where the subject property lies within the range of the comparables for which he has knowledge. So in the context of the scale shown in figure 2.2 property comparisons may fill the whole range from poor to excellent, and more likely they will be nearer the average. It is for the surveyor to establish whether the subject property is better or worse. An added complication is when there are a variety of factors, some of which are better and some worse. In these instances remember to keep it simple. Map out all those factors that are better and those that are the same or worse – imagine them on a scale like figure 2.2. More help is given with this technique later.

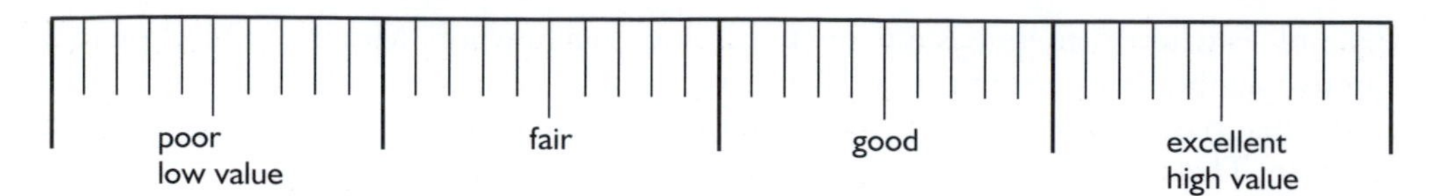

Figure 2.2 Scale of comparison for different properties. The key to successful valuation is correctly placing the property on a relative scale.

2.3.3 An interpretation of the extent and nature of the interest

There is a whole series of specific attributes that can best be described as those issues that affect the client's rights to use the property. These may be either to the advantage or disadvantage of the client and consequently will influence the value of the property generally and the worth to the client. These include:

(a) **Tenure** – the legal right to use the land. This can be freehold or leasehold in England, Wales and Northern Ireland. A feudal arrangement exists in Scotland. The property may be subject to certain forms of ground rent, service charge or chief rent. The impact is that this is an additional cost on the right to use the property. The amount and the conditions surrounding the charge will vary and some will be significant in arriving at the value. There are numerous books on Property Law that will give the legal background and Mackmin (1994) gives a useful interpretation from the surveyor's viewpoint.

(b) **Planning issues** – rights under the planning legislation convey the parameters under which the owners can legitimately use or develop the land and its property, for example residential or commercial occupation. In addition they govern the design and safety standards to which a property has to be constructed, whether for new construction or adaptations of the existing structure or appearance. The right to convert land used for agricultural purposes to residential would usually add significant value as the return is greater. On the other hand a restriction on use may adversely affect value (see chapter 13 for more details).

(c) **Rights of light, way and other easements or covenants** – legal rights defined in the deeds to the original purchase. They can be modified over time and they can be restrictive. Examples could include:

- where a developer and the planning authority agree to an open plan estate that restricts the development of hedges or walls;
- where rights were given allowing a person access to a piece of land over someone else's property. The initial grant might have been given in exchange for some form of consideration to reflect the adverse impact of the access. Subsequent sales will need to reflect that impact.

The overall impact of these restrictions may be to the benefit of an area as a whole. For example a right of way gives access to many but limits the use to an occupier. The benefit may need to be reflected in valuing neighbouring property.

Environmental considerations – this might be another aspect that could influence value. This covers a wide range of factors varying from climatic conditions to previous contamination of the site. Again the situations can be either beneficial or adverse in that they can enhance or detract from the value. Climatic conditions will feature in various parts of the condition section and reference is made to contamination because of the need to ensure correct site treatment to prevent escape of the contaminant (see chapter 11). The latter aspect poses additional costs in the original construction that may well not be capable of being reclaimed through the sale price when compared with other developments without the contamination. In these cases the developer may choose not to build property until profit margins allow.

All of the features described above may not be apparent from a site inspection of the single dwelling. Reliance is often placed on the conveyancer to unearth any matters that may adversely affect price or value. The surveyor should be aware that the conveyancer is depending on him/her to act as the 'eyes and ears' to identify a 'trail of suspicion'. Consider the following example. There are many large blocks of flats in Greater London, some of which have leisure complexes. The right to use the complex can be included within the deeds, in some cases it is not. Some leases make an additional charge for this benefit. The conveyancer will be depending on the surveyor to identify whether there is a leisure complex. This sounds straightforward. But because a lease can extend over more than one block and can include hundreds of flats the task of locating a fitness gym hidden away in a basement might not be that easy. In one particular instance a purchaser knew about a complex and assumed that the right to use it was included in the purchase. In the event, the complex did not appear in the deeds and the surveyor did not mention it. The potential for a legal challenge by a disappointed purchaser would be a distinct possibility. As the customer rarely talks to the surveyor and the surveyor rarely talks to the conveyancer such a breakdown in communication is bound to happen.

2.3.4 *Market analysis*

This highlights all the factors or attributes that make up the proposition by comparison to the current state of the market. The most important aspect is the local market but this should not ignore macro considerations. The general state of the economy must be accounted for as well as indicators of trends within the housing market taken from leading parts of the country such as London. For example, to slow down the economy over the last decade the interest rates were raised. This did have an important influence on the whole property market. The extent and timing were difficult to anticipate. Local conditions such as the closure of key businesses may have an impact upon the local market contrary to conditions at a national level. The ability of the work force to find alternative employment without moving location is a key determinant of the impact. The introduction of a regular analysis of the economy by the RICS is an indicator of the relevance of this topic and increasing expectations of the customer.

Generally the surveyor undertakes the analysis of the market, based on retrospective information. This poses significant problems, as this information does not forecast what will happen in the future. The surveyor has to anticipate what will happen to the property especially with regard to condition. This provides a useful comparison between the effect of condition on value (a tangible) and those factors of a more subjective nature such as micro and macro economic forces and their influence on value (intangibles). Identifying faults in design or structural elements will lead the surveyor through a process resulting in a recommendation to undertake repairs at a cost that can be deducted from an estimation of price. Considering experience of similar conditions, the risk to the property can be fully assessed and the surveyor can be fairly certain of the outcome. It is different with the economy as a whole. The chain of events that can trigger a recession within the property market either nationally or locally can be very complicated. They have been poorly analysed in the past so it is difficult to be as confident. This should not be used as a catch-all excuse. In Corisands Investments Ltd. v. Druce & Co. (1978) the surveyor was found negligent for not allowing for the speculative element in the property market. The judge based his decision on what 'an ordinary competent surveyor' would have done. The test of reasonableness will usually apply. The case of South Australia Asset Management Corporation v. York Montague Ltd. (1996) limited the valuer's liability to the extent of the overvaluation, known as the SAAMCO cap. As a consequence a valuer cannot be held liable for a downturn in values resulting in the lender making a loss, provided the valuer was not asked to anticipate the changes in the market place and there was no evidence at the time of the valuation of a downturn. In simple terms if a valuer considers a property worth £100,000, but at the time of the value the comparable evidence suggests it is worth £50,000 then the valuer can be held liable. However, if the property was worth £100,000, but as a result of a downturn in the property market, it was subsequently worth £50,000 then the valuer would not be liable.

Basic market forces cannot be ignored but these are not neatly packaged as with other commodities. Property is not homogeneous and records of sales experience throughout an area (beyond the limits of an individual firm) are generally poor. Developing a pattern of supply and demand even at the local level is therefore not a precise science. Anticipating future demand for a specific property has to be based on demand at that point in time and the number of similar products available that will be in direct competition. Consequently the clients take an inherent risk with any property transaction in respect of the intangibles. They cannot be sure that they will receive full compensation if deprived of their benefits at a time when the prevailing market conditions differ from those when the initial transaction took place. Lenders usually take out a form of indemnity to cover large loans relative to value.

The key features of the market analysis are summarised in figure 2.3.

No evaluation of the market place would be complete without some consideration of the consumer. In many cases the surveyor may not even be aware of the actual purchaser and is more concerned with the lender. Generally trends develop where

Figure 2.3 A summary of key considerations involved in a market analysis

consumers of like minds and financial circumstances tend to be attracted to particular areas. This has resulted in a variety of consumer-based statistical surveys being undertaken by companies such as Acorn. The base information for these databases is the census although other types of information may support them. The approach is to evaluate socio-economic groupings based on postcode. Companies needing marketing advice to target areas for specific product sales use the information produced. Essentially these surveys will give an indication of the typical lifestyle for someone living in a certain area. Further investigation can reveal what products that particular lifestyle will want to buy and target them with sales campaigns accordingly. These companies are well positioned to expand this data to provide value comparisons.

A further point that is of some significance in the residential market is the availability of finance especially with a highly competitive mortgage market. There was a period when Building Societies had strict limitations on who should receive funds based upon their own limited availability of finance. This put a brake on housing transactions. Once this was removed competitive pressures in the mid-1980s introduced a level of flexibility into the gearing on an individual's income and innovative ways were found to finance the spiralling house price inflation. Some commentators indicated that this availability of finance was fuelling the inflation. The ability to pay was not such a restricting feature in the equation with the 'Yuppie' era of spiralling wage inflation and a willingness of individuals to take on more debt. The philosophy of 'buy today and pay later' regrettably came home to roost. By the early 1990s there was a significant rise in repossessions and bad debts, with negative equity introduced into the homeowners vocabulary: it became clear that the property boom could not be sustained.

2.3.5 Calculations of price, worth, and value

How all the attributes influence price is the key to accurate appraisal and is the subject of much debate. A lack of detailed recording and assessment will mean that the surveyor's interpretation of value may be open to challenge. It can be argued that for commercial property the attributes will specifically relate to the needs of the investor. These are tangible and should lead to commercial gain. This underpins a definition of worth to the investor and there is no reason why this cannot be applied to the residential purchaser or lender. The definition shown in figure 2.4 for investment was given in Chartered Surveying Monthly (RICS 1997). It is set against the author's interpretation for residential worth.

The benefits perceived by purchasers of residential property are translated by the surveyor into the attributes specified in figure 2.4. This is done on the basis of trends. A classic geographical example used in the northern hemisphere is that locations have developed into prime sites because of their position relative to the sun on south-facing slopes as compared with north-facing ones. The perceived benefits for the purchaser may include a longer growing period, lower fuel bills, a more pleasant aspect, etc. The number of benefits is great and varied. One family's benefit (worth) may not be the same as another's. Over time the benefits in a specific area will have grown to such an extent that a trend develops and they become tangible and this is translated as value.

This establishes the relative quality of the property within the market place and then a price can be applied. A monetary value can then be placed on the benefits of owning or lending on that property in that specific area.

Investment	*Residential worth*
Worth is a specific investor's perception of the capital sum that s/he would be prepared to pay (or accept) for the stream of benefits which s/he expects to be produced by the investments.	Worth is a specific home buyers perception of the capital sum or monthly payments that s/he would be prepared to pay for the stream of benefits which s/he expects to be produced by the acquisition of that property.
Price is the actual exchange figure in an open market.	As for investment in commercial property.
Value is an estimate of the price that would be achieved if the property were sold in the market.	As for investment in commercial property

Figure 2.4 Definition of investment and residential worth (Source: RICS 1997).

2.4 Methods of valuation

2.4.1 Introduction

With any form of valuation there has to be a comparison as value is relative. Whether you are valuing fine art or property the surveyor must have a benchmark. From this, good or bad points will be either added or deducted. There is a saying that suggests you should compare 'like with like' and so 'apples should not be compared with pears' follows. However, here lies one problem with comparisons; although only one variety of apples should be compared with another, apples and pears are both fruits that are quite similar in some respects. The same can be said of property. A valuation should start by comparing properties of a similar type, e.g. terraced houses with terraced houses. But what happens when a detached house has to be valued in an area where there are mainly terraced houses and no others to compare it with? A form of comparison is still necessary to achieve a solution. There might not be any detached houses (apples) in the vicinity but are there some similar terraced houses (pears)? This is feasible but requires a significant amount of subjective assessment by the individual surveyor. How the subjectivity can be reduced is described later in this section.

There are two main methods that can be used to process this information namely the Residual Approach and the Comparable Method. The former will only be briefly described as it relates principally to establishing the value of the land. Because the comparable method is the most common approach used for the appraisal of residential dwellings it will be discussed in more detail in the next section.

2.4.2 Comparing transactions

It is important that the surveyor bases a decision on what the market place is saying. A large sample of transactions makes for a large quantity of comparable data. Consequently the surveyor can relate the subject property against a number of others and rank it accordingly. A small number of transactions make this task more difficult. The following list is a number of key features that are used to develop this benchmark. This is not a comprehensive list of all matters that are recorded as specified in the Annex to Guidance Note 3 of the 'Red Book' (RICS 1995). It includes the most common features that surveyors use to make comparisons.

- **The address** – this may seem obvious as a start, but not so obvious is the key information within it. All the locational characteristics are locked within this information and include:
    - quality of neighbourhood
    - view
    - proximity to facilities
    - site conditions (geology, etc.)
    - environmental hazards
    - orientation
    - appearance of the property (sometimes known as kerb appeal).
- **The type and style of property** – i.e. detached, semi-detached, house, bungalow, etc.
- **The age of the property.**
- **The size of accommodation** – measured in square metres and usually an external measurement, but it can also be internal. The important factor is to be consistent.
- **The number of bedrooms** – as opposed to other rooms. This is the key measure for the standard type of property. For more exclusive dwellings the number of reception rooms may be more significant.
- **The number of garages or the facility for parking space.**
- **The tenure of the property** – this may be significant. If leasehold how long is the unexpired term of the lease? What are the costs/liabilities of the arrangements?
- **The general condition.**
- **Any additional features** – especially those that either add or detract from the value.
- **The selling price** – available at that time.
- **The valuation** – based on the comparable information and as prescribed by the customer.

More information will be recorded either on the report or in the site notes but this should give the surveyor a benchmark upon which to assess the value by comparison to other property. It will also act as a filter and should the case prove to be complex, more detailed research can be undertaken.

The experienced practitioner will take all these features for the subject property and consider whether individually they are better or worse than the best comparables (those most similar). For example, how close are the comparables from a location viewpoint, in the same road? In the same area, town or village? The greater the distance from the subject property the more variables that are likely to arise. This will make the comparison more subjective and complex. The best comparisons would be property adjoining or neighbouring the subject one, reducing locational variances and allowing for a focus on more tangible items such as number of bedrooms, size of accommodation and parking allocation. If the apples example were considered, comparison from one variety to another would be by appearance, texture, taste, keeping ability, price, etc. The principle is the same.

The Popular Housing Forum (a body comprising architects, planners and builders) produced a report entitled *Kerb Appeal: The external appearance and site layout of new houses*. This identified some of the characteristics that appeal to purchasers. This illustrates how a surveyor undertaking a valuation for a lender will not always reflect the same priorities as the purchaser (see figure 2.5). This graph identifies the reasons

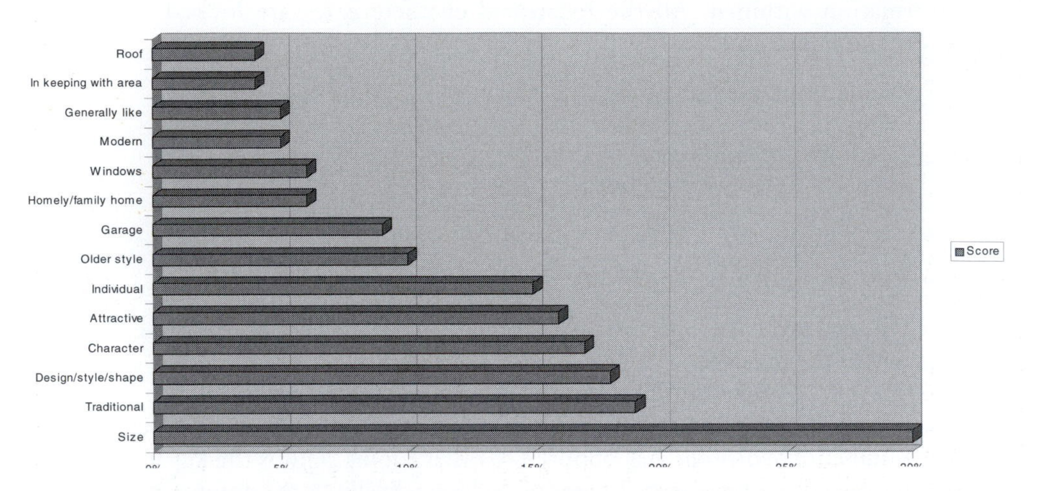

Figure 2.5 Graph showing the 'kerb appeal' of properties (Source: BRMB International: Popular Housing Forum, Kerb Appeal 1998).

for preferring an existing property as compared with a new one. A surveyor would use size as a key attribute and the amenity of a garage for example. However, attributes like character, attractiveness and that 'homely feeling' will be more difficult to quantify and maintain as a sustainable feature. The basic design of a property helps to establish the homely aspect but it is quite often the choice and use of furnishings and decoration, together with general tidiness that actually determines the homely feel. This of course will change with a new occupier.

Fashion can affect the appeal of housing. The same report by Kelly (1998) also listed the 'the ins and outs of housing fashion'. This identifies an important factor that will change with time. Fashion in domestic design is quite likely to change a number of times within the lending term of the standard mortgage. Some of the key influences are shown in figure 2.6. For more information contact the Popular Housing Forum, 9 Upper High Street, Winchester, Hampshire, SO23 8UT.

Fashionable 'in' features	*Features 'out' of fashion*
Broken roof lines	Boxy shapes
Eaves	Futuristic design
Bay windows	Space age design
Dormer windows	Unique
Varied use of materials	Ultra modern
Integral porches	Stuck on features such as porches
Character properties	

Figure 2.6 Fashionable features in domestic house design (Source: Popular Housing Forum 1998).

Situation/ classification	*Movement*	*Damp*	*Timber defects*	*Roofs*	*Thermal insulation*	*Building services*	*Non-traditional dwellings*
Decline for lending purposes	*Structural repairs where the cost exceeds value – foundation failure in need of extensive underpinning and correction of levels within the dwelling to make habitable*	*Water damage has lead to secondary deterioration making property un-inhabitable – water damage from a fire or leak*	*Wood rotting fungi has lead to structural failure – dry rot extending to main structural elements and caused failure as with movement*	*Failure of roof structure to perform leading to structural failure – gale damage or failure of roof covering leading to water penetration as for damp, or failure of roof structure causing main walls to bow (excessive roof spread)*	*Cold bridging on building incapable of remedy, leading to condensation and mould growth that makes property a health risk – block of flats where problem inherent in the design, but cannot be resolved by one occupant*	*Failure of drains leading to structural failure of the main building – as for movement*	*A designated property where costs of repair to an acceptable standard exceed value – a less well known type for which there is not an economical repair scheme*
Further investigation and impracl tical to provide a current value	*Recent movement causing impairment to the use of the building – opening doors and windows restricted*	*Water damage has lead to secondary deterioration with the possibility of dry rot infestation – as for timber*	*Potential for dry rot or a failing of a timber element due to jointing, shrinkage, warping etc. – Timbers in contact with damp materials and mostly concealed*	*Failure of roof structure to perform adequately – a milder yet serious version of decline – roof spread where full impact of damage cannot be assessed or structural calculations necessary to determine extent of repair*	*Inappropriate use of cavity wall insulation or milder version of the decline scenario where a repair might be practical – use of cavity wall insulation in an exposed position making the outer skin too cold and risking frost damage to the bricks*	*As for movement, but caused by drain failure*	*A property where a repair has been undertaken, but it is not known whether it meets the standards specified by an approved scheme*
Further investi-gation required, but quantifiable	*Recent movement of a less serious nature, where a diagnosis can be made from one inspection – cracking needs to be repointed to prevent damp penetration*	*Evidence of damp penetration where the cause can be diagnosed and it does not extend to concealed areas – rising damp to main walls where the floor timbers can be seen in the cellar*	*Defect identifiable – window frames affected by wet rot and needing replace-ment as the only feasible method of repair*	*Defect identifiable and actually failing lead to ingress of water – flat roof in need of re-covering and underside of roof visible so extent of damage apparent*	*Incorrectly installed insulation leading to cold bridging and serious condensation – situations around window or door openings in contact with timber elements*	*Evidence that the services pose a risk to health and safety – obvious defects or old installations for electric wiring*	*A non-traditional dwelling not designated and repairs are typical – spalling of render on a Laing-Easiform exposing the reinforcing*
Defects form part of a programme of routine maintenance	*Old movement that will need regular repointing – thermal movement under a window, where the cracking does not extend below the dpc*	*As above, but the extent of damage is visible and the cost of works minor – defective flashing to a chimney stack and the underside is to a roof void*	*Examples of wet rot on secondary features of the structure – barge boarding*	*Evidence of failure allowing minor water penetration that does not impair habitation – slipped tile*	*Inadequate levels of insulation – insufficient loft insulation*	*Minor defects that do not pose an immediate threat to health and safety – one old plug socket with surface wiring*	*Repairs typical of a traditional dwelling*

Figure 2.7 Table showing how condition affects value.

2.4.3 Comparing condition

Another area of particular significance in this book is the comparison of condition. This leads to even more subjectivity as individuals have different views on the acceptability of cosmetic or minor defects. A newly decorated house will generally look more attractive than one that is well into the maintenance cycle and is in need of painting. However this is a cosmetic feature as decoration does deteriorate with occupation and the normal ageing process. New decoration will carry a premium on the price (provided it is of a reasonable quality and colour) but whether it enhances value is debatable. Decorating is an ongoing cost and if it has been done recently then the cost will be deferred for a further period. This will give benefit to the new owner and so add value. If the new owners fail to redecorate then the added value will be lost therefore decoration is not sustainable.

Where a property has more serious defects they are referred to the lender as conditional if the mortgage is to proceed. The repairs will usually be costed and reflected in the value. Figure 2.7 is an example of a comparable record of how condition affects the valuation process. The comments in each section provide the thought processes that can be adopted by a surveyor.

2.4.4 Comments on the comparable spreadsheet

In figure 2.8 the key benchmark features have been used. Practices differ from region to region, office to office, but this example is probably the most common. It serves the purpose of evaluating the comparison method as a valuation technique. The headings are not tablets of stone and can be extended especially when linked to an integrated database capable of recording all the information on the valuation report or site notes. However, statistical evidence used in the House Price indexes would indicate that there are only a limited number of variables that influence price in the majority of cases. There will always be circumstances when one of the key features does have an overriding impact. For example, the appearance of a property may be so awful that exceptional accommodation internally would be irrelevant as potential viewers were put off before ever getting inside.

Figure 2.8 has limited qualitative information. The norm is to simply list the information. More sophisticated techniques are being developed that ranks the importance of the features. These databases effectively mimic what the surveyor does to value a property. It should be noted that the names of areas and streets have been changed for reasons of confidentiality and to avoid a regional bias.

Each of the categories are described in a little more detail:

- **Address** – The information shown merely gives the physical location. Implicit within that address are all those features that might affect value. For example if coal mining affected the whole area it would clearly need to be mentioned to a client. The surveyor would be so aware of the situation it would be a waste of time to record it every time. The exception might be where the mining was active or there was some particular hazard to that area such as a known fault line. If there was a specific hazard then this would be recorded in a separate comments' field. Location can be ranked as seen in the example Celandine Nook. This area is considered less desirable than Longwood. More sophisticated techniques would score the area

Address	Type	Bedrooms	Age	Size m^2	Tenure	Condition	Garage	Price	Value	Date	Extras	Analysis value m^2
2, Acacia Ave, Longwood, Anytown	SDH	3	45	90	F/H	Good	Space	75000	75000	03/09/97	CH, double glazing	833.33
Good views, faces the south, back garden in shade, but a corner plot, lacks privacy	Typical for the area	2 large 1 small	typical	average	usual	average for this area	usually a garage			On the market for 3 mths – usual for this area	Usual	
53, Acacia Ave, Longwood, Anytown	SDH	3	46	90	F/H	Fair	Garage	80000[a]	77500	10/10/97	CH, double glazing	861.11
On the other side of the street, good views from back garden, better position than 2	Same	Same	Similar	Same	Same	Cosmetically worse than 2	Pre-fab, better than 2	Same		Little movement in the market place in this period	Same	
24, Lilac Road, Longwood, Anytown	SDH	4	40	105	F/H	Good	Garage	92500[b]	90000	15/09/97	CH, double glazing	857.14
A street parallel to Acacia Ave, not as good views, same access to facilities	Same	Extended property	Younger but not significantly same construction	Larger	Same No 53	Same as No 2 better than others	Brick built better than others	Same		Same	Same	
18, Ash Lane, Celandine Nook, Anytown	SDH	3	35	92	F/H	Good	Garage	70000	70000	23/09/97	CH, double glazing	760.87
An area adjacent to Longwood, but higher up the valley side, more exposed and more urban views, similar access to facilities	Same	Same as Acacia	Same comment as Lilac	Similar	Same	Similar	Same as Lilac	Same	Lower price reflects location	Same	Same	

a Higher price reflects better position and garage, but it could have been higher if condition better. b Higher price reflects larger accommodation, better garage, but poorer position.
General comments on Longwood: good area, formerly wooded, some evidence of longstanding movement, local shops and a good school nearby. Access to a motorway, good views of a golf course and valley and to nearby hills, exposed in winter

Figure 2.8 Example of a comparable spreadsheet

relative to other areas. Even within areas, it could be possible to score specific streets. In some cases this scoring might be applied to the actual plot. This only becomes relevant when there is a significant difference between plots. However, it is important to reflect that the plot is a significant part of the value of the property. In the south of the country in certain areas it can account for more value than the actual building.

In all the examples, reference has been made to the precise location and in particular the view of the dwellings. This is merely used as an example to differentiate one attribute from another. It is important as in this case all the properties are in a similar area. Therefore, this is fine-tuning the value but in terms of lending risk it is probably irrelevant. In terms of insurance risk this feature may be more significant as the more exposed areas are likely to be more prone to storm damage or the influence of other climatic affects such as frost. This is a good example of how the differing needs of clients can be significant. In this example view is the key variable (see figure 2.9) which shows the advantage to the eye and conceals the one of exposure, in another area it could be proximity to a railway or tube station, etc. Whatever the feature, the principles of comparison still apply.

The aerial view (figure 2.10) shows the variables of property location. In the lower part of the photograph there is a development of good quality stone dwellings adjacent to a dual carriageway road. To the upper part of the photograph there is supermarket and petrol filling station. Beyond this is a council housing estate. The property fronting the main road is of an older type to those on the private estate and the plots are generally smaller. The value per square metre of these plots will be lower because of the proximity to the main road and the view of the supermarket not being ideal. However the latter point has to be balanced against the convenience factor.

A further point relating to location is occupational mix. It is sensitive but cannot be ignored. In this case there are two forms of housing tenure in close proximity. Those in private occupation and those rented from the local authority. It is not uncommon to find property of differing tenures that are close to each other. In some cases, developers are required to allocate part of a new site for social housing. This can result in families of different socio-economic groups living in the same area. It can also provide accommodation for those not choosing to buy. Whatever the reason this will be interpreted by the market place. This could lead to a reduction in the capital value of the owner occupied property and result in a higher rental value for those that are let. Each case needs to be looked at on its merits; the surveyor needs to be aware of the potential impact and allow for this in the analysis.

- **Type** – In all cases on the comparable sheet the type is the same. Implicit in this is that building costs will have been similar. It would be important to distinguish if the design differed markedly. For example, if one dwelling had a flat roof while another was thatched there might well be an impact on value. Again this aspect may be more significant to the insurance risk than the lending risk, although appearance would be a key feature in assessing the value.
- **Bedrooms** – The number of bedrooms gives an indication of the size of the accommodation and its functionality. (Americans would record the number of fireplaces as this indicates reception rooms, which is of more significance to the customers in

Figure 2.9 The view from this property is both an asset and a liability. The scenic outlook will enhance value but the exposure to wind and rain may affect the durability of building components.

Figure 2.10 The aerial view of this small neighbourhood area shows how location can affect houses differently even though they are relavtively close to each other.

that country.) Within the range of 1–5 bedrooms it should be possible to relate this to the size of accommodation (but there are always exceptions!). Beyond this, any unique or special characteristics may require a more subjective approach. As a rule, a 3-bedroomed house will have a wider market than a 1-bedroom because there are usually more potential purchasers. The 3-bedroomed property is more adaptable and better suited to the average family unit. However, certain locations can match a certain size of accommodation, a student area for example or one near a hospital that may consist of traditional Victorian housing. It is quite likely that a proportion will have been converted to bed-sit or flatted accommodation rather than returning to family units as originally intended. The valuation process is a complex matrix of matching variables.

- **Age** – This can give an insight into the condition and to some extent the design of the property. For example, a property built in the 19th century will not meet the same standard of construction as one built recently. Therefore it would be unusual for it to have the same level of energy efficiency. If it did, this would be considered an advantage especially when compared with properties of a similar age that would almost certainly be energy inefficient.
- **Square meterage** – This indicates the size of the property but the surveyor would need to give thought to how that space has been used. For example a room of 5 m × 5 m with appropriately positioned doors and windows, giving adequate access and natural lighting has more potential than one of 7.5 m × 3.5 m. Such a room would be long and narrow, it would be difficult to fit the furniture in to a traditional layout. Rooms can be too big and feel cold and draughty. A hallway no larger than a passage between doors would give a better appearance, where the stairs lead up to the first floor, as this also gives an impression of more size.
- **Tenure** – There are numerous legal aspects that can influence the value and some very significantly (see section 2.3.3). Tenure is a significant one worth recording as the differentiation between freehold and leasehold is the most important. Other aspects are so wide ranging and exceptional that they would be recorded and interpreted individually. One legal area is planning. For an established area the Planning Authority for significantly large areas defines the policy and therefore close comparables will not vary. The main example is likely to relate to any alterations to the property and whether they comply with current regulations. Breach of planning regulations is likely to be exceptional but significant.
- **Condition** – All-encompassing one-word comments that have limited value to anyone other than the individual who made them. The main use is as a memory jogger. It is important to set a common scale within an organisation so that other surveyors will know when to undertake more research.

    - 'Excellent' comparable to a new house.
    - 'Good' would suggest that there were no repairs other than would be encountered within normal maintenance and that the general appearance was tidy.
    - 'Fair' or some other equivalent word meaning less than average, would need some works. Some of these may be made a condition of the mortgage.
    - 'Poor' would indicate a number of repairs.

- **Garage** – This covers number of garages and parking space. In some areas the ability to park off the street will carry additional value because on-street parking is the norm. In the examples quoted, off-street parking was standard for the area.
- **Price** – The price quoted in the market place at the time of valuation. The surveyor may wish to ask a number of questions about this:

    - How long has the property been on the market?
    - Has the price changed over the period?
    - How many people have looked around the property?

The answers may not always be forthcoming, but if they are then they give an indication of how the customers feel about this property. Clearly if the market is very good then the customers may not have a lot of choice, dependent upon the degree of their need. If the market is very poor then it is likely that only the best at a price will sell. Again these are extremes and the normal situation is more likely.

- **Value** – This is the surveyor's interpretation of all the features put together when compared with similar properties. The reason for the valuation may dictate how this value is recorded (see the definition of value and worth earlier in the chapter). The courts have indicated that there are varying degrees of tolerance that are acceptable. Beaumont v. Humberts (1990) was a case involving a negligent re-instatement valuation. Staughton LJ accepted that:

  a surveyor or valuer may be wrong by a margin of 10% either way without being negligent. That in itself seems, in my uninstructed opinion, a high standard to impose; but it is, as I have said, accepted.

Care must be taken in interpreting such decisions as they will all have been based upon the specific facts of the case. To use the 10% example could be critical on a 95–100% mortgage, especially if comparable evidence exists for a neighbouring property as this could well be a near perfect match. In such cases 10% would be excessive.

The RICS (1995) have given guidance on the procedure for valuing new property, which relates to the number of comparable sales actually on the site and comparison to sales in the vicinity. The surveyor has to determine, on the circumstances of each case, whether she/he has effectively proved the figure reasonable and that it provides a suitable risk for the lender. This may mean that there is a tolerance between the comparable evidence and the price paid in the market place. This reflects the market at that point in time. For example the difference between the best comparable from a month previously may show a 5% increase in value. If the market is fairly active this change may be acceptable and to ignore it would be incorrect. Provided the surveyor records the reasoning behind the decision and this is justified, then the tolerance is acceptable. The reasoning behind the process of fine tuning the valuation is complex. Any differences could be due to a variety of factors ranging from the time of year (the market is usually quieter around Christmas) to the demand for particular properties in popular areas.

- **Date** – The date the valuation was completed should also be the date of the inspection. This is important to establish a reference point. It is unlikely that anything would change between the day inspected and, say, a valuation worked out a day later. If the house did burn down or was damaged in some other way and the valuation was dated after that day then there could be an issue.
- **Extras or Comments** – A catch-all section to cover any additional or overriding factors that will influence the value. Central heating and double-glazing are recorded here because without them the value would be significantly affected. In respect of central heating this is probably by as much as the cost of the installation.
- **Analysis** – To illustrate the comparable method, a simple analysis has been included but there are many variations. This example highlights some of the difficulties and the subjective nature of the analysis. All the figures vary. That is because the figure

has to reflect not just the building, but also the immediate and extended environment, condition and all the other features. In practice that means that the difference between 2 and 53 Acacia increases by £27.78 per m^2 (3.3%), or £2500 in capital terms. This reflects that No. 53 has a garage but No. 2 does not. Nos 2 Acacia and 24 Lilac also differ in size. The difference in price per square metre is £23.81 (2.8%). This has removed the gross size differential and focused on the key differences of the brick garage and the marginally poorer location. The capital sum is £2500 (2.7%). This would not cover the cost of the average brick garage. Therefore, it may be appropriate to analyse this as £4000 (4.4%) enhancement for the garage. This still would not cover the cost of a brick garage but it may be all that the area will stand in value terms. This has to be counterbalanced by £1500 (1.6%) depreciation for the location. Surveyors in different parts of the country may argue with this but it must be remembered that all these figures are relative and therefore percentages may be more appropriate as shown. The variances can only be arrived at by analysing previous examples and a key feature is to link that analysis to price and to keep doing that. Failure to monitor the market would lead to a self-perpetuating situation whereby the surveyor would have a stable or even declining set of figures. See figure 2.11 for examples of property in the spreadsheet area.

2.4.5 Residual approach

The traditional way that developers establish the value of land is to start with the sale price of the properties that they intend to build. This is multiplied by the number of units to give the gross value of the site. All the costs are then deducted. The remainder is the price that the developer should be able to pay for the basic site. More detailed examples of this approach are given in works such as *Valuation and Sale of Residential Property* by David Mackmin (1994).

Figure 2.11 Houses typical to those included in the example spreadsheet in figure 2.8.

Although the land element is an integral part of the building, the principle of the residual approach should not be lost. The features that are attached to it at a point in time are because the essence of valuation of property relates to how the land has improved this. In reality it is the land that will remain long beyond the works that are undertaken on it. The proof of this is seen throughout time. Buildings have been continually erected and then demolished as they have become obsolete and others have been built in their place that are more appropriate to the time and needs of the occupier.

A garden can be landscaped to produce a masterpiece. Although the underlying architecture will remain, a couple of growing seasons without any maintenance will render the improvements worthless. Another example of a feature that is related to the land could be restrictions on use such as planning conditions. A condition that the dwelling should retain certain historic features could have a significant impact on the value of the land and buildings and any changes or improvements that could be made. There are many other features that fall within this category and may restrict or enhance current value.

2.5 Effect of condition on value

2.5.1 A review of factors that affect value

This section will review, by reference to recent research and practical experience, how clients perceive the benefits that accrue from expenditure on their property. In trying to understand the impact of condition upon value it is important to set this within the context of other issues that affect value.

Interpreting the value of the benefits to an individual property tends to come from practical application. Very little has actually been written about the process in the UK probably because the variety of benefits is so great they are not well recorded. If they are, they are jealously guarded because these factors form the basis of the surveyor's art and his business. Where property is similar (i.e. a row of terrace houses) then there may be few attributes that differentiate one property from another. The sunny side of the street may be known to be of significance to locals. Where there are only small back yards then orientation to the sun is very important. It is probably of less worth to the occupier who has a larger garden where the shadows do not extend over the whole plot. Conversely local knowledge about a former resident may also be sufficient to give the house a stigma that will make it unsaleable. The key learning point here is that benefits are of differing value to differing individuals and this may be related to location or other attributes.

2.5.2 Home improvements and value

In May 1997, 1200 questionnaires were sent to estate agents throughout the UK (Which 1997) asking whether 20 popular home improvements would influence the price and saleability of a home. The findings are shown in figure 2.12. A literal interpretation of these findings is inadvisable. This is because the appreciation in value of a property is more likely to be associated with an improvement in the general property market than as a direct result of a single home improvement. House prices have on average more

How home improvements can affect value a lot	*How home improvements can make a house easier to sell*
Will increase the value a lot	**Will make selling much easier**
1. Two-storey extension	1. Central heating
2. Central heating	2. New kitchen
3. Garage	3. Garage
4. New kitchen	4. Off-road parking
Will increase the vaue a little	5. Double glazing on a modern house
5. Conservatory	6. New white bathroom suite
6. Off-road parking	7. En-suite bathroom
7. En-suite bathroom	**Will make selling a little easier**
8. Loft conversion	8. Paint exterior woodwork
9. Double glazing on a modern house	9. Landscape gardens
10. New white bathroom suite	10. Conservatory
11. Landscaped garden	11. Two-storey extension
12. Original fireplaces	12. Original fireplaces
Will make no difference	13. Redecorate in neutral shades
13. Replace sash windows with PVCu ones	14. Loft conversions
14. Redecoration in neutral shades	15. Hanging baskets and window boxes
15. Swimming pool	**Will make no difference**
16. Paint exterior woodwork	16. Roof insulation
17. Roof insulation	17. Replace sash windows with PVCu ones
18. Redecorate in distinctive shades	18. Redecorate in distinctive shades
19. Hanging baskets and window boxes	19. Knock through rooms
20. Knock through lounges	20. Swimming pools

Figure 2.12 Table showing whether home improvements affect the value of property (Source: *Which* 1997. Published by the Consumers Association, 2 Marylebone Road, London NW1. For more information phone 0800 252100).

than doubled within the last 15 years (even allowing for the recession in the late 1980s to early 1990s). Consequently £1000 spent on central heating back in 1983, on a house worth £50,000 would have been lost within the overall appreciation of value. The average value could now be in the region of £100,000. This inflationary action has been active since the post-war building boom. For example, a 4-bedroomed detached house in South Yorkshire newly built in the mid 1960s was originally bought for £4,500. In 1997 it was sold for £105,000 (2233% increase!). Future predictions are for a better-managed economy and for a more stable level of inflation. Consequently, it would be unwise to rely too heavily on the market to overcome errors in judgement on home improvements. The following guidance may be helpful.

A home improvement should be:

- **of benefit to the actual owner and should give worth to that individual.** If the improvement can be replicated cheaply then it is unlikely to add value. An example would be if instead of building an extension to a house the same amount of accommodation could be acquired by buying another house more cheaply. Alternatively if a neighbouring property has had its windows replaced by PVCu double glazed windows and those of the subject property are single glazed and in poor repair, then the value of that property is likely to be depreciated by the cost of the replacement windows. Situations are rarely as clear cut as this but this gives an indication of the principle.

- **of good quality and the appearance should be in keeping with the design of the actual property and of those around it.** The classic flat-roofed two-storey extension on a 1930s semi-detached house may give benefit to the owner but its appearance is not up to modern-day expectations. Therefore its added value will be minimised. The addition of a pitched roof will be expensive but will certainly enhance the appearance. It could also reduce maintenance costs and therefore will add some element of value. Whether this will equate to the money spent on the improvement will have to be judged in the individual circumstances. Another aspect would be where the extension greatly increases the size of the accommodation to the extent that it is not characteristic of the type and socio-economic grouping of the person who might want to live in that area. Then this will not add value, merely benefit the individual owner. The same applies to the detached property positioned in an area of terraced houses.
- **should provide additional functionality.** One example given by the Consumers Association (*Which* 1997) shows a conservatory on a reasonable quality bungalow owned by an elderly couple. They considered the main benefits were 'a pleasant place to have their dinner and entertain friends'. The value of the property was put in the region of £145,000. This represented an increase in value (partly due to inflation) of £15,000 which covered the cost of the investment (11.5%). If this had been a £30,000 terraced house, expenditure of £10,000 (33%) or more on a conservatory forms a high percentage of the base value. It is unlikely to give that degree of added value especially if the durability of the conservatory is considered. A brick-built dwelling has a minimum life expectancy of 60 years, whereas secondary features such as softwood window frames have probably a life expectancy of 10 years. A typical conservatory falls somewhere between the two. Consequently the terraced house owner has to look carefully at what level of functionality can be achieved by the type of improvement. A summary of guidance to the surveyor or owner is shown in figure 2.13.

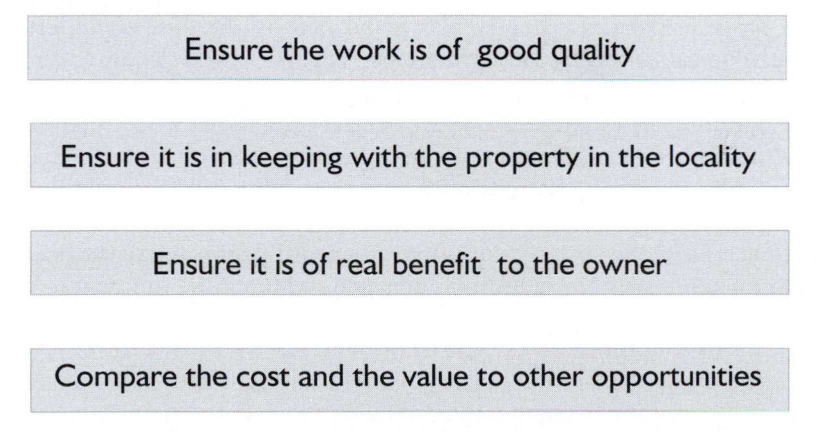

Figure 2.13 Summary of guidance when assessing the relative value of home improvements.

2.5.3 Condition and value

The main purpose of this section is to concentrate on the influence that condition has on value. Firstly, a distinction can be drawn between home improvements and condition. In some cases they can be the same as with the example of replacement windows. But in one important respect they are different. The works to retain the condition are essential if the value of the property is to be maintained. Provided they are done correctly (i.e. sustainably) then they will be of benefit to the owner. The owner may not agree with this especially if repairs were unexpected and the money spent on them was allocated for other matters which gave a more tangible return, e.g. a family holiday or new car.

The effect of repair works on value can be as unpredictable as home improvements. One thing is for certain – a property allowed to fall into disrepair will lose value compared with a better maintained one. It is this latter point that is important and relates back to the definition of 'value' from the dictionary – a fair equivalent, the degree of this quality (see section 2.2.1). In undertaking the appraisal of a residential dwelling, there are three levels of condition that need evaluation (see also figure 2.7):

(a) the defect that is so serious it will, or has caused major structural failure (e.g. foundation failure);
(b) the defect that is serious and could, if left unattended, lead to a partial structural failure. An example would be defective roof covering that is allowing water penetration. That could lead to rot occurring in structural timbers. In other words those defects that render the dwelling not 'wind and water tight';
(c) the defect that is not serious, mainly cosmetic and would only affect appearance if left unattended.

Example (a) is so serious that the value of the dwelling would be seriously eroded if left unattended resulting in a total loss with demolition being the only option. In such cases it is only worthwhile expending money on the property if the cost of the works do not exceed its completed and fully repaired value, excluding the land.

In respect of (b) the same criteria applies. In this case it is more likely that the nature of these repairs is such that they currently affect individual elements of the property. Therefore the value of the property will be eroded by the cost of undertaking the repair relative to better repaired ones. If all the properties in the locality have the same defect then the comparison on value has to spread to other localities to find comparables that are in better repair (see figure 2.14). In market situations where there is pressure on supply then this cost equation could be reduced as individuals will have differing views on how to undertake the repair, or how to pay for it. Ultimately the ability to pay influences value.

In the final area (c) the costs of works may not be fully reflected in the price or value as they may be indicative of on-going maintenance and therefore inherent in the pricing of that type and age of property.

The impact of condition can also be a complex matrix of two or more attributes, which affect value. For example condensation, which has a perceptible impact upon the occupier, can be a result of a combination of factors including location (colder region), the quality of finish within the dwelling (level of insulation in the dwelling), the type of property (non-traditional type) and the use and layout of the accommodation (high moisture production).

How all of this should be reported depends upon the needs of the particular client. The lender will want to know of high-risk situations that could influence the security of the loan. They will want these elements reduced if at all possible, for example the damp staining to a rafter, indicative of a leak to part of the roof that can only get worse and could ultimately lead to dry or wet rot. The potential for structural failure needs to be assessed and reported accordingly. The occupier may choose to review a report in a different light dependent upon the perception of benefits that will accrue from the property. If the leak is not actually affecting the occupier's use of the property (i.e. the water is not dripping onto the bed) then they may wish to defer an expenditure in preference to another which will produce a more immediate and perceivable benefit, such as redecoration.

The folly of this decision is apparent to those who view the property in a more commercial way. Consequently it is important in all cases to consider the needs of your client and where practical ensure that advice is given in the right context and prioritised if appropriate. The following section gives an example of categorising the condition of a property that is being considered for a loan.

2.5.4 Distinction between lender and purchaser

Figure 2.7 shows the assessment of condition relative to value. It attempts to classify certain types of defect that would have a significant impact upon the saleability of the property and the decision to lend money. Situations where the surveyor is unable to give a value are rare, but they do occur. Special attention should be given to avoiding the need for unnecessary requests for specialist reports. Where the surveyor can make an accurate diagnosis on site and the full extent of the repair can be identified then the impact upon the value should be determined.

Figure 2.14 Properties in one street in a similar condition.

The purpose of figure 2.7 is to give practical examples for guidance. Situations will always vary and care should be taken in interpreting the guidance. Individual lenders may have specific instructions that will override this less formal type of advice.

The situations relate to sections within this book and the classification to actions that need to be considered when acting for a lender. Again there may be specific instructions that apply to individual lenders that will override this guidance.

Definitions

The following is a summary of the terms used in figure 2.7:

- **Decline for lending purposes** – In these cases the defects are so serious that they question the future stability or durability of the dwelling. There may be occasions where only a comprehensive refurbishment programme could provide a solution.
- **Further investigation and impractical to provide a current value** – The extent of the defects are quite serious and cannot be fully diagnosed so it is not possible to quantify the damage and the cost of the works. The surveyor therefore needs more information to rank the property relative to its comparables. Figure 2.15 shows a tree immediately adjacent to a cottage of some age. The property is displaying signs of movement that may or may not be associated with the tree. This is an example of a case where further investigation would be sensible. The vendor would be well advised to obtain independent reports before sale!
- **Further investigation required but quantifiable** – The surveyor can diagnose the extent of the defect and therefore quantify the likely cost so the needs of the lender can be satisfied. The purchaser will need to be kept fully aware of the implications of the costs of the necessary work.

Figure 2.15 The proximity of the tree to the cottage should be of clear concern to any surveyor and a justifiable reason for recommending further investigations.

- **Defects form part of a programme of routine maintenance** – This helps complete the picture of the type and general condition of the property. Probably of limited significance to the majority of lenders but helpful to the purchaser and to the future saleability of the dwelling. May indicate areas that could give rise to an insurance claim in the future.

It can be seen from figure 2.7 that in respect of the lender the key areas of concern can be categorised as follows:

- dampness
- movement
- timber failure
- health and safety
- cost and its impact upon value.

The significance of the defects is whether they impair the quality of life of the occupants. Some are related to the same cause, so they form a matrix that will combine to produce a plan of action for the surveyor. Questions must be asked whether the defects pose a risk to the health and safety of the occupants. If so, is this imminent or an acceptable future risk and therefore what is the impact upon value ? How much will it cost to repair? The significance of this latter point will be how does this rank against other property within the same area. The defect may be common to all property and be an acceptable measure supported by qualified contractors. A good example of this has become the insertion of a damp proof course in a 1900s terraced house. Plenty of contractors are capable of undertaking the work with minimum inconvenience and the costs are well known, so a quantifiable impact upon value (see figure 2.7).

The purchaser needs different information to the lender. The lender is concerned that the loan can be repaid and the property has some influence on this, but it is mainly the purchaser that provides the basis for the security. However, the purchasers have needs from the property as a home that may be impaired if the defects affect their use of the property. It is important to draw a distinction between those who view the property in a more commercial way, consequently it is important in all cases to consider the needs of your client and where practical ensure that advice is given in the right context and prioritised. Figure 2.16 shows a property leaning at a considerable angle, built on sand; it may well provide reasonable accommodation for the occupier and they may have taken internal measures to ensure everything is on the level. However, saleability is restricted and the commercial view must be taken if a loan is required. Would a prospective buyer prefer this house or one genuinely on the level. What reduction in price would they expect to compensate them for the movement?

Achieving a balance between genuinely persuading the client against purchasing a property that is not suited to them is an art in itself. The key is that even a dwelling in the worst condition can be rectified by the right client. Therefore understanding your client is the most significant issue.

Figure 2.16 No it isn't the way this photograph has been printed! The front fence posts are vertical, it is the house that is leaning over!

2.6 Valuing in the UK

The principles of valuation apply equally with England, Scotland, Wales and Northern Ireland. The legal process may be slightly different in Scotland and there are varying forms of tenure including feudal and payments such as chief rents. However, for the purposes of providing a value they should be treated in the same way. The question should be how does the variable impact upon the saleability and consequently the value. Comparison to similar situations should provide a resolution.

References

Beaumont v. *Humberts* (1990). 2 EGLR 166.

Corisands Investments Ltd v. *Druce & Co* (1978). EG 315, 318.

Kelly, R. (1998). 'Designer housing scares off customers'. *The Times*, 3.10.98, London.

Mackmin, D. (1994). *Valuation and Sale of Residential Property*. Routledge, London.

Rees, W. H. (1992). *Valuations Principles into Practice*, 4th edn. Estates Gazette, London.

RICS (1995). *Appraisal and Valuation Manual*. RICS, London.

RICS (1997). CSM def. of investment.

Roberts v. *Hampson* (1989). 2 All ER 504.

Smith v. *Bush* (1990). 1 AC 831.

South Australia Asset Management Corporation v. *York Montague Ltd* (1996). 2 EGLR 93

Which 1997. Special Report: Home Improvements. October 1997, Consumers Association. London.

Part II

The survey and identifying the problem

Chapter 3

Carrying out the survey

3.1 Introduction

The type and extent of inspections and surveys have been well described by professional institutions and the courts. This chapter will outline good practice rather than re-interpret any particular standard survey.

To simplify discussion in this chapter the term 'survey' has been adopted in its broadest sense. It is taken as covering both mortgage valuations and Homebuyers surveys as it identifies good practice rather than trying to further explain any particular standard activity. More precise descriptions of both types of activity are given in the RICS's *Appraisal and Valuation Manual* (RICS 1996).

The term 'surveyor' has been adopted and applies to those professional people who see themselves as either 'valuers' or 'surveyors'.

This chapter looks at how to carry out a survey from first principles. For many experienced practitioners this may appear too introductory but it has been included for two reasons:

- the book is designed for a broad readership including cognate and non-cognate entrants into the profession. For these readers introductory material is necessary;
- for experienced professionals it will be useful to 'check' current practice against what is deemed to be acceptable practice.

Any successful survey will consist of two stages:

- preparatory work
- carrying out the actual survey itself.

Both of these will be looked at in more detail.

3.2 Preparatory work

A survey develops in stages and begins long before the surveyor actually enters the dwelling. These stages are described below.

3.2.1 Office preparation

This could be considered as an informal 'desk-top' study. This stage is very important for new surveyors and those who are working in a new geographical area. Once the address of the property is known there are a number of quick checks that can produce some useful information. These can include:

- asking colleagues in the office if they are familiar with the area or the type of properties;
- referring to old and current Ordnance Survey maps for the area. For several urban areas, maps from the early 1900s have been published commercially. A few minutes spent looking at these can give a quick insight into the history of the site. This might be specifically useful in respect of building movement and contaminated land issues as the position of old quarries, pits, ponds, streams, etc. could be revealed;
- looking at reports produced by the surveying organisation from the files of similar properties in the area;
- check any general files or special databases that the office maintains or subscribes to.

3.2.2 Equipment

Not having the proper equipment on a survey is no defence against an action of negligence. The list of surveying equipment will vary depending on personal preferences, the organisation the surveyor works for and the sort of work to be tackled. Assuming commissions equivalent to Homebuyers surveys will be carried out, a typical list could include:

- measuring tape. A 5–7.5 metre retractable steel tape is suitable for most jobs but occasionally a 20–30 metre fabric tape will be needed;
- an electronic moisture meter with spare batteries. This should be calibrated every time it is used;
- a spirit level – a small-hand held level is useful for checking alignment of door frames, window sills, etc. A 1 metre 'bricklayers' level' will be essential to check the levels on floors, the verticality of walls, etc.
- a plumb bob and line – a quick and effective way of checking the verticality of walls;
- equipment for simple 'opening up' such as:
  - a robust claw hammer;
  - a large flat-head screw driver;
  - a 'wrecking' or crow bar (450 mm long);
  - bolster or cold chisel;
  - two large and two small drainage inspection keys;
- a bradawl or other suitable probe;
- a surveyors' ladder – four sections, minimum 3 metres long. Others that retract or fold are useful but check whether they are suitable. They have more moving parts and so for some people they can be difficult to operate;
- powerful inspection torch and spare batteries;
- suitable clipboard and paper for notes and sketches;

- suitable camera (with flash) and spare film. Some surveyors prefer to use Polaroid for instant records of significant features;
- binoculars (× 8–10 magnification);
- health and safety equipment:

  - personal attack alarm and spare battery;
  - mobile or detachable car phone;
  - first aid kit;
  - safety helmet;
  - face mask with disposable filters for loft inspections, etc. These must be a suitable specification for the dust that can be expected;
  - safety goggles;
  - pair of protective gloves for lifting inspection chamber covers, etc. Many organisations are now recommending disposable gloves. This is because once the gloves have been used if they are put back into the survey tool kit they can contaminate other tools;
  - disposable rubber gloves for dirty or unhealthy locations;
  - appropriate steel top capped wellingtons or other suitable safety footwear.

This may seem a lot of equipment but all of it could be used on a typical Homebuyers survey. It does not need to be carried around room to room but it should be available if required.

3.2.3 Slow drive around the area

A cursory inspection of the neighbourhood can reveal useful information. This could include:

- house types and general condition, i.e. how many roof coverings have been replaced, structural movement to other properties, replacement windows, etc.;
- type and number of trees in the area;
- local environmental hazards or nuisances such as factories, scrap yards, traffic rat runs, etc.

This can be gleaned from the car during a slower than normal drive to the property.

3.2.4 A long walk up the path

During the short time it takes to park a car outside a property, up the front path and knock at the door, a sharp-eyed surveyor can pick up a significant amount of useful information:

- the type and general condition of the house;
- relevant garden features such as proximity of large hedges and trees, etc.;
- any neighbouring owner issues such as boundary wall problems, possible nuisances, trees the other side of the boundaries, etc.

By the time the front door is opened, the surveyor should have a clear mental agenda of issues and 'trails of suspicion' that need to be followed and investigated.

3.3 Relationship with the vendor

A surveyor may be just one more person in a long line of visitors that have trodden a path through the dwelling. If the property has been on the market for some time then the vendor can be jaded by the whole experience. Although there is no contractual relationship between the surveyor and the vendor (although this may change in the future), a positive relationship can reveal a lot of useful information about the dwelling. It is surprising how forthcoming a proud owner can be even about the deficiencies in their home. On the other hand an over-attentive owner can get in the way of a surveyor making it difficult to approach the task in a methodical and systematic way. The following tips may help:

- If possible talk to the owner personally to arrange the time of the survey. During this brief conversation the following issues can be discussed:
    - how long the survey will take and which parts of the dwelling that will need to be inspected. This will allow them to move stored items, open access hatches, etc.
    - ask general details about the property, e.g. its type, size, age, etc. These are useful facts that can help inform the office-based preparations;
    - inquire whether any work has been carried out to the property such as extensions, alterations, repairs, etc. If yes ask them to sort out any documentation they may have. This could include guarantees, planning and building regulation permissions, etc.;
    - whether there are any bats in the property (see section 6.4.3).
- Turn up on time with proof of identity. Remember, the person that gives you access might not be the person who you originally talked to (see section 3.5.4 on personal safety).
- After arriving:
    - explain the purpose of the survey. Not all vendors understand the complexities of house sale and purchase procedures;
    - outline your approach to the survey, how long it will take, what rooms and spaces you will need to inspect, etc.;
    - ask the owner to take 5 minutes to show you around the property. It gives a quick introduction to the house and gives an opportunity to ask gentle but probing questions. Arrangements can be made about inspecting areas that are occupied, i.e. sleeping occupants in bedrooms, etc.;
    - once this initial and brief walkabout is complete, politely inform the owner that you want to carry out the rest of the survey on your own.
- During the survey take great care not to damage the property in any way. Typical examples have included:
    - marking wall and ceiling surfaces when assembling ladders for loft inspections, etc.;

- soiling carpets with muddy boots, dog faeces, etc;
- disturbing decorations when unscrewing access hatches, using moisture meters, etc.;
- knocking ornaments off mantelpieces, shelves, etc.

3.4 Carrying out the survey

3.4.1 Procedure

Experienced surveyors will have their own approach to carrying out a survey (Wilde 1996, p.19). The method described below is just one possible alternative. The main rule is that it must be methodical and systematic. The same routine should be followed in every survey. This can help create an impression of competence if challenged in court. A typical procedure could include:

- Work around the dwelling internally:
    - inspect the loft space first (if there is one);
    - inspect the rooms on the uppermost floor by working around that entire floor in a clockwise direction;
    - finish on the landing and inspect the stairs down to the next floor;
    - follow the same process on each floor down to the lowest one;
    - inspect any cellar or sub floor void if accessible.
- In each room, inspect the various elements in the following sequence:
    - ceilings;
    - walls (including skirtings);
    - floor;
    - windows and doors;
    - heating;
    - electricity;
    - plumbing;
    - other amenities (e.g. toilets, basins, sinks, etc.);
    - fittings and fixtures (e.g. cupboards, fitted wardrobes, fireplaces, etc.);
    - any unusual or special features.
- External inspections. It is a good idea to go outside towards the end of the survey to avoid the problem of tramping muddy boots around the house. This should include:
    - the main elevations including all secondary elements (doors, windows, etc.);
    - observable roof surfaces including chimneys stacks, etc. On flat roofs lower than 3 metres this could include a ladder inspection. Some roof surfaces can be inspected from the windows of upper rooms. Rainwater goods should be assessed at this point;
    - any significant garden features such as retaining walls, etc;
    - all outbuildings;
    - boundaries, fences and gates;
    - any special features such as rights of way, etc.;

- drainage inspections, etc.

This systematic approach is fine in theory but some defects may show signs in more than just one space (Wilde 1996, p.18). Therefore, the surveyor will occasionally have to inspect some areas a number of times to properly follow a particular 'trail'. Once the survey has been completed, always inform the owner especially if you are leaving.

3.4.2 Inspecting flats and maisonettes

The different types of standard surveys usually contain a clear description of what parts of a flat or a maisonette should be inspected. Although this may vary between organisations the key features include:

- **externally** – the exterior of the whole property that is accessible from the common areas. This is to gauge the general state of repair and includes the normal rules governing flat roofs. The extent of this assessment was described by Holden (1998a, p.26) who suggested that the surveyor must '... take reasonable steps to establish any specific problems which the legal advisors should raise with the management company.' Outbuildings should also be inspected in a superficial way apart from those used for leisure activities. These are normally excluded from most surveys;
- **internally** – the interior of the actual dwelling, the communal area from the front entrance door of the whole building to the flat entrance door and the remainder of the staircase. This will include any roof spaces that are accessible from within the actual property. The same rule also applies to floor voids and cellars.

3.4.3 Accessibility issues

The *Homebuyer Practice Notes* (RICS 1997) defines the term 'accessible' as '... visible and readily available for examination from the ground and floor levels, without risk of causing damage to the property or injury to the surveyor.' This provides a good starting point for articulating precisely how far a surveyor should go during a survey. Whatever the type of commission, the client is entitled to know what the surveyor will or will not look at. The sections identified below outline practical examples that defines the term 'accessible' more accurately:

- **inspecting timber floors** – see section 6.2.5;
- **inspecting drainage systems** – see section 10.14.5;
- **inspecting roof spaces, lofts and voids** – see section 7.2;
- **inspecting flat roofs** – see section 7.7;
- **testing services** – see section 10.1.4;
- **moving furniture** – most occupied dwellings are furnished and this will limit the extent of a survey. The RICS (1997) advise that '... furniture ... and other contents are not moved or lifted.' Looking at this from a customer's perspective, where there are small movable items most people would be disappointed if a surveyor refused to move them. Holden (1998a, p.25) points out the difference between doing the minimum required and '... the little bit extra which you may feel will serve the best interests of the client ...'. Therefore, it would be reasonable to move the following:

- easy chairs, sofas, beds, etc. where there is enough space, they are on castors and easy for one person to push aside;
- small coffee tables, magazine racks, dining room chairs, standard lamps, waste baskets, etc.;
- a few ornaments and other possessions (say 3 or 4 items) to allow the furniture to be moved.

In all cases the owner's permission must be obtained before anything is moved. Larger items such as sideboards, dressing tables, wardrobes, fridges, freezers, washing machines, televisions, etc. would be too heavy to move and so should be left in place. It is not possible to be more precise as each case must be judged on its own merits. A surveyor should always apply the test of reasonableness.

3.5 Health and safety

Over the last few years, safety in the construction industry has become more important. The *Construction, Design and Management (CDM) Regulations* (HMSO 1994) has made health and safety central to all new construction, repair and maintenance operations. This section will look at health and safety issues relevant to those carrying out appraisal of residential property.

3.5.1 Reasons for a health and safety policy

There are a number of reasons why all working environments should be as safe as possible:

- **moral** – unsafe conditions will have an impact on the health, well-being and life expectancy of the individuals involved. This will in turn affect family life and ultimately have an impact on the whole community;
- **economic** – unsafe conditions will result in economic loss through:
  - working days lost;
  - the cost of investigating and administering claims and incidents;
- **legal** – legislation has created a criminal liability on the company, the manager and the employee where the legal requirements are contravened.

The main legal instrument is the Health and Safety at Work Act 1974 (HMSO 1974). Under section 2(1) it states: 'It shall be the duty of every employer to ensure so far as is reasonably practicable, the health and safety and welfare at work of all employees.' Briefly it creates the following duties:

- the provision and maintenance of safe plant and systems of work;
- adequate arrangements to ensure safe use, handling, storage of equipment, goods, etc.;
- appropriate information, instruction, training and supervision of duties, etc.;
- provision of a safe place of work including safe entry and exit;
- provision of a healthy working environment.

This duty not only falls on the employer, the employee must adopt a common sense approach including:

- to take reasonable care;
- to co-operate with the employer;
- not to mis-use safety equipment.

These are important points for the individual surveyor. Following an accident, if the investigation reveals that the surveyor did not take sufficient care then his or her legal rights may be undermined. This is important for younger surveyors. Occasionally youthful 'bravado' leads to silly risks being taken. For example, putting a chair on a table is not a safe way of getting into a loft. A surveyor does not have to fall from a great height to sustain an injury that will ruin a working life.

3.5.2 Planning for health and safety

Because surveyors inspect so many different types of property in a variety of conditions, it is impossible to identify all the possible risks that may have to be faced. Therefore all surveyors, whatever their position in a company, should be able to assess the risk of any particular activity. This process involves:

- identifying those areas of work that could pose a risk;
- listing the controls that should be put in place to eliminate or reduce the risk.

It will be impossible to eliminate all risks but a measure of the extent of the precautions required can be calculated by a test of 'reasonableness' and 'practicality'. This test balances the risk (defined by its severity and frequency) against the cost of eliminating that risk (in terms of money, time and effort). In other words what is the chance of a serious accident happening and how often is it likely to happen? If it is likely to be frequent then it is worth spending time, effort and money on reducing the risk.

A key feature of the working practices of a surveyor is that the place of work is often someone else's home or building. This affects the measures that can be taken to ensure a safe environment. Therefore it is essential that any working practices are underpinned by a clear health and safety policy. This should form the basis of an ideology that is at the core of all induction courses and training initiatives. A typical approach could be:

- identify all the areas of work;
- identify the potential hazards;
- for each hazard match the severity of the potential outcome (i.e. from death down to minor or nil impact on health) to the frequency of the event;
- for each of these events review what procedures exist to minimise the risk. The aim is to reduce the risk to a low level (i.e. minor or nil injury);
- put a regular review procedure in place so the procedures can be assessed from time to time;
- maintain an accident and incident record and evaluate each incident against the current procedures.

Using ladders

One of the greatest risks that surveyors face is associated with the use of sectional ladders when gaining access to loft spaces. Although this sounds elementary, the authors suspect that few surveyors are formally taught how to use ladders properly. The main issues are highlighted below.

The equipment – most surveyors use sectional aluminium ladders consisting of three or four sections that extend to between 3 and 4 metres in length. Most models slot together and are secured into position by specially fitted bolts and wing nuts. These should conform to the appropriate standards and be complete.

Use of the ladder – there are a few key rules:

- Each time the ladder is used, the surveyor must check that it is in good condition and has not been damaged when other people may have been using it. All the sections should be properly slotted together and all the bolts and nuts properly tightened.
- Where possible the ladders should be safely positioned using a 1/4 ratio. A 4-metre-high ladder will need to be set 1 metre away from the wall (see figure 3.1). This may pose particular difficulties where loft access is concerned. Often it is very difficult assembling a sectional ladder on a small landing. Because two sections fail to reach the hatch but three sections extend into the actual loft space the surveyor always has to face the dilemma of how to open the hatch itself. There are two options:
  - Prop the ladder against an adjacent wall so that the loft access hatch can be removed. Then an additional section can be added to the ladder extending it into the loft.
  - The ladder can be propped against the reveal of the hatch opening so the surveyor can climb up, remove the hatch, climb down and reset the ladder and enter the roof space but the reveal has to be deep enough to support the ends of the ladder.

  The choice will depend on which is the most safe.

- The ladder should be stood on a firm and even base. It is best not to wedge one side of the ladder on sloping ground. Try to level up the ground first. Where on soft ground it is best to put down a support board (see figure 3.2).
- Before climbing the ladder make sure that the underside of footwear is free from soil, etc. Be very careful if the rungs become wet, icy or greasy.
- Wear suitable footwear. Training shoes, plimsols or sandals are not appropriate.
- If you intend to get off the ladder (to inspect a flat roof or get into a loft space) it should ideally extend about 1 metre above the eaves or the ceiling level. This will make getting on and off the ladder safer.
- When on a ladder, never over-reach in an effort to inspect a specific feature. Get down and reset the ladder.
- If possible carry any equipment in your pockets when climbing a ladder. If your hands are carrying a torch, damp meter, tape and clipboard it will be difficult to steady yourself and it will be easy to drop one of them on someone's head!
- Where the ladders have to be placed on public footpaths, roads, etc. the base should be properly secured to prevent accidents to both the surveyor and the public. Although placing a sandbag over the bottom rung might help, having another person at the bottom 'footing' the ladder is the ideal solution but this may be difficult to organise on most standard surveys.

These precautions may seem like stating the obvious but falling from a ladder or being hit by one that has fallen over hurts!

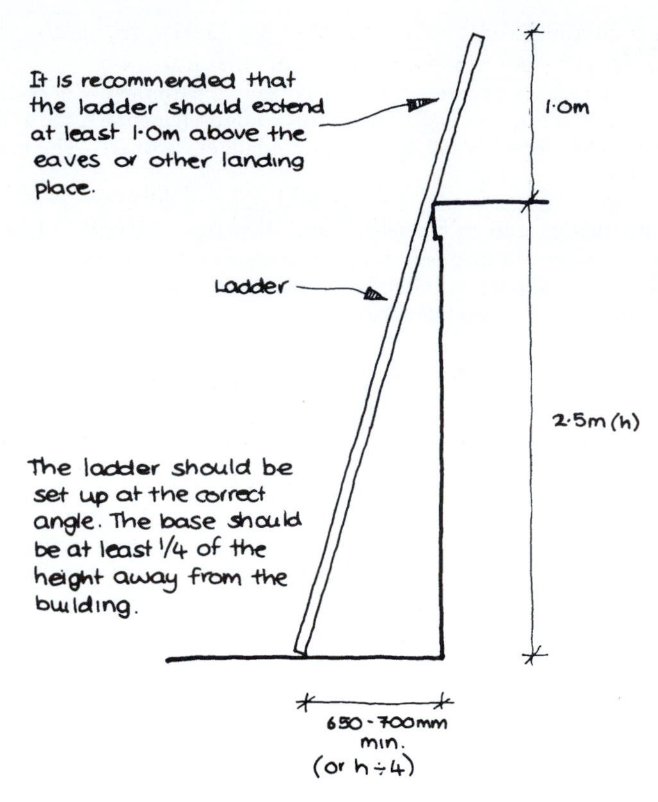

Figure 3.1 Correct position of a ladder when inspecting a flat roof.

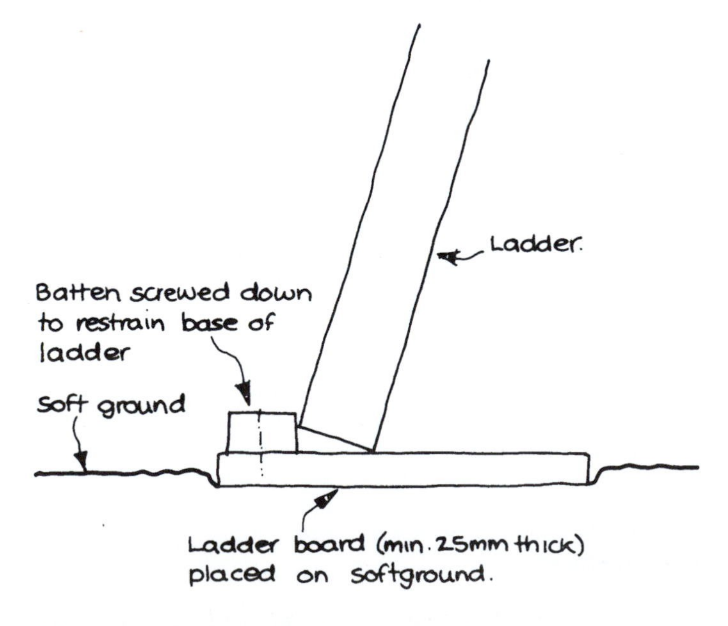

Figure 3.2 Use of ladder board at base of ladder when on soft ground.

Establishing such a philosophy will enable individual surveyors to carry out 'on the job' risk assessments as potential dangers present themselves.

3.5.3 Safe working practices

As previously mentioned, the surveyor's working environment will be as variable as the properties inspected. Despite this it is possible to identify some key areas and the attendant risks.

The office environment

All aspects of a surveyor's job must be evaluated and there can be many risks in the office environment. For example, particular emphasis needs to be given to the kitchen area and the use of electrical machinery. All equipment should be properly maintained by a reputable company under a formal service agreement.

Handling equipment

A high proportion of injuries and lost working days are due to injuries caused when handling equipment or objects. This could include:

- lifting heavy objects such as ladders, inspection chamber covers and furniture;
- less obvious tasks such as twisting to get something off the back seat of a car or reaching into the boot;
- getting a heavy box of paper from a high shelf in the office.

The key factor is to take great care when handling all types of equipment and to make sure individuals do not have to overstretch or lift objects that are too heavy.

Use of specific equipment

There are two types of equipment:

- that which helps a surveyor to do the job more effectively such as damp meter, ladders, measuring tapes, inspection chamber lifting tools, etc.;
- that which gives the surveyor additional protection such as safety helmet, protective face mask, steel toe-capped shoes or boots.

In both cases it is important that they are tools that are up to the task, for example a face mask that will stop the appropriate particle size and shoes that will stop a nail in a floorboard from injuring a foot. The specification of the equipment should be checked carefully.

3.5.4 Personal safety

Due to the solitary nature of the work, surveyors can be vulnerable to personal attack and injury. The best advice on how these risks can be minimised has come out of the work of the Suzy Lamplugh Trust. This is the national charity for personal safety and takes a positive approach to the aggression and violence that professionals may have to face. They have collaborated on a number of publications aimed at people who work alone in other people's homes. Much of the advice is targeted at health service workers and is relevant for the surveyor. This has been summarised below (Bibby 1995).

Carrying out a survey in an occupied premise

As a general rule when a surveyor is in someone else's home it is important to remember that it is their home and their territory that is being invaded. Therefore they are in command and this should be challenged as little as possible.

Before you go on the survey think about the following:

- Do you have to go alone? This may be difficult when resources are scarce and workloads are heavy but two people are much safer than one. In many cases, accompanied surveys can be part of a new entrant's supervision procedure or justified by formal quality assurance requirements. If it presents difficulties then two-person surveys could be restricted to areas where the risk is considered the greatest.
- Before leaving your workplace, make sure somebody knows:
    - where you are going and what your plans are;
    - what time you expect to finish the survey and return to work or arrive home;
    - arrange to check in with someone in the office to confirm you have returned. You should always check in even if you go straight home. A phone call from the safety of your armchair may stop your colleagues worrying about you.
- Always take a mobile phone with you that is properly charged and carry your personal attack alarm.
- Go in daylight wherever possible. Organise your day so you are travelling in darkness rather than surveying buildings in a fading light. In the winter this may mean completing the survey by mid-afternoon.
- When you arrive at the property:
    - think about the location of the dwelling. Is it a tower block, down a country lane or on a one way street? What dangers are associated with the location?
    - park in a position where it will be easy to leave quickly. Always lock your car and do not leave valuables or bags, etc. in full view on the seat;
    - when the door is answered say who you are, why you are there and show them your ID or business card;
    - check who you are talking to and make sure it is the same person you arranged access with. If not, consider carefully whether you should go in;
    - let them know how long you will be, where in the house you will need to go and then wait to be invited in. Be assertive but remember you are in someone else's territory.
- When carrying out the survey:
    - get to know the layout of the house so you can get out quickly if you need to;

- take only essential things into the house that can be carried around with you. Do not leave possessions in several different rooms in the house. This may delay your departure if you need to leave in a hurry;
- do not take documents into the property you don't want the occupant to see. Be careful when using a personal recorder to dictate notes. Owners may be hurt and offended by even the most objective assessments of their home;
- different people live in different ways. Try not to react to the condition of the dwelling itself even if it is smelly, very untidy or plain dirty. Do not be judgemental and if you are, do not let your feelings show;
- if there are dogs or other animals that appear threatening, politely ask for them to be kept out of the way. Many surveyors have been badly bitten by large dogs who 'wouldn't hurt a fly';
- if you accidentally go into bedrooms, bathrooms and WCs where occupants are sleeping or changing, apologise and leave the room quickly;
- if any of the occupants appears drunk, is aggressive or over-attentive then say that you have an urgent appointment and leave straight away. Do not look back, get in your car and drive away calmly.

The Suzy Lamplugh Trust advises that you should do what is necessary to protect yourself. Do not worry how this may appear to others and do not worry about failure. Trust your instincts and act accordingly.

Keeping it in context

It is not the intention to fuel the anxieties of newly qualified surveyors. The fear of crime and dangerous incidents is sometimes out of all proportion to the likelihood of them occurring. But unless surveyors are aware of the dangers and plan for them they will always be vulnerable. The potential for violence may depend on a whole range of factors. These could be nothing to do with the surveyor; the person may have had a bad day and the surveyor was the last straw. The reason for the house being sold might be marriage break-up, bankruptcy or bereavement and the surveyor might be the personification of what they see as a personal tragedy. Therefore surveyors must always remain alert to the possible dangers. If these guidelines are followed every time they will become integrated into the normal inspection process and carried out unconsciously.

3.5.5 Dangers posed by the building during the survey

Each property has to be taken on its own merits but the development of a survey routine (see section 3.4.1) will help. Safety awareness has to be a top priority when entering the dwelling, room or outbuilding. The RICS (1991) state that the surveyor should be aware of the ordinary risks of the job and '... obvious signs of a need for extra care, shown for example by the disrepair or age of the building, should heighten the surveyor's awareness of his (sic) own safety' (Reville 1998, p.175). This was illustrated in Rae v. Mars UK (1990). In this case, a surveyor fell into a deep pit in a dark, unlit store-room of a factory he was inspecting. The building owner was found liable but the court also decided that the surveyor had not been '... quite as careful as he could have been'. This was because he had a torch at the time but it wasn't switched on. Consequently the surveyor's damages were reduced (Reville 1998, p.176). The following examples outline potentially dangerous situations that surveyors need to aware of:

- Does the flooring in the building seem in good condition or is it likely to collapse because of rot or beetle infestation? The key indicator here would be the general condition of the building. Are there any floorboards missing or other openings in the floor?
- Are the electrical installations safe? Are there any exposed wires? Do the heating appliances appear in good condition or do scorch marks or flue stains suggest that combustion products could threaten health? Is there a smell of gas?
- Will a flat roof take the weight of a person? Look out for uneven surfaces, ponding of water and any other signs that may suggest poor condition. Special care should be taken if the surface is covered with snow.
- Is it safe to go in the loft space? Do the ceiling joists appear strong enough to support a person (min. 38 × 72 mm at 600 mm centres as a suggested minimum). Are they overloaded by stored items? Is the loft space boarding properly secured or is it a haphazard collection of different types of timber panels that are likely to collapse or tip up when stood on? Will it be safe to walk across the back of the ceiling joists, i.e. are they covered in insulation and are the roofing members close enough to walk steadily on?
- Is the building structurally stable? Are the walls, ceilings, etc. likely to collapse? Will the roof covering come loose and fall on to people below? Such situations could be made worse if there is a strong wind blowing or high snow loads.
- Are the outbuildings safe to go in? Are the gardens safe to walk around? The main dangers could include rubbish or vegetation covering various holes, excavations, inspection chambers, cesspools, etc. Are there any dangerous animals loose? Is there any fast moving or deep water?

These are just a few of the possible dangers that could face a surveyor on an inspection. Experience will result in many of these checks becoming automatic but unless safety is seen as a key consideration from the earliest stages then a lack of awareness could lead to increased risk.

3.5.6 Contamination, debris and unsafe atmospheres

To some extent these are linked to the points made above but include special dangers:

- used and unprotected syringes;
- surfaces that may be contaminated with chemicals or other hazardous substances;
- areas containing asbestos – see section 11.5.2 for more information;
- presence of rats, pigeons, cockroaches, fleas, etc. that could spread diseases and illness.

The proper use of protective clothing and equipment can protect against many of these risks. For example a face mask is essential when in a space infested by pigeons. A handy can of flea spray will save hours of embarrassing scratching!

3.5.7 *High-risk locations*

This will include locations where there are higher than average risks. These might include:

- building sites;
- farms;
- areas subject to landslip and subsidence;
- some inner city areas;
- workshops where semi-industrial processes are undertaken.

When carrying out work in these type of locations, surveyors should be specially attentive.

3.5.8 *Travel and vehicle safety*

Most surveyors spend a considerable amount of time in the car travelling between their office and the properties to be inspected. There are a number of precautions that should be taken:

- make sure the car is well maintained and roadworthy. In addition to the formal servicing, read the car maintenance manual and stick to the routine daily, weekly and monthly checks;
- ensure that there is enough fuel in the car to get to the destination;
- carry an emergency car kit and join a motoring breakdown association;
- carry an appropriate map so that you will not need to stop to ask for directions;
- the journeys should be planned so that you do not get tired. Take regular breaks;
- always follow the highway code and stick to the speed limits.

3.6 Taking site notes

Depending on personal style and preference, surveyors will either make written notes and sketches to record their impressions of a property or dictate their thoughts into a personal recorder. Opinion is divided on the most suitable method and is discussed in the case of Watts v. Morrow (1991). Although this case related to a structural survey the fact that the surveyor used a personal recorder was a matter of concern to the judge. The particular worry was that the surveyor dictated his **report** directly into the machine not just his site notes. Once back at the office, the tapes were passed straight to the secretary who typed them up. The report was amended and then sent to the client. This resulted in the report being '... strong on immediate detail, and I regretfully have to say, negligently weak on reflective thought', according to the judge.

This ability to ponder over a recently carried out survey is an important part of the process. To be able to look at the dwelling as a whole is vital. Seeing how the different elements interact and draw reasoned conclusions is what surveyors are paid for. The absence of any site notes makes this much more difficult. In the Watts case, this lack of written notes also meant that the surveyor had nothing to highlight a 'trail of suspicion'. He had trouble recalling some aspects of the property which was put down to not having a written record of the survey. This attitude is supported in the *HSV Practice Notes* (RICS 1997), under DHS section B it states that 'There is no objection to the use of machines for recording site notes'. Here the emphasis appears to be on 'site notes'.

In Bere v. Slades (1989) a valuer admitted that he had no recollection of a valuation he had carried out other than from his site notes. The court decided in favour of the valuer. This was probably helped by the positive impression of competence promoted by his good record keeping.

The importance of making sketches was mentioned in the judgement of Fryer v. Bunney (1982). Here the surveyor had checked a property with a moisture meter and reported that no dampness readings had been registered. In the event there was indeed dampness that could have been detected with the moisture meter if it was used more effectively. Murdoch and Murrells (1995, p.94) concluded that 'The moral of this story must be for surveyors to prepare a very simple sketch plan of a property as it is surveyed, marking roughly where a protimeter has been applied'.

A slightly different view has been put forward by Wilde (1996). In a personal review by an experienced building surveyor, he suggested that there may be nothing wrong in dictating the report directly into a personal recorder during the survey. This is because it would often be a first draft soon to be revised by the surveyor in the office and so giving an opportunity for reflection.

In conclusion, if there are no specific organisational quality assurance procedures, the following record of a survey should be made:

- clearly handwritten site notes with enough sub-headings to indicate which parts of the building the notes relate to. Sketches can add to the descriptive value of the record;
- the notes should be legible and set out in a logical order. Ideally they should be in pencil in wet weather and re-written in the dry if necessary;
- if a personal recorder is used then the tapes should be transcribed as soon as possible. The transcribed notes should be used when compiling the report.

The site notes should be filed along with the rest of the client information for as long as practicably possible.

3.7 Time taken to do the survey

The time taken to complete a survey is the sort of question that is so difficult to answer that in an ideal world it is not worth asking. It will depend on:

- the type of commission;
- the size and complexity of the property;
- the age and condition of the building;
- the knowledge and experience of the surveyor;
- ease of access and even the co-operative nature of the owner.

One commission may go smoothly while another takes up a considerable amount of time. Notwithstanding these influences, at a time when surveyors are being asked to increase their productivity, establishing indicative time scales may well provide useful benchmarks for this important professional service.

Mortgage valuations – this has been accepted as an inspection that '... is, of necessity a limited one' according to Judge Ian Kennedy QC in Roberts v. Hampson (1989). The time taken '... should not take longer than 20–30 minutes' (Lloyd v. Butler 1990). This average time depends on the problems identified during the inspection. In the Hampson case the judge stated, 'It is inherent in any standard fee work that some cases will colloquially be "winners" and others "losers", from the professional man's point of view. The fact that in an individual case he may need to spend two or three times as long as he would have expected, or as the fee structure would have contemplated, is something that he must accept. ... If, in a particular case, the proper valuation of a £19,000 house needs two hours work that is what the surveyor must devote to it.' This principle was somewhat moderated by the statements in Lloyd v. Butler (1990) where it was stated that not all trails of suspicion need to be followed. Although many valuers and surveyors may wince at the two hour 'loser' not many complain about the 'winners' where the straightforward nature of many properties result in inspection times of less than 20–30 minutes. Despite this variance, most commentators would consider 30–45 minutes to be an average for this type of inspection.

Homebuyers survey and valuation – although there is no explicit legal discussion of the time period taken to do an average HBSV, the comments of Fawcus J in Hacker v. Thomas Deal & Co (1991) about the time taken to carry out a building survey may be useful. Commissioned to survey a very large house in Belgravia, London the surveyor started at 9.00 am and had finished by about 1.30–2.00 pm. The judge appeared satisfied that this was an appropriate length of time and stated that '... one can readily understand why he could not have been involved in any other survey (that day)'. Based on the assumption that the HBSV 'envisages an inspection more limited in scope than a structural survey ...' (Murdoch and Murrells 1995) it would seem reasonable to expect the average Homebuyers survey to take about 2–2.5 hours.

3.8 Referring to specialists

3.8.1 Introduction

The inability to fully investigate the true nature of a defect during a standard survey should not be used as an excuse for surveyors to refer the matter to someone else. The client pays a fee for a survey that tells them about the property they are interested in buying. If the report results in a series of referrals to other specialists for further investigations then it is not surprising that the client is unhappy. Not only will these additional inspections cost more money, they may delay and ultimately frustrate the whole property transaction. There is a balance to be struck here – on the one hand giving the client clear and focused advice that they need and expect while on the other ensuring that the surveyor does not exceed knowledge and skills boundaries. If this happens not only would the client be given the wrong advice but the surveyor would be open to legal challenge.

This is mentioned in the RICS Practice Statement that describes the Homebuyer survey (RICS 1997). Under competence and responsibilities (PS 11.7) it states that '... reports should include caveats and recommendations for further investigation only when the surveyor feels unable to reach the necessary conclusions with reasonable confidence'. This view was reinforced by Holden (1998a) when clarifying the service offered by the HSV. He states that excessive recommendations for further investigations is '... alien to the concept of the HSV service ... and a dereliction of responsibility'. But referring to specialists is a delicate judgement. It can be best illustrated by using a medical analogy. When a patient goes to the doctors, a General Practitioner (GP) will investigate their signs and symptoms and in the majority of cases, an effective diagnosis will be made. If the cause of the patient's illness is not clear then the GP may carry out further tests (blood and urine tests for example) that may provide the missing information. Another option is to refer the matter to another medical practitioner who has a specialist skill. This is because no GP can have the full range of knowledge required to properly assess the human condition. The GP also has the confidence that the specialists that she sends her patients to are (on the whole) well trained and are indeed 'specialists' in their field.

In the surveying profession no 'general practice' surveyor will have the full range of skills required to fully assess the condition of all the dwellings they encounter. That is where the analogy with the surveying profession ends. Surveyors cannot have the confidence in the 'specialists' being so well trained as in the medical profession. There are few other trade associations or organisations that have a clearly prescribed competence level. In many cases, referrals in survey reports can result in clients being sent to individuals who have **less** skill than the surveyor who made the referral in the first place. It is like a GP when she is unable to diagnose the cause of an illness, sending the patient to a pharmacist. Although pharmacists are well qualified and able to give sound advice to their customers, they are not as qualified as doctors. They are also selling a product. This last point is important. Take the damp proofing business as an example. The vast majority of prospective home owners who are referred to specialist damp proofers are being sent to a contractor who is selling a product. Generally, they are not paid a fee to provide an objective assessment. They submit a report that describes the condition of the property together with an estimate and a hope that work will come their way. This is not to say that damp proof specialists are dishonest and unscrupulous. Far from it. The British Wood Preserving and Damp Course Association have clear standards and sponsor reputable NVQs in surveying for dampness. The problem is that the further investigation of the defect and the commercial interest of the organisation are not adequately separated. This will always leave the client in danger of exploitation.

Holden (1998a) tried to clarify what the RICS and ISVA hope to achieve through the standard approach of the Homebuyer survey. If a surveyor is confident that remedial works of some kind are required and the only doubt is the extent then the client should be advised to obtain a quotation. If on the other hand the surveyor has a 'real difficulty reaching a conclusion about the nature of the trouble' a recommendation for further investigation should be given. So for example, if a dwelling is clearly affected by rising dampness, the client should be referred to a contractor to obtain a quote for the remedial work. Alternatively, if a dwelling is clearly damp but the signs and symptoms offer no definite explanation of what is causing it, the client should be referred to a 'specialist' that can carry out the appropriate diagnostic work. This may or may not be a 'contractor'.

3.8.2 Referring clients to specific contractors

Properly recommending further investigations is only the first part of the process from the client's point of view. The question of who should do the work is the next challenge. Many surveyors are reluctant to recommend specific contractors because of a concern that they will be unable to guarantee the work carried out. As a consequence many feel they might be held partly responsible for any defective work done.

This area was explored in a question and answer article about the Homebuyer survey (Holden 1998b). The issue of whether a surveyor should recommend specific contractors or specialists when asked by a client to do so received the following response:

> It is proper to recommend particular contractors. Indeed most clients would expect you to have the relevant local knowledge and could reasonably take offence if you declined to do so.

Two precautions are recommended:

- offer at least two real choices so accusations of favouritism can be avoided;
- include a brief disclaimer that makes it clear that the surveyor cannot accept responsibility for the work of any third party that may have been recommended.

If the client asks the surveyor to obtain quotations on their behalf then it is recommended that the contractors send the quotes directly to the client. This will limit the surveyors involvement in the process and emphasise the contractual relationship remains between the contractor and client.

Where the surveyor is prepared to offer this sort of advice there is still the problem of establishing which contractors are reputable. In their consultation paper *Combating Cowboy Builders* the Department of Environment outlined the problems facing domestic consumers when selecting a builder without professional advice (DETR 1998). People usually rely on one of two methods to select a builder:

- Yellow Pages or adverts from a local paper; or
- word of mouth.

Both cases are fraught with difficulty as neither source guarantees competence. Monitoring the effectiveness of contractors would be enormous job. The DETR estimates that there are 163,000 builders operating in this country with many more falling below the VAT threshold. Although surveyors are better placed to navigate their way through this minefield, establishing a list of reliable builders and specialists will still be a challenge. Here are a few tips:

- Evaluate the performance of builders used by other parts of a multi-disciplinary surveying practice and build up an approved list.
- Always use builders that are part of a recognised trade association especially those that have insurance-backed guarantees.
- Work closely with local authorities in the area. Many have a good deal of information on local builders through renovation grant work, care and repair agencies, etc. Often they might be prepared to share these lists.

- Link up with private sector validation schemes such as Green Flag and the Automobile Association. These have very strictly run emergency repair schemes that offer a high degree of vetting of approved builders.

Establishing positive links with a group of reliable contractors will give great benefit for all concerned:

- the client will have the peace of mind in knowing that the chosen builder has some form of established track record;
- the surveyor will be able to offer a more complete service to their clients. This may lead to 'follow-on' commissions for organising alteration, extensions and other repair and refurbishment work to the client's property once they have moved in;
- the recommended builders will receive a steady stream of work. Hopefully this will encourage them to value the relationship and ensure that a good service will be provided.

Even when such a system works well there will be a few occasions when things go wrong. In these instances the surveyor might find themselves drawn in to mediating role between the client and their builder. Although this will be time consuming and contractually ambiguous it could be a price worth paying for the enhanced service it provides.

References

Bere v. *Slades* (1989). 2 EGLR 160.

Bibby, P. (1995). *Personal Safety for Health Care Workers*. Arena, Aldershot.

DETR (1998). *Combating Cowboy Builders: A Consultation Paper*. Department of the Environment, Transport and the Regions, London.

Fryer v. *Bunney* (1982). 263 EG 158.

Hacker v. *Thomas Deal & Co* (1991). 2 EGLR 161.

HMSO (1974). *The Health and Safety at Work Act*. HMSO.

HMSO (1994). *The Construction (Design and Management) Regulations*. HMSO.

Holden, G. (1998a). 'New Homebuyer: more of your questions answered'. *Chartered Surveyors Monthly*, vol. 7, number 8, May. RICS, London.

Holden, G. (1998b). 'New Homebuyer: more questions answered'. *Chartered Surveyors Monthly*, 8(2), October. RICS, London.

Lloyd v. *Butler* (1990). 2 EGLR 155.

Murdoch, J., Murrells, P. (1995). *Law of Surveys and Valuations*. Estates Gazette, London.

Rae v. *Mars (UK) Ltd* (1990). 03 EG 80.

Reville, J. (1998). 'Surveying safely – rights and responsibilities'. *Structural Survey*, vol. 16, number 4, pp. 172–176. MCB University Press.

RICS (1991). *Surveying Safely – A Personal Commitment*. Royal Institution of Chartered Surveyors, London.

RICS (1996). *Appraisal and Valuation Manual*. Royal Institution of Chartered Surveyors, London.

RICS (1997). *Homebuyer Survey and Valuation*. HSV Practice Notes. RICS Business Services, London.

Roberts v. *Hampson* (1989). 2 All ER 504. (1988), 2 ECLR 181.

Watts v. *Morrow* (1991). 4 All ER 937, 2 EGLR 152, CA.

Wilde, R. (1996). 'Making notes on a survey'. *Structural Survey*, vol. 14, number 2, pp.18–21. MCB University Press.

Chapter 4

Building movement

4.1 Introduction

Most surveyors and valuers get very nervous when they come across a building that has been damaged by building movement. This often results in recommendations for further specialist investigations especially by less experienced or less confident surveyors. This is despite the studies by the BRE (1990, p.2) which have shown that the majority of houses underpinned following the dry summer of 1976 were only exhibiting minimal signs of damage. This high level of underpinning activity still continues. According to the Association of British Insurers (1998) in 1998 there were nearly 40,000 individual claims costing in the region of £400 million.

The reasons for this over-reaction are complex:

- All mortgage valuations and Homebuyers reports are 'point in time' inspections. In many cases it is not possible to give a balanced professional view on just one visit only.
- Many surveyors simply do not have enough experience of assessing damaged buildings.
- Surveyors feel they do not have the time or resources to carry out a full diagnosis;
- The expectation of the building owner has risen and building damage is no longer tolerated (Evans 1998, p.18). Owners will often push very hard for underpinning schemes or other expensive repair work to go ahead.

The role of the surveyor in the process is crucial. Many subsidence claims are triggered by the vendor when a potential purchaser receives an adverse report. Therefore surveyors have a clear responsibility to make sure that their judgements are measured so that:

- the lender and purchaser can be properly advised;
- unnecessary and wasteful repair works are avoided, and:
- the liability of the professional involved is protected.

4.2 Incidence of movement

The dry summer of 1976 resulted in thousands of buildings developing cracks for the first time. Statistics show that the vast majority of the resulting underpinning claims were concentrated on the southeast and surrounding regions. The BRE (1990, p.2) put forward a number of reasons for this:

- the main soil type is clay that is prone to shrinkage during dry spells;
- the level of rainfall is lower in this part of the country; and
- there were a considerable number of trees in the urban environment.

Despite the greater risks, building movement can occur in all regions for a variety of reasons. Surveyors must develop an objective procedure that can help them assess the consequences of any building damage.

4.3 Assessing the cause of damage

4.3.1 Extent of inspections

In the ever-changing world of mortgage valuations, it is difficult to draw a confident division between where the inspection ends and another survey begins. The first task of a surveyor is to notice that the building has been actually damaged. In this context 'damage' is not only cracking, it can also include distortion, sticking doors and windows, disrupted services and lack of weathertightness.

Once the damage has been noticed then the decision taken by the surveyor will depend on the type of commission being carried out:

Mortgage valuations – although the principle of 'following the trail' has been firmly established, the 'trail' in a mortgage valuation may be shorter. With experience, many surveyors will be able to judge quickly when a building is so significantly damaged that further investigations and referrals will be required.

Homebuyers survey – because this is seen as a 'survey' and expected to be more extensive, the 'trail' of diagnosis will have to be followed to a greater length and be better considered. The surveyor has more opportunity to look at the damage to a building and set it in context. A more informed view may help to give balanced advice and so avoid unnecessary referrals to specialists.

Whatever the type of commission, it is essential that all surveyors understand the 'process' of diagnosing the cause of building movement. This is so that:

- early in their careers surveyors can follow a step-by-step approach that will give them a greater confidence in their decision making until they develop a more intuitive level of skill;
- when even experienced surveyors come up against an unfamiliar or novel form of building damage, they have a clearly defined approach that will help them give appropriate advice and minimise potential liability.

The most popular approaches to this process have been promoted by the BRE (1990) and the ISE (1994). A combination of the two provides the basis for a robust, objective diagnostic process for surveyors to follow.

4.3.2 Diagnostic process

Any structure can be damaged in a number of different ways and this damage will provide the surveyor with information on which an assessment can be based. The approach consists of a number of distinct stages.

Step one – triggers for 'following the trail'

The following approach should be triggered only if evidence of building damage is noticed. This is not restricted to cracking but could include:

- out of alignment – sloping sills, lintels, masonry coursing and floors, etc.;
- windows and doors that don't open properly;
- environmental clues such as fissures in the ground, damage to adjacent properties, etc.

Once a 'trigger' is noticed then any surveyor must 'follow the trail' of the defect. The most effective way of achieving this is to carry out a *present condition* survey following several clearly prescribed stages.

4.3.4 Step two – the present condition survey

The BRE (1990, p.3) suggests that the first stage involves the objective collection of information that helps define the extent of the damage.

SKETCH THE EXTENT OF THE DAMAGE

A quick proportional sketch of each elevation should be carried out that clearly indicates the position and direction of the cracks. These need only be simple line drawings but must show:

- the type of crack – whether it shows tensile, compressive or shear characteristics (see figure 4.1);
- the width of the crack and whether it tapers along its length;
- how the cracks are distributed around the building;
- whether the cracks extend through the width of the wall and affect the internal surfaces.

Figure 4.2 illustrates part of a typical sketch that might be produced. A simple floor and site plan should also be included showing the positions of any drains, trees, slopes, etc.

Not all masonry affected by building movement will crack. Much of the housing stock before World War II was constructed using softer sand/lime mortar. In this situation it is common to see masonry courses 'flowing' with the movement rather than cracking. This is still evidence of movement and so should be noted.

DETERMINE THE AGE OF THE DAMAGE

Knowing the approximate age of the cracks can give important clues as to whether the movement is still active. The external cracked surfaces can give a clue. If the edges are well defined and sharp then the cracks could be recent. The exposed faces of the masonry within the depth of the crack should be lighter in colour than the weathered external surface. If the crack has moss, grass or even an elder bush growing out of it then it might be safe to assume it has been there a long time!

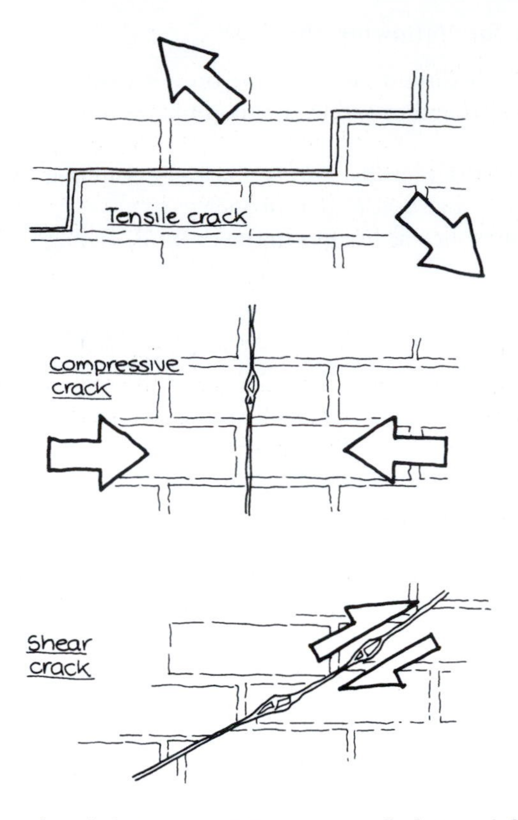

Figure 4.1 Different types of crack that can occur in masonry walls that result from tensile, compressive and shear forces.

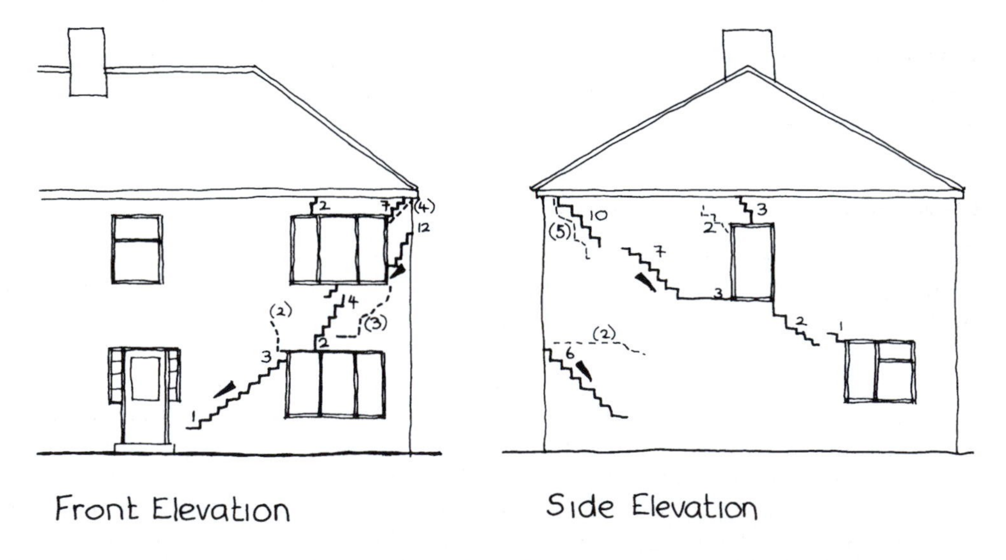

Figure 4.2 Typical 'present condition' sketch of a dwelling that a surveyor should make once building damage is noticed in a property.

Establishing the age of cracks

Asking the occupants is one way of establishing the age of a crack but beware! If you ask a vendor how long a crack in their lounge wall has been there you will probably get an evasive answer that ranges between decades to centuries. On the other hand, if you avoid any mention of cracks but ask when they last decorated they will only be too willing to tell how recently it had been done. Never underestimate the usefulness of subtle questioning of the vendor!

The age of the internal decorations and any repaired external finishes (such as repointing) may also give an insight into the age of the movement. If the cracking has not affected internal decorations that are five or more years old it could indicate that the movement has stabilised (ISE 1994, p.45).

IS THE BUILDING DISTORTED?

If a building is damaged by ground movements then part of it will rotate. This can result in walls leaning over, floors tilting and lintels and sills sloping away from the horizontal. Measuring the extent of these distortions precisely is a very sophisticated process but most surveyors will be able to come to a broad view. For a Homebuyers survey this could be by:

- using a bricklayer's spirit level (1.0 m long) to gauge the verticality of walls and slope of floors;
- placing a smaller spirit level (say 250 mm long) on sills, underside of lintels and door heads to see if they are sloping;
- dropping a simple plumb bob and line out of a convenient window.

This sort of information will be very broad brush. If considered in isolation it could easily lead to incorrect conclusions but when matched with the other evidence it could add depth to the analysis.

DEFINE THE SERVICEABILITY OF THE BUILDING

The functional performance of the various elements of a building will be affected by building movement. Describing the extent of this loss of performance will assist in the diagnostic process. The features to look out are as follows:

- **doors and windows sticking** – The door and window openings in any building represent breaks in the structural continuity of the walls and will quickly distort. The frames in the openings will 'rack' or be pushed out of shape. Any opening casement windows and doors will 'bind' against their frames. In extreme cases they will be incapable of being opened at all. Often building owners will regularly plane down the frames so the doors and windows can still be used. After a number of years strange shapes will often result (see figure 4.3).

Figure 4.3 Distorted window opening in a damaged building. The top sash has been regularly adjusted so that it will close. What will the carpenter do when she gets down to the glass?

- **cracked window panes** – High levels of movement can induce such stresses in window frames that the glass panes will actually crack. This is especially notable with metal frames.
- **draughts and rainwater penetration** – Where the cracks are wide enough on exposed elevations, wind-driven rain can penetrate to the internal surfaces. The gaps that appear around window and door frames can be very vulnerable especially where the waterproofing mastic sealant has become disrupted.

CONSTRUCTIONAL DETAILS

Some types of construction will inherently be more vulnerable to movement by their very nature. Making a note of the particular characteristics will enable the surveyor to rule out potential causes and so simplify the diagnostic process. Examples include cavity walls and materials prone to expansion and/or contraction such as calcium silicate bricks, etc. Each of these are so closely associated with clearly defined forms of building damage that prognosis may be quite simple.

AGE OF THE PROPERTY

The ISE (1994, p.37) point out that new houses should be treated with extra caution. This is because normal settlement and drying out cracking is often mistaken for subsidence problems. The definition of 'new' in this case is usually applied to dwelling less than 5 years old. Even though cracking might be more commonly encountered in these newer properties, any cracks 2–5 mm and over should still be referred for further investigation. Cracks smaller than this should be considered 'normal' unless there is any other evidence to the contrary.

Step three – classifying damage

Once all the relevant information has been gathered together the analysis can begin. The BRE wanted to develop an approach that encouraged assessors to remain objective. This can give professionals a common language, criteria that are clearly understood and would lead to a shared understanding of any particular level of damage to a building. To achieve this, the BRE developed a system of classifying the into three main categories (BRE 1990, p.3):

- **aesthetic** – This damage affects the decorative finish of the building and has little effect if any on its functional performance.
- **serviceability** – The effects are so pronounced that the performance of different elements will be affected. For example the jamming of doors and windows; wind and rain entering through cracks in walls and the fracturing of some service pipes.
- **stability** – In this category, parts of the building will be so badly damaged that stability might well be threatened.

Although this can help a surveyor begin to define the level of damage, the categories remain too broad. To make this approach more useable these definitions were refined to produce six different categories of damage (see figure 4.4). This enables a wide variety of cases to be easily classified. The most important feature about this technique is the written description of the levels of damage. This includes the 'ease of repair' as well as descriptions of the damage itself. This encourages any surveyor to look at factors other than just the cracking. This can avoid surveyors relying on subjective personal views and 'gut reactions'.

The way to use this table effectively is to look at the damage to a building and ask the question 'If the movement has stopped and is unlikely to continue, what work would I have to organise to repair the damage?'. Look at the descriptions carefully and classify the damage that can be seen. Broadly speaking this falls under two headings:

- for damage in categories 0, 1 and 2, the repair works are mainly cosmetic involving crack filling, redecoration and repointing;
- categories 3, 4 and 5 include more significant repairs ranging from the rebuilding of small sections of masonry through to partial or total rebuilding of the whole building.

A few other points should be noted when using this table:

- the classification is not based on crack width alone. They are given for guidance only. Damage to buildings is a much broader concept than just the cracking;
- the assessment must be based on the visible damage at the time of the survey and not how it might progress. It is vital to avoid any subjective speculation at this point.

CAUSES OF DAMAGE

Only when the objective classification of the damage has been completed can consideration be given to what could be causing it. As a general rule there are two different types of movement that cause damage to buildings:

Category of damage	*Description of typical damage (Ease of repair in italic type)*	*Approximate crack width (mm)*
0	Hairline cracks of less than about 0.1 mm are classed as negligible.	Up to 0.1 mm*
1	*Fine cracks that can be easily filled during normal decoration.* Perhaps isolated slight fracturing in the building. Cracks rarely visible in external brickwork.	Up to 1 mm*
2	*Cracks easily filled. Re-decoration probably required. Recurrent cracks can be masked by suitable linings.* Cracks not necessarily visible externally; *some external repointing may be required to ensure weathertightness.* Doors and windows may stick slightly.	Up to 5 mm*
3	*The cracks require some opening up by a mason. Repointing of external brickwork and possibly a small amount of brickwork to be replaced.* Doors and windows sticking. Service pipes may fracture. Weather-tightness often impaired.	5 to 15 mm (or a number up to 3 mm)*
4	*Extensive repair work involving breaking-out and replacing sections of walls, especially over doors and windows.* Window and door frames distorted, floors sloping noticeably. Walls leaning or bulging noticeably some loss of bearing in beams. Service pipes disrupted.	15 to 25 mm* but depends on number of cracks
5	*This requires a major rebuilding job involving partial or complete re-building.* Beams lose bearing, walls lean badly and require shoring. Windows are broken with distortion. Danger of instability.	Usually greater than 25 mm* but depends on number of cracks

* Crack width is just one factor in assessing the category of damage and should not be used on its own.

Figure 4.4 Table showing the classification of visible damage to buildings (Reproduced courtesy of BRE).

- movements that originate in the building structure itself, such as thermal expansion and contraction, moisture movements, cavity wall tie failure, chemical changes such as sulfate attack, poor design and workmanship, etc.;
- movements associated with problems in the ground beneath the building. These can include subsidence caused by variation in soil types and over-loadings, mining activities, settlement of filled ground, shrinkage and swelling of clay, leaking drains, etc.

Relating this back to figure 4.4, as a general rule damage associated with movements of the building structure rarely exceeds the damage described in category 2. Movements that originate beneath the building can result in damage from all categories.

This initial distinction begins to be useful. Two diagnostic principles can be stated:

- Unless there is clear evidence that the observed damage will not progress to higher than category 2, then it is not worth finding out the cause. Further investigations are usually unwarranted and all that is required is the restoration of the appearance. The key factor here is judging whether the movement is progressive. This is discussed in more detail below.
- If the damage is in category 3, 4 or 5 then this is clear evidence that foundation movement is the most probable cause. Because this can be progressive and significant, further investigations are required before the client can be properly advised. Even if the foundation movement has ceased never to move again, it is likely that the building will be so significantly weakened that remedial works will still be necessary.

Step four – assessing whether damage is progressive

This initial judgement is a useful starting point but the next stage is vital. Small cracks have a habit of growing into large ones! Many surveyors have walked away from minor plaster cracks only to return a few months later to find that the cracks have matured into category 5 'whoppers'. To assess whether it is progressive, two questions have to be asked:

- Is the damage due to foundation movement?
- If yes, will it get worse?

This difficult decision can be made easier by looking for the characteristics that set foundation movement apart from other causes. These are summarised below:

- **Cracks affect both the internal and external faces of walls in close proximity.** This is especially significant in cavity walls. Other causes such as wall tie failure and expansion problems affect the outer leaf only.
- **The damage extends down below the dpc and into the ground.** When support beneath part of a foundation is removed or weakened the building will virtually 'break' and rotate around what could be thought of as a 'hinge' point. The cracks usually begin from this point and are well below the dpc level (see figure 4.5). If the cracking had been caused by superstructure deficiencies then the cracks would only appear above the dpc. This is because:

  - the brickwork below the dpc is fully restrained by the ground;
  - the dpc acts as a 'slip plane' allowing the wall above it to move independently.

- **The cracks are normally tapered, being wider at one end than the other.** Because the building will tend to rotate, the cracks will be wider at one end than the other. For most cases, the cracks will be wider at the top apart from where the ground 'sags' in the middle of a building causing wider cracks towards the bottom (see figure 4.6) (BRE 1991, p.14).
- **The cracks are normally concentrated on one part of the building.** Most ground problems are relatively isolated resulting in damage to one part of the building (mining subsidence being a notable exception). Other types of superstructure defects can cause damage that is distributed around a number of elevations of the dwelling.
- **Cracks are normally diagonal in direction.** Other causes of damage can cause horizontal cracks (cavity wall tie failure and sulfate attack) or vertical cracks (moisture and thermal movements).
- **Floors and walls tilt and window and door openings distort and jam.** This is caused by the rotation of the building.
- **Cracks tend to travel around openings.** The ISE (1994, p.43) state that openings attract cracks because they find it easier to travel around the openings than through the adjacent masonry panels. Door and window frames will twist out of alignment within their openings and any mastic sealant will be disrupted.
- **The point of maximum damage can be represented by a single large crack or a series of smaller, closely spaced ones or an 'array' of cracks.** This depends on the stress levels in the masonry structure and the way that material reacts to it.

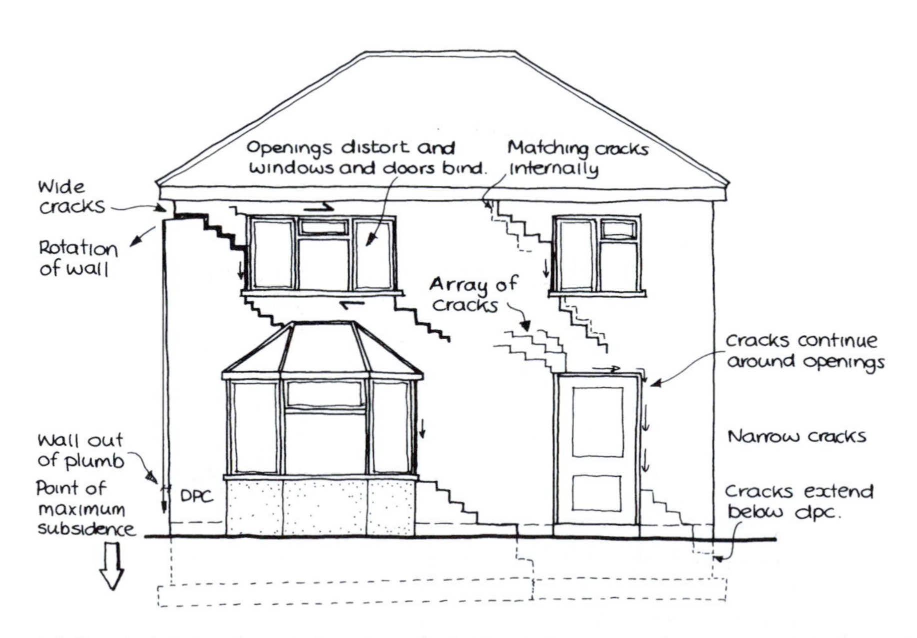

Figure 4.5 Sketch showing characteristic signs of subsidence damage to a domestic property (reproduced courtesy of Structural Engineers Trading Organisation).

These signs must be seen as typical symptoms. Because buildings are so different in the way they react to ground movements some cases will not easily fit into this these standard descriptions.

Step five – triggers for further investigation and referrals

Two courses of action are now available:

- If the damage has occurred over a short space of time and can be categorised as category 3 or higher, it is not necessary to carry out any further evaluations. This is because the damage is so significant the next stage involves the investigation of the subsoil conditions and other factors. A referral to a specialist is entirely appropriate.
- If the damage is category 2 or below, a surveyor's judgement is needed. Even if the evaluation of the physical symptoms suggests foundation movement, referral for further investigation is not automatically justified.

In the latter situation the building and surrounding area should be assessed for any 'risk' indicators that could suggest the likelihood of continuing foundation movement will be high. The following 'risk' indicators account for 80–90% of the subsidence cases in this country.

NEARBY TREES AND SHRUBS

This factor alone accounts for 50–60% of all subsidence claims and is normally associated with shrinkage of soils and especially clays (ISE 1994, p.37). More detailed consideration is contained in section 4.6 but the most important issues are:

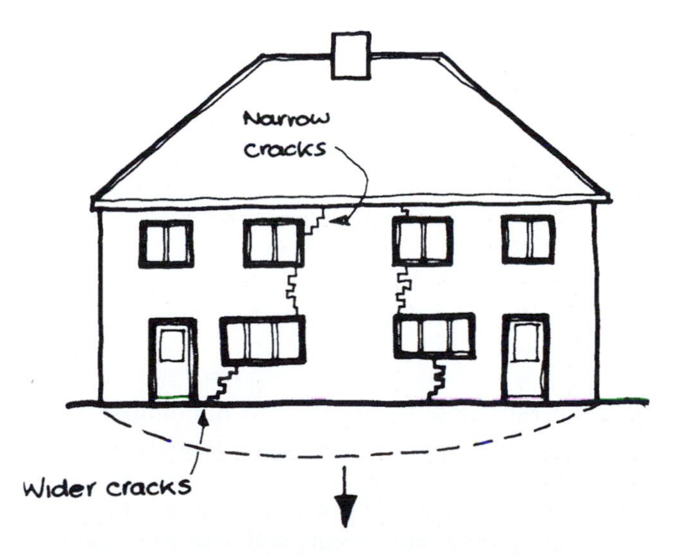

Figure 4.6 Sagging of a building in the middle. It is important to note that the cracks are wider towards the bottom of the building in this case.

- Are there any trees within 20 metres of the dwelling?
- Are there any hedges or rows of large shrubs within 10 metres?
- Is the subsoil in the area likely to be shrinkable?
- Are there any drains running close to the building that could have been damaged by tree roots?
- Are there any tree stumps or other evidence that trees or shrubs have been removed in the past?

A positive answer to any one of these could lead to a referral for further investigation before a client could be properly advised.

PROXIMITY OF SLOPING GROUND

There is a natural tendency for subsoils to move from higher to lower levels. If this happens the foundations will be affected as well as other external features (i.e. retaining walls, paths, drives, etc.). This is a problem that will need the advice of a professional with considerable geotechnical knowledge and experience because the causes and solutions are very complex. Signs that suggest problems include:

- trees that are inclined down the slope of the land;
- 'fissures' or large 'cracks' in the surface of the ground;
- slumped areas of ground often corresponding to the fissures described above;
- garden walls and other external constructional features that are being pushed over or showing other signs of distress.

This is likely to be more active during the winter months than in the summer because of the increase in moisture in the ground.

POSSIBILITY OF LEAKING DRAINS

Drains can cause subsidence in two ways:

- by softening the subsoil (especially clays) allowing the building to 'sink';
- finer elements in the subsoil can be 'washed out' and/or washed into the drains leaving voids that allow subsidence to occur.

Determining whether drains are leaking is a very time-consuming and expensive business (see chapter 10). At the initial stage, the following factors may help to establish the likelihood of problems:

- Are there any drain runs close to the affected building? Look out for lines of inspection chambers, rodding eyes, gullies, etc. If there are no drains within 5 metres then it is unlikely to be a cause.
- Are the drains rigid or flexibly jointed? Older rigid jointed drains are very susceptible to leaks as even slight ground movements will lead to cracking. Generally, pre-1970 drains tend to be of the rigid type while those after usually have flexible joints. Therefore, properties older than 25–30 years are seen as having a greater risk.

- Are there any trees or shrubs planted directly over or close to a drain run that is close to the dwelling? Roots can penetrate even the smallest of cracks in the search for moisture. This can lead to damage and further leakage.
- Do the drains run beneath vehicular access points? Drains can be damaged by cars or heavier vehicles if they are too shallow or have not been constructed well enough.
- What is the condition of the inspection chamber? Lifting the lid of a nearby chamber can give an insight into the condition of the drain system. For example can you see any soil or other granular material running down the drain that could mean a leak further up the run? Are tree or shrub roots in the chamber? Is the benching and brickwork in good condition or is it spalling away? All these signs could increase suspicion that the drains are in poor condition. Detailed guidance on assessing the condition of drains is contained in chapter 10.

SUSCEPTIBLE SOILS

The very nature of some subsoils will mean that ground movement is more likely in some areas than others. Most problems are caused by shrinkable clays especially when trees and large shrubs are close by. An increasing number of houses have been built on former landfill sites that also pose a high level of risk. Site inspections alone cannot usually determine soil types so good local knowledge is essential. This emphasises the need for surveyors to keep up-to-date local information.

4.4 Calling in the specialists

The process of assessing building movement is summarised in figure 4.7. If there are risk indicators present, a recommendation that further investigations are made by a specialist is not only appropriate but serves the client well. Once the decision has been made, to whom should the client go? The question is controversial.

Over the last few years, building surveyors have campaigned to be recognised as an appropriate specialist that can carry out further investigations. A publication of the ISE (1994, p.33) questions this. They suggest that all the different types of surveyors are appropriate 'experts' for the initial appraisal. When a subsidence problem occurs only those that '... are capable of accepting responsibility for ... foundation design, ground engineering and/or specification of remedial works including site inspections during the contract ...' should be commissioned to carry out further investigations. In a joint publication, the Institution of Civil Engineers and the Building Research Establishment (Freeman et al. 1994, p.49) considered the issue. They commented that the 'necessary skills to do this does not fall neatly into any one of the professional disciplines within the building industry.' Consequently further investigations into subsidence damage should be organised by an appropriate professional experienced in that sort of work regardless of their designation. It is worthwhile noting that many lending institutions have clearly defined criteria for who should be used in these situations.

4.5 Reporting on building damage

The preceding sections outlined an assessment framework against which building movement can be judged. This can be summarised as follows:

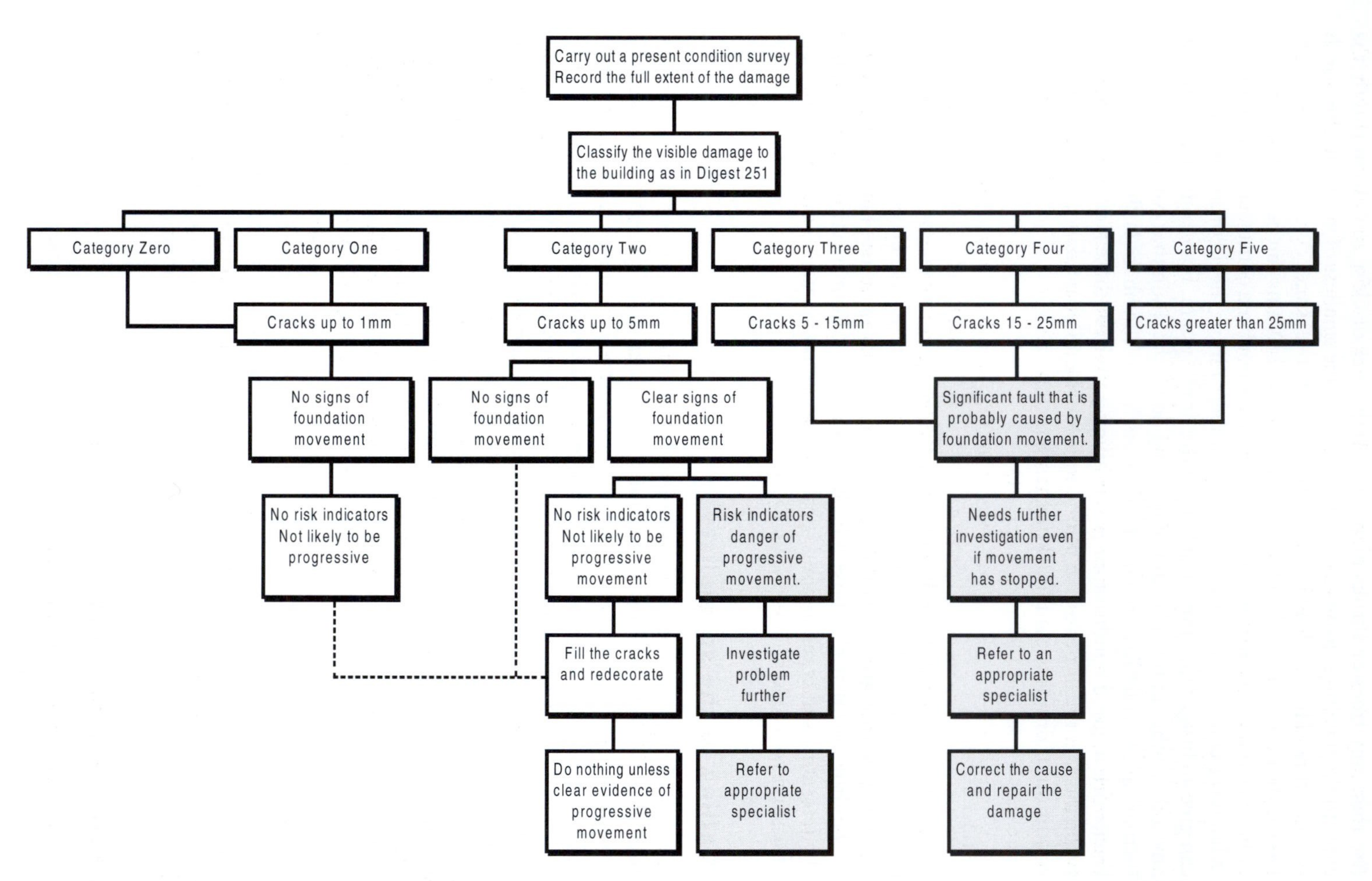

Figure 4.7 Summary of the process of assessment of building movement. This table identifies the main stages in the process. The shaded boxes represent those situations when a referral to a specialist may be required (reproduced courtesy of BRE).

- If the damage observed is in category 3, 4 or 5 (cracks of 5 mm or over) then the movement probably originates in the ground. At this magnitude the dwelling is so significantly damaged that further investigation is required even if the movement has stopped. Referral for further investigation is appropriate.
- If the damage is category 2 or less (i.e. cracks of 2–5 mm) and:
    - the damage appears recent (less than 5 years);
    - the pattern of the damage suggests foundation failure;
    - there are 'risk' indicators present;

  then referral to a specialist for further investigations is probably appropriate.
- If the damage is category 2 or less, it is older than 5 years and there are no 'risk' indicators then referral to a specialist might not be appropriate. Restoration of appearance is all that may be required.
- If the damage is in the categories of 0 or 1 (that is less than 2 mm) and no risk indicators are present then it is probably of no consequence. No further action is required and the damage can be dealt with during normal decoration works.

This guidance is very broad and has to be applied in context for each and every case. There are always exceptions to the rules. If surveyors are going to adopt a more positive approach to assessing building damage, this framework will provide a robust basis for good practice.

4.6 Trees and buildings

4.6.1 Introduction

Trees in urban areas can enhance the quality of the environment in many different ways. They can also do a great deal of damage to nearby buildings if they are too close and are not managed properly. The ISE (1994, p.37) claim that 50–60% of subsidence claims are associated with trees. The BRE (1990, p.3) state that a tree must be at least its mature height away from any building. If they are in a group then this should be increased to 1.5 times the mature height. This is sensible advice because the damage caused during drought conditions can be considerable. But if this rule is blindly followed then most trees in urban settings would have to be chopped down. In some streets it is difficult to find a tree that is not within its mature height of a building. Evans (1998) considers the recent 'drought' that was formally declared at an end in the summer of 1998 by the Environment Agency, resulted in the wide-scale persecution and premature death of many trees. One of Evans' greatest causes for concern is the 'safe distances' that have been widely publicised. He claims that they are 'alarmist' and represent a '... fundamental misinterpretation of the root spread data produced by the Kew tree root survey'. Consequently current guidance has to be sensibly moderated and applied by the surveyor.

4.6.2 Trees and their effects on buildings

Trees can cause a number of problems:

- the moisture extraction of the root systems can cause shrinkage in susceptible soils;
- the pressure of a growing mass of roots can cause retaining walls to collapse;
- roots can enter even the smallest cracks in drainage pipes causing further disruption and blockage;
- leaves can block gutters causing them to overflow. The resulting dampness can lead to the establishment of wet and dry rot in the building;
- trees can fall on buildings (and people!) and damage them.

Some trees can also lead to soiling of pavements, cars, etc. through the deposit of sticky liquid and fruits. Building owners are faced with the additional costs of clearing up leaves, branches and twigs, etc.

4.6.3 Factors to investigate

To avoid over-reaction, when assessing the effect of trees on a property, a number of key issues have to be evaluated;

- **step one** – the nature of the subsoil;
- **step two** – the type and maturity of the tree;
- **step three** – how close it is to the building.

These will be looked at in more detail below.

Step one – nature of the subsoil

Shrinkable, cohesive soils are the most at risk. The extraction of moisture by tree roots causes the clay to desiccate and shrink. If this occurs too close to a building, especially one with shallow foundations then subsidence could result. On the other hand, with non-cohesive soils (sands, gravels, etc.) trees are unlikely to have any effect. This emphasises the importance of knowing about local subsoil conditions.

Step two – type and maturity of the tree

Not all trees pose the same problems. Research has shown (BRE 1985) that some trees have a higher rate of moisture extraction than others. In these cases and where the subsoil has shrinkable characteristics, greater caution must be exercised. The most dangerous trees include:

- ash
- beech
- elm
- horse chestnut
- lime
- maple
- oak
- popular
- sycamore
- willow

This list may vary between different commentators and lending authorities so it is important to check any particular requirements. There are thousands of different tree varieties in this country and correct identification is the vital first stage in properly advising a client. Like any complex process it is important to follow a step-by-step approach to tree identification. There are many different sources of information but Barrett (1981) put forward a method that would be the most helpful to the less experienced surveyor. This involves looking at three different characteristics of the tree:

- the leaf
- crown and trunk shapes
- bark types

IDENTIFYING LEAVES

Because most trees will be identified by their leaves, this section will look at this aspect only. There are several stages:

- **categorising into generic leaf group shapes** – all conifers have either needle or scale-like leaves. Broad-leafed trees can have simple or compound leaves. These different types are illustrated in figures 4.8 and 4.9 and represent the first stage in the identification process;
- **classifying individual leaf shapes** – once it has been initially grouped then the leaf should be further classified according to the different categories shown in figure 4.10. This then reduces the range of possibilities down to a few manageable varieties. Only high water-demand trees have been included in this particular table. Barrett (1981) has put forward a method that would be the most helpful to the less experienced surveyor.

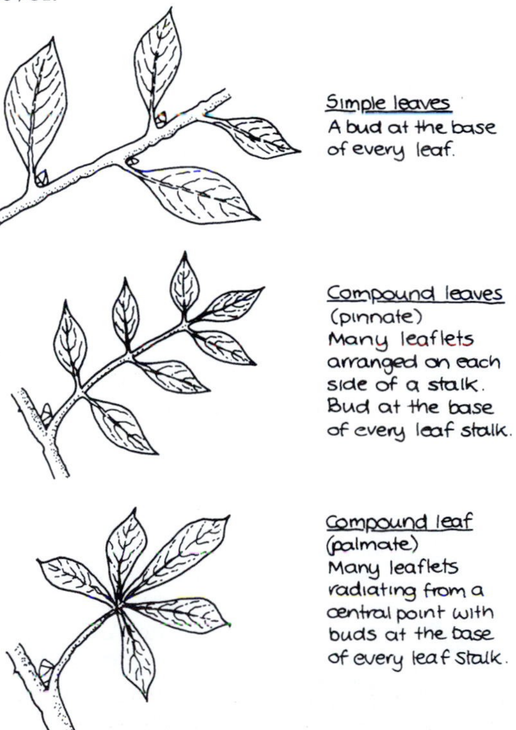

Figure 4.8 Sketches of generic leaf shape groups for deciduous trees (reproduced and adapted from *Usborne Guide to Trees of Britain and Europe*).

MATURITY OF THE TREE

Trees like any other living things go through a number of stages in their life cycle. It is important to try and identify this because it can give an indication of what growth has occurred and what can be expected. This can influence the advice given to the client. There are three main cycles:

- **First cycle** – This is generally the first third of the tree's life where growth is the most vigorous. It pushes out roots at a substantial rate and can cause a lot of damage to buildings during this stage.
- **Second cycle** – The tree tends to consolidate its position reaching an equilibrium with its environment. If there is a substantial dry period then the root system will rejuvenate and expand to find new sources of moisture.
- **Third cycle** – This tends to be a period when the root system contracts until the tree dies. There is little chance of damaging growth during this stage.

Identifying which cycle a tree is in can be estimated by measuring a tree's height and girth of the trunk. Although the girth dimension might be easier for a surveyor to measure with a tape, noting the height might be a little more problematic as climbing is not recommended! There are a number of techniques that could be used:

- Hollis (1986) suggests using a card simply marked out with a number of lines representing angle measurements from the horizontal. Sighting along this card, the surveyor walks away from the tree until the top of the tree corresponds to the 45 degree angle. As long as the distance from the tree is measured, the height is a matter of simple trigonometry.

Leaf Group Shape	Description	Typical tree type
	Single, very narrow leaves attached individually to the twig	• Douglas Fir • Western Hemlock • Norway Spruce • Yew
	Needles in bunches or clusters growing from woody knobs on the twig	• European Larch • Cedar of Lebanon • Atlas Cedar
	Stiff or soft and flexible needles in pairs, 3s or 5s.	• Scots Pine
	Feather-like arrangement of needles in flat rows	• Swamp Cypress • Dawn Redwood • Coast Redwood
	Tiny or large scale-like leaves that overlap and cover twig.	• Leyland Cypress • Western Red Cedar

Figure 4.9 Table of the most common coniferous leaf shapes (reproduced and adapted from *Usborne Guide to Trees of Britain and Europe*).

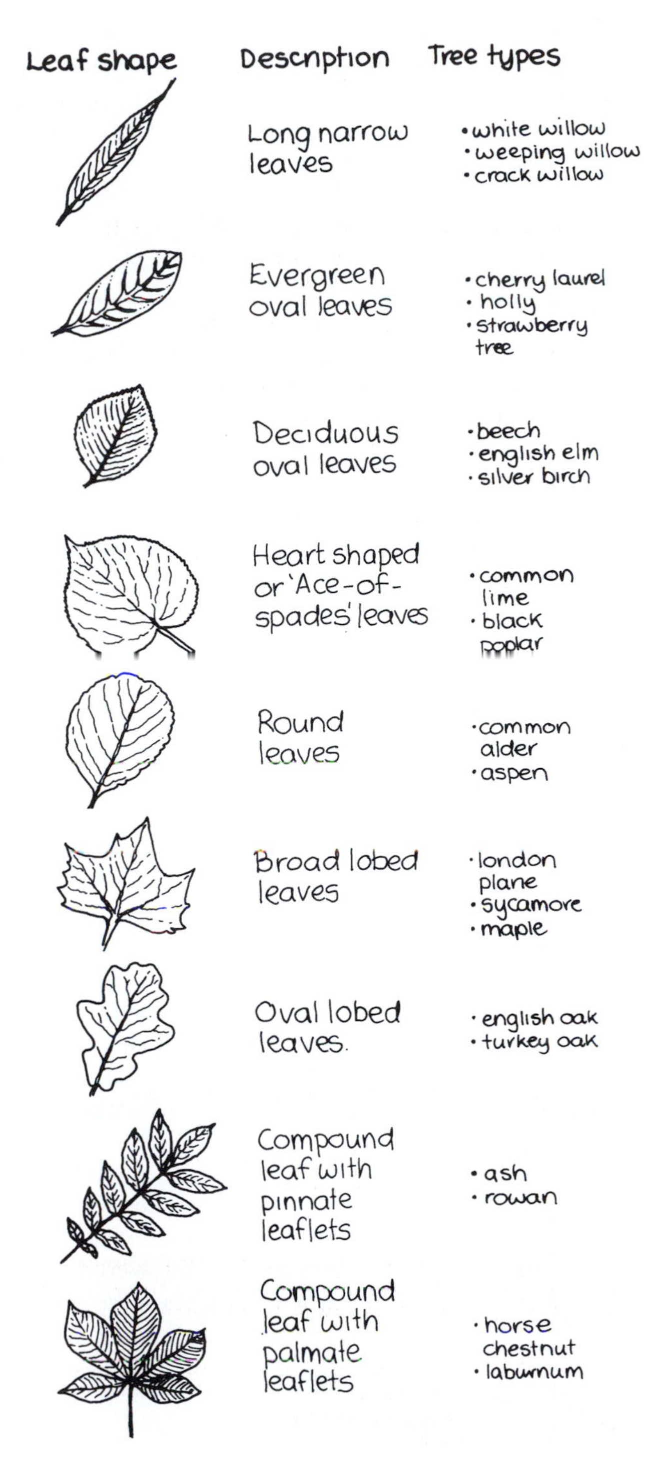

Leaf shape	Description	Tree types
	Long narrow leaves	• white willow • weeping willow • crack willow
	Evergreen oval leaves	• cherry laurel • holly • strawberry tree
	Deciduous oval leaves	• beech • english elm • silver birch
	Heart shaped or 'Ace-of-spades' leaves	• common lime • black poplar
	Round leaves	• common alder • aspen
	Broad lobed leaves	• london plane • sycamore • maple
	Oval lobed leaves.	• english oak • turkey oak
	Compound leaf with pinnate leaflets	• ash • rowan
	Compound leaf with palmate leaflets	• horse chestnut • laburnum

Figure 4.10 Table of the most common deciduous leaf shapes (reproduced and adapted from *Usborne Guide to Trees of Britain and Europe*).

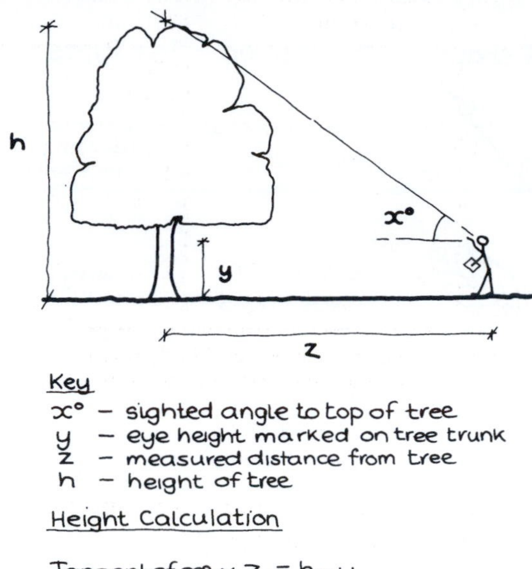

Figure 4.11 Estimating the height of trees.

- Another more conventional approach is to use a traditional surveying instrument called an Abney level. This measures of the angle of inclination and therefore results in a more accurate estimate of the height of the tree (see figure 4.11). This method is better suited when the size of the garden is limited.

Using the approximate height and girth information in figure 4.12, a rough estimation of a tree's life cycle can be made.

Step three – proximity to the building

Once the type and life cycle of the tree have been identified, the distance from the building is the next issue. The BRE have applied a general rule of:

distance away from building = 1 × mature height of the tree.

This advice has been modified over the years. For example, a study carried out at Kew Gardens in London suggested that halving the safe distances for the various trees on shrinkable clay is only likely to result in a 10% increase in damage to buildings (Cutler and Richardson 1989). A large insurance company has discovered that 90% of damage to property on shrinkable soils occurred when the trees were within the following distances:

- high water demand trees, 20 m
- medium water demand trees, 10 m

Tree type	Mature height(m)	Girth (m)
Ash	23	varies
Beech	20	up to 6
Elm	25	4–6
Horse chestnut	20	max 2.8
Lime	21–24	4.5–7.5
Maple	21	varies
Oak	24	3–10
Poplar	28	2.5
Sycamore	21–24	1.5
Willow	24	6

Note: These sizes will vary according to species and location. The figures have been based on an average taken from a number of sources. This assumes an urban environment where mature height will be restricted.

Figure 4.12 Height and girth of some common trees.

Therefore, these two distances could represent a compromise between minimising the risk to the building and protecting the tree.

ORNAMENTAL TREES AND SHRUBS

Many property owners seeking privacy and seclusion plant a variety of ornamental shrubs and hedges close to their dwellings. One of the most common is the Leyland Cypress or 'leylandii'. If they are well managed and clipped they may not present a problem. Because these trees have a mature height of up to 25 metres, uncontrolled growth can lead to high levels of water extraction especially if there are 20–30 of them along a boundary! On shrinkable clay, anything over 4 metres high within 4 metres of the building should be seen as posing a threat.

4.6.4 Reporting on trees

Advising clients on what to do about trees would be easy if it was just a matter of cutting down those that were too close to the building. In many cases if a tree is removed, moisture will return to the subsoil causing it to swell. This would then lead to further damage to the building through the heaving up of the ground beneath. This can be more damaging than subsoil shrinkage and can continue for many years. Consequently care must be taken over what advice is given. A decision framework is outlined below but this must be set against the clear and preferred guidance that many lending institutions and other clients give.

Where the subsoil is known to be a non-shrinkable, non-cohesive type

There is no significant risk from trees unless:

- the branches are either touching or soon likely to touch the building during wind action. Pruning and tree management might be needed;

- the tree is in poor condition where parts of it are in danger of collapse;
- where drain runs pass very close to the base of the tree and/or there is evidence of root growth in the drainage system.

Where the subsoil is known to be a shrinkable cohesive type or the true nature of the soil is unknown

There are no significant risks from trees unless:

- high water demand trees are within 20 metres of the building;
- medium water demand trees are within 10 metres of the building;
- ornamental shrubs/hedges 4 metres or taller are within 4 metres of the building;
- the branches are either touching or likely to touch the building during wind action. Pruning and tree management might be needed;
- the tree is in poor condition where parts of it are in danger of collapse;
- where drain runs pass very close to the base of the tree and/or there is evidence of root growth in the drainage system.

Referral to specialists

Where trees do pose a danger to buildings or their drainage systems, further investigations are often appropriate. Because of the inter-related nature of these problems, the following specialists might be required:

- a structural engineer or building surveyor to deal with the building matters;
- an arboriculturalist or tree specialist to advise on tree management/removal;
- a drainage contractor to test/inspect the drainage system.

This level of investigation could be very expensive and so advice should be carefully given.

Liability to other building owners

If the damage has been caused by trees on neighbouring properties, the insurance company may seek to recover the costs of the remedial work from the neighbouring owner. Although this is a complex area of law, when advising on trees the surveyor must also look beyond the boundaries of the property under consideration. For example are there trees on the public highway or neighbouring gardens far enough away? Are there trees in the property being surveyed too close to the neighbour's house? These matters should be brought to the attention of the client.

4.7 Other forms of building movement

So far the focus of this chapter was identifying foundation movement because it is a defect that leads to the very damaging consequences. This section will briefly outline some of the other causes of building movement that may also lead to damage. These all originate within the building structure itself.

4.7.1 Lateral instability

Section 4.3.2 stated that movements that originate in the superstructure of a building rarely give rise to damage that is greater than category 2. Lack of lateral stability to the walls of a dwelling is one of the many exceptions to that rule! For new buildings, the Building Regulations (HMSO 1991) demand that the walls are securely tied back to the rest of the structure (figure 4.13). This is to ensure that relatively thin walls do not buckle under normal loading conditions.

Invariably older buildings were not provided with this lateral restraint especially to gable and flank walls. Where joists run from the front to back of a property, structural integrity is further undermined by the presence of stairwells that represent structural discontinuity (see figure 4.14). In the worst cases, the wall can begin to 'bow' outwards to form a large 'belly' or bulge in the wall. If this goes beyond a certain point, instability and partial collapse can result. The level of instability depends on:

- the strength of the bond between the bricks and mortar;
- the level of restraint offered by the rest of the building structure;
- the slenderness ratio of the wall. That is the relationship between the height of the wall and its thickness. The higher and thinner the wall is, the less load it can carry.

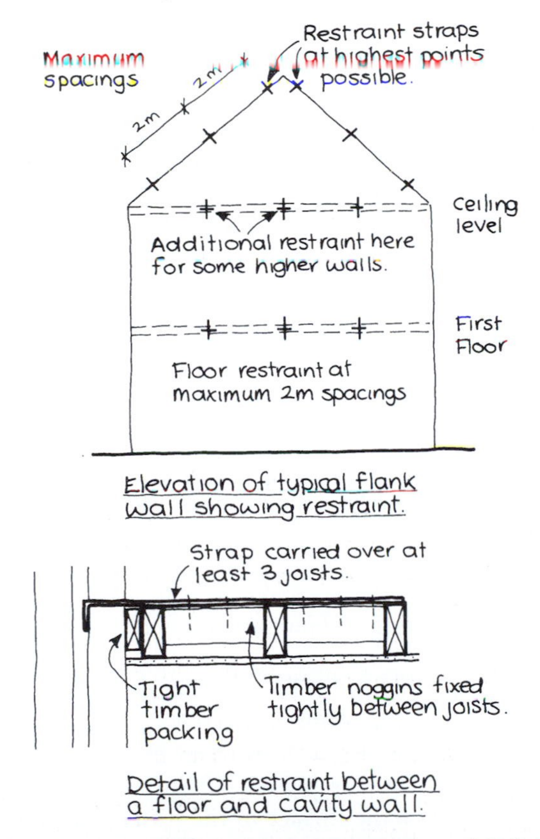

Figure 4.13 Restraint of walls as prescribed in the Building Regulations (1991) for new houses (Crown copyright. Reproduced with the permission of the Contoller of Her Majesty's Stationery Office).

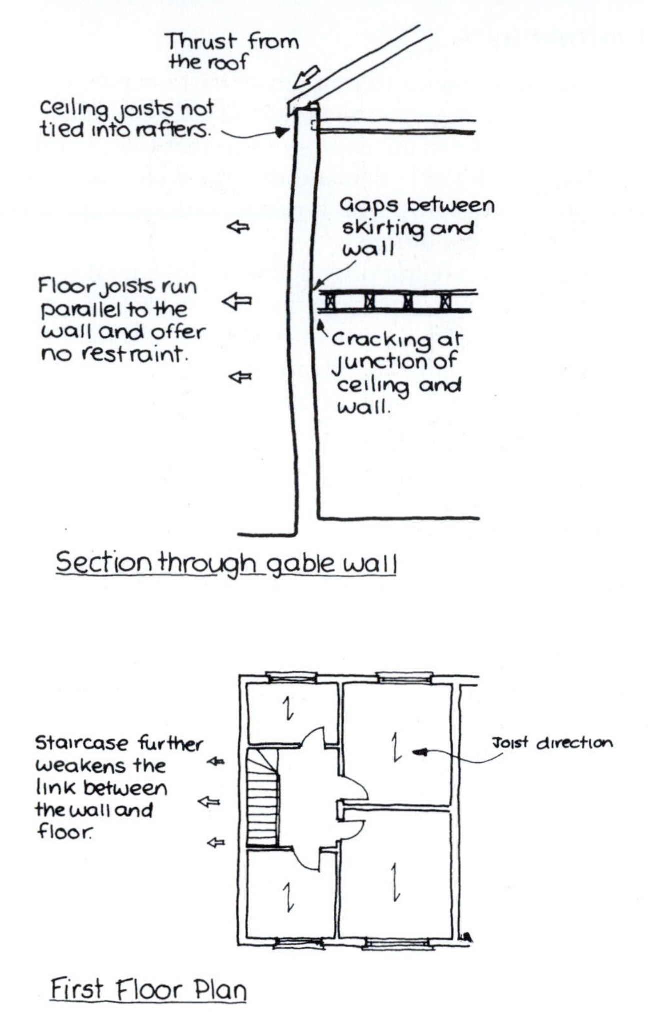

Figure 4.14 A typical situation where lateral instability to a gable wall can occur.

The balance between these factors will determine the amount of distortion.

Recognition

Bulging walls can be difficult to spot. Here are a few tips:

- All walls can be affected by this problem but it is most common with gable and flank walls so these will be the prime suspects.
- Visually sight down the line of a wall from an external corner. This may reveal parts of the wall that are considerably out of true.
- Using a bricklayer's level on the surface of the wall can help but it is unlikely that it will be placed in the position of maximum bulging. It can also be misleading as it will not extend over a large enough area of the wall.

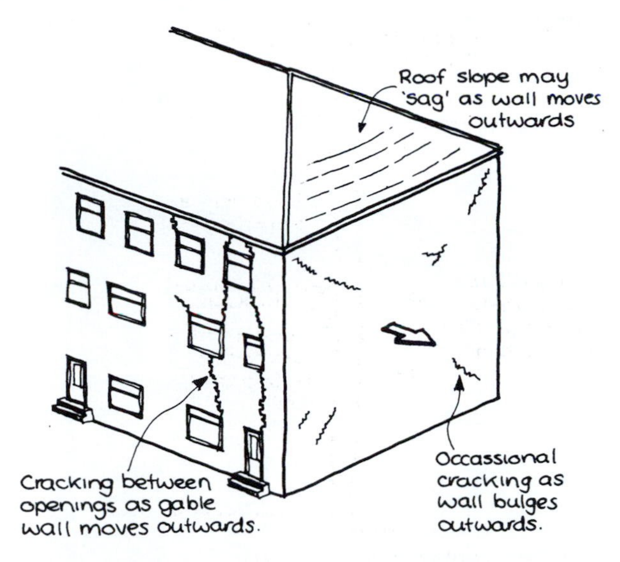

Figure 4.15 Typical cracking patterns associated with lateral instability to gable and flank walls.

- Dropping a plumb line from either the top of a 3-metre ladder or from an adjacent window can give a rough but effective measure of the alignment of the wall. This level of investigation may go beyond the scope of mortgage valuations but would be appropriate for a Homebuyers survey.
- The cracking pattern associated with bulging walls can be varied. Depending on circumstances, there may even be a total absence of cracking especially with weaker lime mortars. Try to imagine a large balloon that is being inflated in a building. The cracking that would result can be similar to that caused by bulging walls (see figure 4.15).
- Internally, this movement can be characterised by:
  - cracking at ceiling/external wall junctions and at internal/external wall junctions;
  - gaps between the external wall and adjacent flooring.

The affected wall might also show evidence of previous remedial repair attempts. Restraint plates and associated tie bars secured to floor and roof timbers might be evident (see figure 4.16). These may indicate an attempt to resolve the defect but the adequacy must be assessed. Local builders often install their own 'traditional' solutions without professional guidance. In these cases, ask the following questions:

- When was the work carried out? Was it properly designed and installed by a competent person?
- Is there any evidence of further movement since the measures were installed?

If there are negative responses to these questions, then further investigations may still be required. More specific advice on the suitability of retro-fitted lateral restraints has been published by the BRE (1997).

Figure 4.16 External restraint plates to these terrace properties clearly indicates the possibility of lateral instability.

Assessment and advice

Official guidance on how much a wall can be out of line before it becomes structurally unstable is contradictory. The BRE (1989) suggests that a two-storey wall should not have been more than 20 mm out of plumb at the time of its construction. For a single-storey building this equates to 10 mm. Anything less than this is of no concern. Parkinson (1996) suggests the following:

- where the wall is less than one sixth of its width out of plumb (35 mm for a one brick thick wall), no remedial work may be required;
- where the wall is between one sixth and one half out of plumb (35–100 mm for a one brick wall) then remedial tying back to the main structure will be required;
- where the wall is more than half of its width out of plumb then stability is threatened and rebuilding is likely to be necessary.

Because the latter two categories involve specialist knowledge, referral to a building surveyor or a structural engineer may be appropriate.

4.7.2 Cavity wall tie failure

Cavity wall tie failure is a modern phenomenon and is more common in some parts of the country than others. Cavity walls have been used since the early 1900s in many northern and western regions but were not widely used in the south until after the Second World War. Local knowledge is very important here as construction methods vary even within regions.

The cause

Cavity wall tie failure has been well described by other authors so it is only briefly summarised here:

- Older, thick wrought and cast iron ties were originally used to tie the leaves of the walls together.
- These were often inadequately protected; many were merely dipped in liquid bitumen before being built in to the mortar course.
- In exposed areas long periods of dampness have led to corrosion of the wall tie. Because rusting iron can expand by up to seven times its original size, these ties can literally 'lift' up the outer skin of the wall resulting in a horizontal 'gap' or 'crack'.
- In some regions this effect was intensified through the use of aggregates that contained high levels of sulfates. When combined with moisture, it produces an acidic environment that accelerates the corrosion process.

Recognition

This affects cavity walls. This sounds obvious but many surveyors have attributed cracking in a solid wall to wall tie failure! It is often unmistakable. Straight horizontal cracking at 450 mm spaces (or every four or five courses) is obvious especially on rendered cavity walls. Typical signs and symptoms are illustrated in figure 4.17 and described below:

- The horizontal cracking often increases in frequency in the higher areas of the wall. This is because there is less self weight of the wall at upper storey levels allowing the pressure of the expanding wall ties to overcome the weight of the wall.
- Look out for recently repointed bed joints. Building owners will often repoint cracks to prevent wind blown rain from entering their homes.
- Cavity wall tie failure usually occurs on the most exposed elevations (west and southwest facing).
- The cumulative effects can result in the expanding outer skin of the wall lifting up the edge of the roof structure. This produces a 'pagoda effect' as the verge tiles rise up.
- Where there is a room in the roof, the pressure of the expanding wall can be overcome by the substantial weight of the roof structure and the uppermost floor. This prevents the upward movement and forces the external skin to bulge outwards because it has nowhere else to go. This can often be mistaken for lateral instability or bulging caused by sulfate attack.

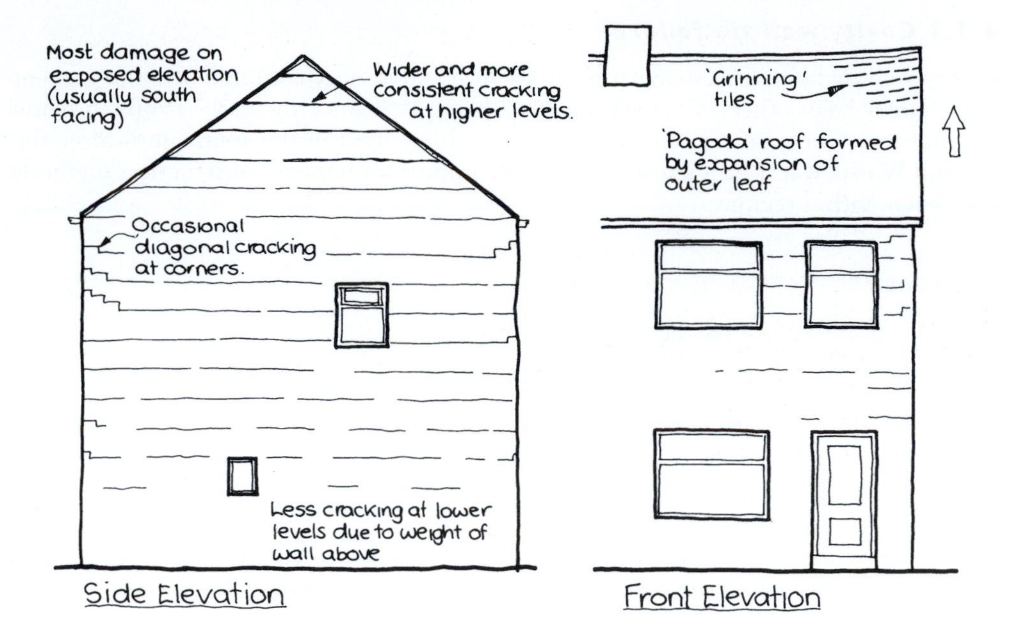

Figure 4.17 Typical signs of cavity wall tie failure.

- Small diagonal cracks can occur around window openings or at the corners of the building where the expansion can be partially restrained.
- The cracking only affects the outer skin. It is rare that it extends to the inner skin.
- Look at similar properties, do they show signs of wall tie failure or evidence of repair?

In the worst cases the corrosion can be so bad that the ties break or snap. This will leave the outer skin unstable and vulnerable to possible collapse especially on windy days.

Some properties will show evidence of repair. There are two main methods:

- a complete brick is taken out and replaced so that the old wall tie can be completely removed and a new one inserted (figure 4.18);
- the old wall tie is isolated from the wall by grinding away the portion built into the external skin. A new wall tie is installed through a hole drilled immediately above or below.

Where this has been done, a check should be made whether all elevations have been included, also that the work has been carried out by a reputable contractor with an insurance-backed guarantee.

Assessment and advice

The risk of cavity wall tie failure must be assessed in a balanced way. Here are a few tips:

- pre-war cavity walls are probably the most at risk;

Figure 4.18 A dwelling that has had its cavity wall ties replaced. The technique probably involved the complete removal of the defective wall ties. The guarantee for this work would have to be checked and should be backed by an appropriate insurance scheme.

- regions that are known to have used aggressive mortars must be surveyed carefully;
- performance is the still the best indicator. If there is no evidence of cracking or partial repair/repointing on the most exposed elevations then there is not likely to be a problem.

If these symptoms are noticed then further investigations are required. Although there are specialist contractors and installers in most regions of the country, care must be taken about to whom your client is referred.

4.7.3 Sulfate attack

This is the term that is applied to the chemical reaction between sulfates and the various constituents of cement. It can lead to the eventual breakdown of the mortar joint. Three ingredients are necessary for sulfate attack to begin:

- sulfates – these are usually present in the bricks themselves or can be introduced by combustion gases in a brick chimney;
- tricalcium aluminates (a normal part of cements);
- water.

In most cases the sulfates and cements are always present therefore the controlling factor is water.

Recognition

Because of the influence of water, sulfate attack can be expected in the following locations:

- exposed parapet walls where both sides are exposed to driving rain;
- exposed brick elevations (south facing);
- chimneys and other flues.

Sulfate attack can also affect cellar walls and other retaining structures. The main visual signs to look for are:

- Expansion of the mortar especially the bed joints can cause considerable vertical expansion giving rise to symptoms of 'pagoda roofs' and/or bulging brickwork similar to cavity wall tie failure. The distinguishing feature being that sulfate attack affects every bed joint while cavity wall tie failure affects every fourth or fifth bed joint only.
- Close inspection of the mortar joints reveals:
    - fine cracks along the length of every mortar joint. This is especially noticeable on rendered walls where the horizontal cracking becomes obvious;
    - whitish appearance to the mortar joints especially where they meet the brickwork;
    - in very wet conditions, the mortar may be reduced to a soft powder or a 'mush';
    - the edges of the bricks may spall.
- Chimneys and parapet walls may lean away from the direction of the prevailing wind. This is because the damper, more exposed side allows greater rates of expansion than the drier side (see figure 4.19).
- Because the expansion occurs in all the mortar, vertical mortar joints will be affected too. This can cause the wall to increase in size horizontally causing the wall to slip on its damp proof course. Oversailing of the dpc is often noticeable at the corners of the building.

Advice

The type of advice will vary depending on the severity of the attack. One of the key factors will be the amount of disruption to the brickwork. Figure 4.20 attempts to quantify the level of damage and relate this to possible advice to clients.

4.7.4 Expansion of masonry

All building materials will change size after they have been built into a structure. This will be due to the influence of thermal or moisture changes. For example, concrete and calcium silicate bricks are formed by a wet process and will shrink after manufacture. Clay bricks are fired in a kiln and so will tend to swell and expand. If this movement is not accommodated the stresses that build up in the walls can result in cracking. The level of damage is generally at the lower end of the spectrum and is rarely more than an aesthetic issue. This cracking is sometimes confused with other more important causes of movement so a brief review of some of the main symptoms will be useful:

- Cracking due to expansion is often consistent in width and can be distinguished from foundation movement which tends to be wider at one end than the other.

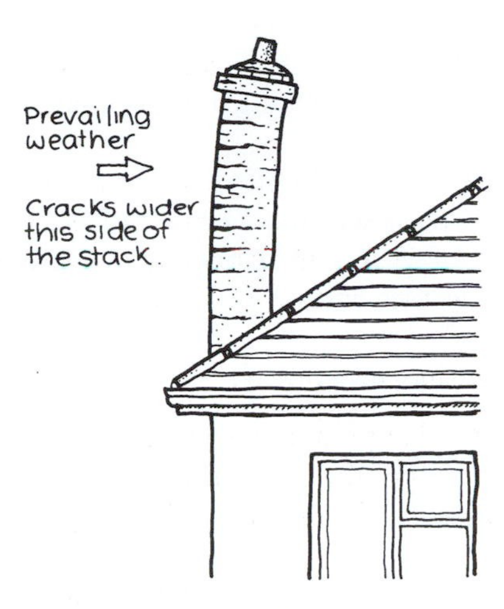

Figure 4.19 Typical damage to a chimney caused by sulfate attack.

Condition descriptor	*Description of damage*	*Repair work/ advice to client*
Good (satisfactory repair)	There is little sign of deterioration apart from the breakdown of small areas of pointing in vulnerable areas (e.g. to parapet walls, chimney stacks, etc.).	There is no clear evidence of sulfate attack. Usually there is no need for any particular repairs apart from normal maintenance work.
Fair (minor maintenance)	Although a number of mortar joints are showing signs of damage the majority are sound. The only evidence of disruption is at the corners of parapets, stacks, etc. where the elements may be cracked but stable.	In these cases, sulfate attack may not be advanced. It may be appropriate to: • rebuild isolated areas where they are damaged (e.g. corners or parapets, top of stacks, etc.); • protect exposed areas by adjusting critical detailing to prevent percolating water, etc.; • repoint remaining brickwork to min. depth of 20 mm depth with sulfate resisting cement.
Poor (significant or urgent repair needed)	Most mortar joints are badly affected and are cracked and friable. There is extensive disruption to the wall with some parts unstable (i.e. corners of parapets, stacks leaning, etc.).	Where sulfate attack has progressed to this level the only real option is to rebuild the affected area including the following features: • low sulfate content bricks; • sulfate resisting cement; • critical detailing adjusted to protect the element from excessive exposure to dampness. In some cases this could mean re-cladding the affected area to reduce the level of exposure.

Figure 4.20 Assessment of damage caused by sulfate attack. This table links condition with client advice.

- The cracks are often a straight line (although they can be diagonal!) and will often crack the masonry unit as well as the mortar joint.
- Where the wall is expanding, the whole building can 'slip' on the physical dpc. This is because the wall below is restrained by the ground while the structure above can move more freely.
- Because of this slippage effect, cracking due to expansion rarely extends below the dpc (see figures 4.21 and 4.22).

Ideally buildings must be designed and constructed to allow for any expected change in dimensions. Movement joints should be present in most walls with spacings broadly in accordance with the following rules:

- clay bricks – not more than every 12 m;
- calcium silicate bricks – 7.5–9 m at least. Special provision needs to be made to restrain these bricks at openings, returns, corners, etc. as well;
- concrete masonry – about every 6 m.

It is important to note the type of brick used in a dwelling because this can help the diagnosis of any cracking.

4.7.5 Roof spread

Chapter 7 looks at the problems of pitched roofs in more depth. If the feet of the rafters of a traditionally built roof is not properly restrained then the rafters can push outwards. This can move the wallplate and the top courses of the wall. Visually this will cause the top of the wall to lean over and could possibly result in horizontal cracking at high level (see figure 7.2).

Figure 4.21 Cracking to a masonry elevation caused by expansion of brickwork. These semi-detached properties are built of calcium silicate bricks that are vulnerable to expansion defects.

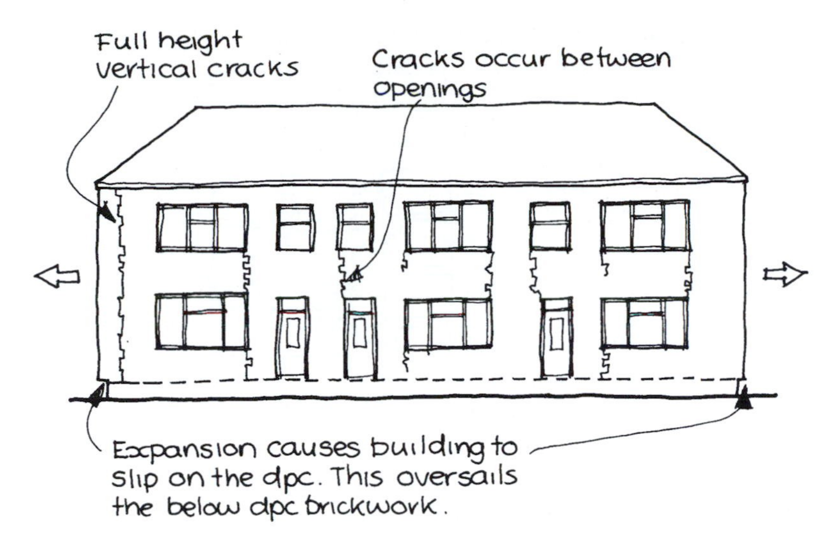

Figure 4.22 Sketch showing typical damage caused by expansion of masonry walls.

4.7.6 Embedded timbers

In traditional construction it was common to build large timbers into the thickness of external walls to provide a form of stiffness to the construction. They were often used as fixing grounds for internal boarding. In some walls, these timbers can be as close as 100 mm to the external surface of the wall. On exposed elevations, dry and wet rot can result causing the timber to shrink and distort. The weight of the walling above will cause the timber to compress. This will induce stresses on the external face where similar movements are prevented. Header bricks and through stones are often broken causing the external face to 'bow' outwards.

Visually this defect will manifest itself as a bowing or bulging in the brickwork and will be difficult to distinguish from that caused by lateral instability. In some cases the internal face of the wall can remain vertical while the external face bows outwards (see figure 4.23). The same assessment criteria should apply to this defect as with lateral instability.

4.7.7 Stone walls

Stone was once a popular building material but because of its high cost it soon became limited to the best quality work. Stone external walls can be more common in some parts of the country depending on the availability of the material from local quarries. Even where it was more available it was often used only on front elevations to save money. Although many of the structural principles are the same as for brickwork there are a few essential differences that need to be emphasised.

Types of stone walling

Before typical defects of stone walling are discussed, the different types of construction will be quickly reviewed.

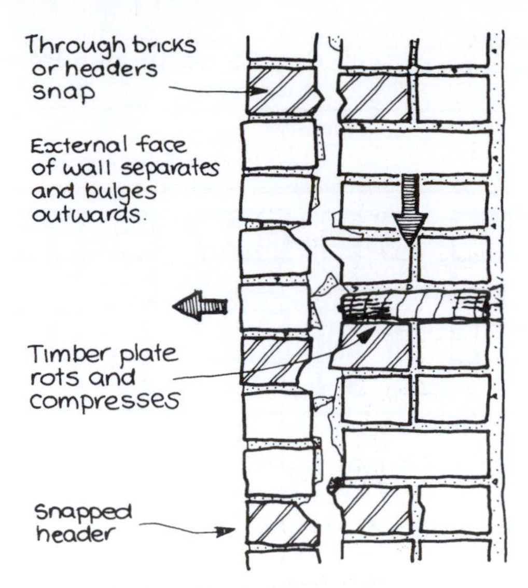

Figure 4.23 Damage caused to a solid wall by the failure of an embedded timber.

- **Types of building stone** – these include:
  - **igneous rocks** – the main type is granite;
  - **sedimentary rocks** – these include the sandstones and limestones that form the bulk of stone that is used in house construction;
  - **metamorphic rocks** – examples are slates and marbles.
- **Classification of walling** – stonework is classified by the way the walls are built. These consist of two generic types with variations as described below.

RUBBLE WORK

These are stones that have been quarry dressed and can range from the cheapest roughest stone work used on boundary walls to better quality work. The approach is similar to building brick walls but because the stones are generally irregular great skill is needed to make sure the wall is properly constructed. Stability is usually ensured if enough 'bonders' (stones that reach beyond the middle of the wall) and 'throughs' (stones that extend the full thickness of the wall) are used (see figure 4.24). The gap between the internal and external faces are usually filled with inferior materials (i.e. rubble). The category includes:

- **random rubble** – including 'uncoursed' and 'built to courses';
- **squared rubble** – which has several different types, 'uncoursed', 'built to courses' and 'regular coursed';
- **miscellaneous** – a variety of different types that are often unique to specific regions including 'polygonal walling', 'flint walling' and 'Lake District masonry'.

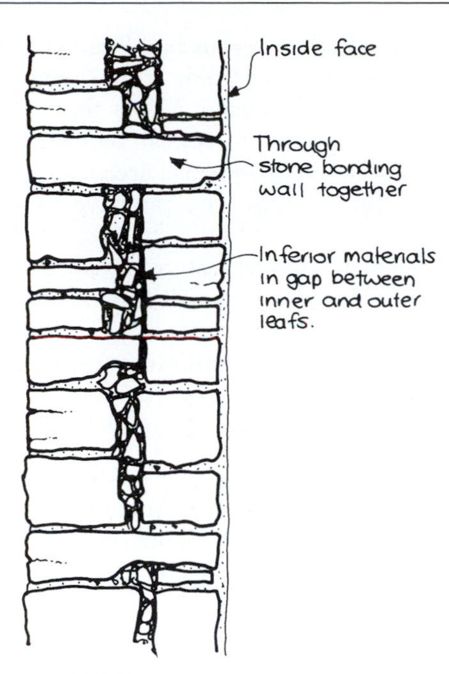

Figure 4.24 Section through typical stone wall.

ASHLAR

This class includes stones that have been accurately dressed so fine bed and end joints can be formed. The face appearance of ashlar can vary. Sometimes it can resemble Flemish bond in brick walls, arranged in courses that alternate between thick and thin, sometimes set in courses that diminish in thickness from the base upwards. Most ashlar walls are termed 'compound walls' because they can be backed with brickwork or rubble construction to reduce the costs of the construction.

Defects in stonework

Because stone is a natural material, its physical properties may vary even between blocks cut from the same quarry face. Despite this there are a number of agents of deterioration that are common to all different types of stone.

- **Incorrect bedding** – all sedimentary rocks were laid down in beds and if the stones are not laid with their bedding planes running horizontally they could be vulnerable to damage. This is especially true where stones have been 'face bedded' or edged bedded. Andrews et al. (1994) suggest an analogy with the pages of a book. If the book is laid flat it can take a lot of weight without any adverse effect. If the book is placed upright and pressure is applied from above the pages bend outwards and the book soon collapses. The same principle applies to stone. Water can travel more easily down the vertical sections and salt and frost action can cause the outermost layers to spall away (see figure 4.25).

- **Salt crystallisation** – like all other porous building materials, salts in solution can soak into the stone. These salts may come from the atmosphere, other building materials, the stone itself or the subsoil. When the wall dries out the salts will crystallise and the resulting internal forces can damage the stone.
- **Effects of acid rain** – rainwater in many urban areas can be slightly acidic because of dissolved carbon and sulfur dioxides. This can have a severe effect on some walls especially limestones.
- **Expansion of embedded metals** – steel cramps have been used on ashlar stonework for centuries. When these corrode they will expand and cause the stone to crack and fall away.
- **Frost attack** – like in brickwork, when saturated stonework freezes the ice crystals can impose sufficient pressure on the stone forcing it to spall away. Stones with smaller pore structure tend to be more vulnerable to this effect.
- **Structural instability**:
    - cracking – stone walls are affected by structural movement in similar ways to equivalent brick walls. One of the main differences is that because the individual stones are usually stronger than their brick equivalents and laid in a more haphazard way, stone walls will tend to crack differently. Cracks will follow mortar joints much more readily and because of the random coursing the nature of the cracking will be much harder to analyse;
    - bowing and bulging – stone walls can be affected by lateral instability in the same way as brick walls (see section 4.7.1). Because of the way stone walls are

Figure 4.25 Face bedded masonry wall. The outer layers of the stone can easily be 'peeled' away by the action of salts and frost. In this photograph the stones that have deteriorated the most are face bedded.

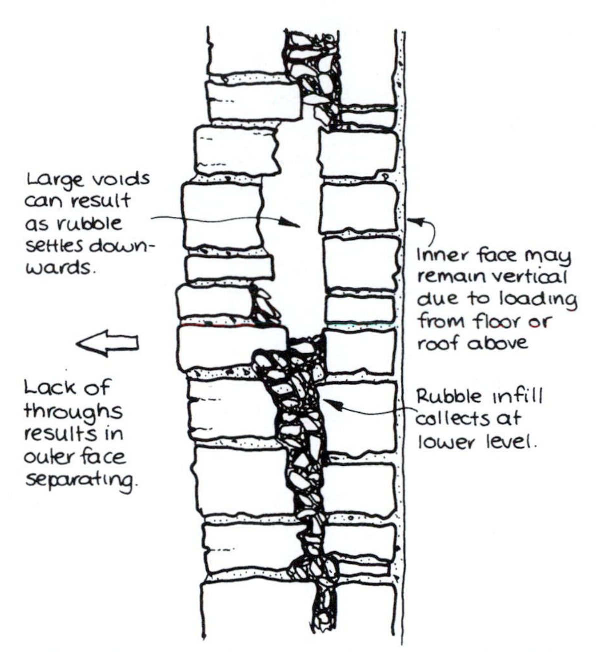

Figure 4.26 Separation of stone leaves can lead to the bulging of the external face of the wall leaving the internal side vertical.

constructed deficiencies in the way the internal and external faces are bonded together can lead to a separation of the wall. A typical example is shown in figure 4.26.

4.7.8 Stability of bays and enclosed porches

The bays and porches of dwellings are often built on shallower foundations than the rest of the building using very differential constructional methods. They may have single-skin walls that are poorly rendered. Some may be framed with timber or possibly be of composite construction. Older stone bays may be formed of what are in effect large stone blocks that are highly embellished and very expensive to replace. The intermediate floor of two-storey bays can be supported by cantilevers of the floor of the main house. Whatever the method there will be little structural connection between the bay and the main house.

These deficiencies will often result in differential movement, cracking, and further deterioration due to penetrating dampness. One of the key problems is where there is a timber bressumer or beam that spans the bay opening (see figure 4.27). Water penetration can often lead to timber decay and structural failure.

Key inspection indicators

- Does the bay/porch appear to be constructed to the same standard as the rest of the house? Intermediate columns and piers should be checked for slenderness in particular.

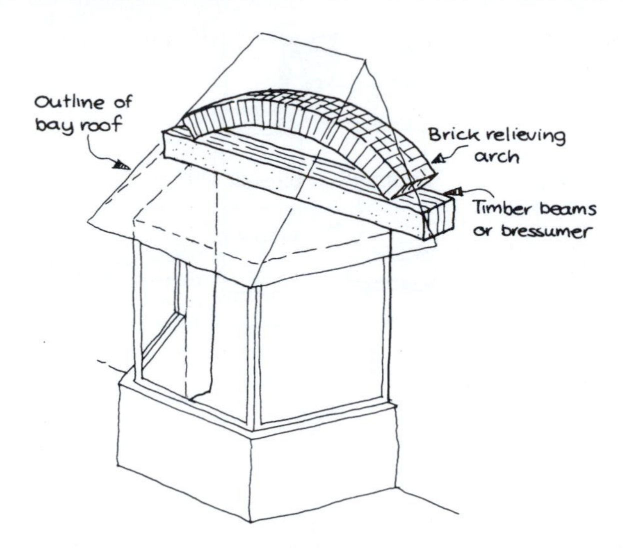

Figure 4.27 Diagram of timber 'bresummer' beams over a typical bay window opening.

- Are there signs of cracking or pulling away at the junction of the feature and the main house?
- Are there signs of water penetration into the bay? Staining to the ceiling is a typical sign.
- Have the windows of the bay been replaced? Often windows to bays are load-bearing and unless replacements are properly constructed structural movement may occur.

4.7.9 Assessing concrete structures

Although concrete structures are normally associated with larger industrial and commercial property, a significant proportion of residential properties may include major concrete components. Even relatively low rise blocks (say 3–4 storeys) can have intermediate concrete floors and balconies. Some buildings can have a complete concrete frame where the brickwork acts as a cladding with no load-bearing function. In these cases it is important that the surveyor properly assesses these components because where concrete elements are defective the repair costs can be high.

Traditionally concrete is seen as a durable, inert and long-lasting material. Its popularity has steadily increased and it is now used all over the world. But like all other materials it is susceptible to deterioration especially that caused by chemical changes.

Nature of concrete

To properly assess the condition of concrete structures, it will be useful to review a few of the main factors that influence concrete durability:

- The permeability of concrete will affect its durability, low permeability being more durable than one that is highly porous.

- Human factors can affect concrete such as poor design and construction practices, poor quality materials, etc.
- Most concrete is reinforced with steel. This steel is protected from corrosion in two ways (see figure 4.28):
  - concrete cover provides a mechanical barrier preventing water and oxygen affecting the steel;
  - the chemical reaction between the concrete and steel creates a highly alkaline environment around the reinforcement protecting it from corrosion (called passivation).

If there are deficiencies with any of these aspects then the long-term durability of the concrete could be in doubt.

Corrosion of reinforcement

One of the key factors involved in the breakdown of concrete is the corrosion of the reinforcement. Because steel expands during the corrosion process, it can cause the concrete cover to crack, spall or even delaminate (see figure 4.29). Corrosion occurs when moisture and oxygen are allowed to penetrate through the concrete cover to the depth of the reinforcement. A number of processes may help to fuel this:

- **Carbonation** – this is where carbon dioxide in the air acts on cement products. When moisture is present, calcium carbonate is formed allowing a volume reduction in the concrete. It also 'carbonates' the concrete reducing its alkalinity to a level that allows the corrosion of steel (often called depassivation). Where carbonation penetrates the depth of the concrete cover, corrosion of the reinforcement will often follow. Although the rate of carbonation will depend on a number of inter-related factors, good quality concrete is the best protection. For example, concrete

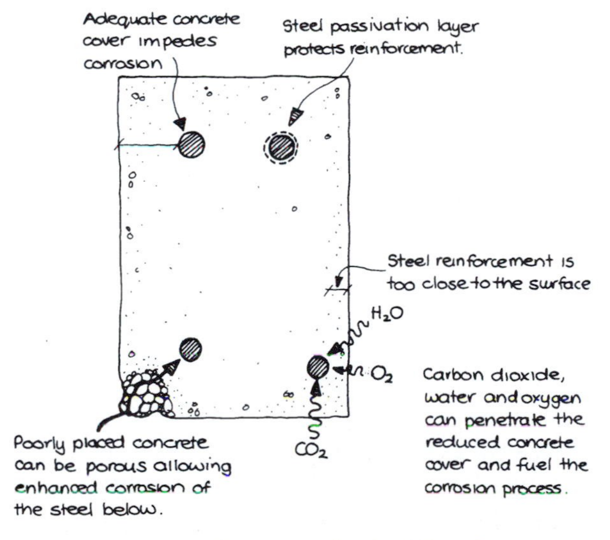

Figure 4.28 Diagram showing the main influences on the durability of concrete.

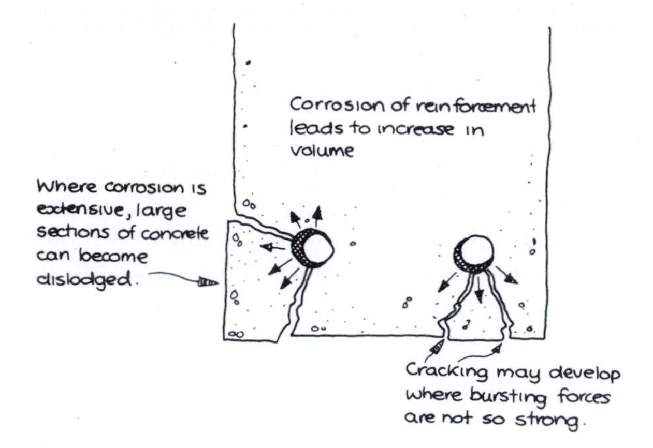

Figure 4.29 The affect of corrosion of reinforcement in concrete sections.

that is poorly designed and placed can carbonate to a depth of 25 mm in less than 10 years whereas better quality materials may take 50 years to carbonate to a depth of only 5–10 mm.

- **Chloride intrusion** – where high concentrations of chloride ions exist in concrete the protective film on the reinforcement can be broken down. Corrosion of the steel can often result causing expansion and subsequent cracking and spalling. Typical sources of chlorides include:
  - contamination from the aggregates and water used in the original concrete mix;
  - de-icing salts used on nearby roads and pavements;
  - calcium chloride that used to be used as an accelerating agent in concrete mixes;
  - highly polluted environments.
- **Alkali–aggregate reaction** – this is a complex chemical reaction where the alkaline solutions present in cement react with certain forms of silica in the aggregate to produce an alkali silicate gel. The gel can absorb water, expand and cause spalling and cracking of the concrete cover.

Testing

Where concrete has begun to show signs of deterioration, a number of specialist tests are usually carried out. These include the following:

- the most common test for carbonation is by spraying a phenolphthalein solution on a freshly fractured surface of concrete. This reacts with free calcium hydroxide present in uncarbonated cement to produce a pink colour. Where it has carbonated there is no reaction and it remains clear;
- cover measurement using a magnetic field;
- permeation characteristics test to see how easily water penetrates the material;
- a variety of laboratory-based chemical tests that establish the precise nature of the concrete;
- strength tests including rebound hammer, penetration resistance tests, etc.

Condition	*Description of condition*	*Advice to client*
Good (satisfactory repair)	The concrete appears in a good condition. There is little evidence of cracking or spalling. There are few rust stains and where they occur they are not associated with cracks. Any cracking is hairline. There are few (if any) surface defects like honeycombing, exposed aggregate, etc.	Based on this visual survey the concrete elements appear to be in good condition. Although there is no apparent need for further investigations, if the client wants to be sure about the future serviceability of the concrete, then a specialist should be appointed. This should be a person who can assess the condition of the concrete and advise and organise any necessary repairs.
Fair (minor maintenance)	The concrete is affected by a number of noticeable defects. There are a few cracks up to 1 mm in width that seem recent. There is a small amount of spalling in a few locations where the occasional reinforcement bar can be seen. There are surface imperfections that suggest poor quality concrete in a few areas.	This visual survey identifies deficiencies that could suggest the concrete is affected by a number of defects. Although the basic structure appears sound a specialist should be appointed to carry out further investigations.This should be a person who can assess the condition of the concrete and advise on any necessary repairs. This may result in some repair works to protect the future durability of the structure.
Poor (significant or urgent repair needed)	The concrete is in a very poor condition. Many surfaces are affected by cracking and spalling. In a number of areas this has revealed heavy corrosion of the reinforcement below. The surface imperfections suggest poor quality concrete	The visual survey has revealed the concrete structure to be in a very poor condition. Although there has been no specialist assessment it is likely that a considerable amount of money needs to be spent to ensure the structure continues to fulfil its function. It is important to appoint a specialist who can assess the true condition and advise and organise the necessary repair work.

Figure 4.30 Table linking the condition of concrete to client advice.

All of these tests are beyond the scope of a typical survey. They would also be outside the experience and skill level of the surveyor. The key is to try and visually assess the concrete structure and consider whether a specialist examination should be recommended. Figure 4.30 attempts to link the condition of concrete surfaces with client advice.

Repair

The repair of defective concrete structures is a specialist operation that can range from the making good of isolated spalled areas to the reforming of large sections of defective concrete. It can include epoxy resins, polymer latex and polyester compounds. Where carbonation is involved the whole surface of the concrete may have to be coated with a protective and decorative covering. Depending on the extent of the work it can be very expensive and have an impact on value. Should a specialist investigation be required it is essential that the client should be referred to a professional person who can investigate the condition of the concrete and specify and organise an appropriate repair scheme.

References

Andrews, C., Young, M., Tonge, K. (1994). *Stonecleaning. A Guide for Practitioners*. Historic Scotland, Edinburgh.

Barrett, M. (1981). *Usborne Guide to Trees of Britain and Europe*. Usborne Publishing, London.

BRE (1985). 'The influence of trees on house foundations in clay soils'. *Digest 298*. Building Research Establishment, Garston, Watford.

BRE (1989). 'Simple measurement and monitoring of movement in low rise building, Part 2: settlement, heave and out-of-plumb'. *Digest 344*. Building Research Establishment, Garston, Watford.

BRE (1990). 'Assessment of damage in low rise buildings'. *Digest 251*. Building Research Establishment, Garston, Watford.

BRE (1991). 'Foundation movement and remedial underpinning in low rise buildings'. *BRE Report BR 184*, Building Research Establishment, Garston, Watford.

BRE (1997). 'Connecting walls and floors: design and performance'. *Good Building Guide*, 29 Part 2. Building Research Establishment, Garston, Watford

Cutler, D.F., Richardson, I.B.K. (1989). *Tree Roots and Buildings*, 2nd edn. Longman, Harlow.

Evans, D. (1998). 'The root of all evil?'. *The Valuer*, pp.18–19. November/December 1998.

Freeman, T.J., Littlejohn, G.S., Driscoll, R.M.C. (1994). *Has Your House Got Cracks? A Guide to Subsidence and Heave of Buildings on Clay*. Thomas Telford, London.

Hollis, M., Gibson, C. (1986). 'Surveying buildings'. *Surveying Publications*. London.

HMSO (1991). *Building Regulations. Part A, Structure*. HMSO. London

ISE (1994). *Subsidence of Low Rise Buildings*. Institution of Structural Engineers, London.

NES (1996). *National Home Energy Rating 'Home rater' Program Manual*. National Energy Services, Milton Keynes.

Parkinson, G., Shaw, G., Beck, J.K., Knowles, D. (1996). *Appraisal and Repair of Masonry*. Thomas Telford, London.

Chapter 5

Dampness

5.1 Introduction

5.1.1 What is dampness?

The climate of the UK can be described as 'warm and wet'. The influence of the sea and the weather systems that follow the Gulf Stream make sure that dampness is the dominant element of our climate. People living in the west of the country will be acutely aware of this (Lacy 1977, p.24). As a consequence, the exclusion of moisture is one of the biggest challenges that any building has to face. The BRE estimates that over two million houses in the UK suffer from 'severe' levels of dampness with another two million affected to a lesser extent (Garrett and Nowak 1991).

5.1.2 The effects of dampness

Dampness in dwellings has two main effects:

- It damages the health of the occupants (Ormandy and Burridge 1988, p.272):
  - through mould growth – which is closely associated with damp dwellings. Mould is a fungus that spreads by putting out spores. These microscopic particles are classed as 'allergens' and can cause an allergic reaction in the respiratory tracts of a significant number of people. Asthma sufferers are particularly vulnerable;
  - house dust mites – these microscopic insects are close relatives of ticks and spiders and live on the dead skin cells that all humans shed. They can be found in carpets, soft toys, soft furnishings, etc. but are most common in mattresses and bedding. They thrive in warm, moist conditions. The more damp a dwelling, the larger the population of house dust mites. The droppings of the house dust mite are also classed as 'allergens'. This will affect sensitive respiratory tracts in the same way as mould spores, asthmatics again being vulnerable.
- Physical effects on the building fabric include:
  - the spoiling of decorations;
  - the breakdown of plaster finishes;
  - corrosion of embedded metal components;
  - wet and dry rot in timber elements.

Therefore the financial loss resulting from dampness-related defects can be considerable.

5.2 Measuring dampness – different methods

Any building owner that has suffered from dampness problems will be familiar with a succession of different types of surveyors prodding a variety of moisture meters into the affected parts of their homes. This method of detecting moisture is by far the most common but is not the only approach. Even when they are used, the results must be interpreted with care. The alternatives are described below.

5.2.1 The senses

This includes:

- sight – very underrated. Most forms of dampness have distinct visual characteristics that can help diagnose the cause;
- smell – although this is a very subjective measure, a 'fusty' smell can alert a sensitive surveying nose to dampness problems in a property;
- touch – this can be very misleading. Touching a cold or cool surface can give the impression of dampness when the material might actually be dry, so this is very unreliable.

5.2.2 The moisture meter

There are many different types of moisture meters although one manufacturer dominates the market. Essentially there are two types:

- **capacitance meters** – these meters have a flat face that is pressed against the surface. It registers moisture content at or near the surface. The problem with this type is that a rough surface can affect the accuracy of the readings. Not many of these are used in practice;
- **conductivity meter** – these probes, usually pins, are pressed against the surface to be tested and the electrical resistance between the probes is measured. The more moisture present, the higher the reading that is registered. Older versions had a needle gauge but newer meters usually incorporate digital readouts, various coloured lights and even audible buzzers!

Interpreting the readings

Different building materials hold moisture in different ways. For example, porous, lightweight materials (timber, etc.) are generally able to hold a high proportion of their weight in moisture. Heavier, more dense and less porous materials can hold proportionally less. That is why a brick will be very wet when it has a 5% moisture content while timber can be considered 'dry' at 12%.

Because a conductivity meter is a fairly crude instrument, it cannot accurately measure the precise moisture content of a wide variety of materials. As a consequence it is calibrated to measure the moisture content of wood only. Timber is a consistent material and the probes can usually be pushed below the surface. The percentage shown is an accurate measure of the real moisture content. For brick and plaster and all other materials, the results are not so reliable. Therefore most manufacturers state that the moisture readings are only relative measures. Sometimes called 'wood moisture equivalent' percentages, they can give an indication of how wet a surface is but not an absolute measure. Therefore it is better to use the various gradations of coloured lights as a measure of how damp an element is. One manufacturer gives the following guide:

- **green** – readings in this band can be considered 'air dry' and pose no particular threat;
- **amber or yellow** – this suggests that there is excess moisture present and it is coming from sources other than the air. If readings in this category persist then further investigations may be required;
- **red** – a serious moisture condition exists. There is a clear source of excess moisture and further investigations are definitely required with the possibility of remedial action being very high (Oxley and Gobert 1994).

Therefore applying the principle of 'follow the trail', where red or yellow readings are recorded a surveyor must investigate the problem further.

Other restrictions on use

Moisture meters are not only limited by their inability to measure precise moisture contents of a variety of materials but also by other limitations:

- The way some materials gain and lose moisture can 'fool' meters. Where the pore structure of a material is very small, moisture can be lost from the surface layers leaving the heart of the material still saturated. Conversely, the same surface layers may become very wet (i.e. condensation) and cause high readings while the underlying material may be relatively dry.
- The physical nature of the contact between the pins of the meter and the material can affect the readings. On hard surfaces especially it is essential that the pins are properly seated on the surface.
- High levels of salts in the material being measured can affect the readings. This is because the electrical resistance is lowered by salt crystals and can result in high readings. Although the presence of salts can be a problem in itself it does not necessarily mean that the material is damp (see figure 5.1).

These limitations clearly demonstrate that the instruments should be seen as tools that help the surveyor. They do not give an instant diagnosis and the information they provide must always be set in context with other signs and symptoms. More information on good operating practice of moisture meters has been published by TRADA (1991).

Using a moisture meter on a survey

In all surveys, the moisture meter is an essential tool. Despite the limitations it can help surveyors identify damp areas that normally would not be apparent visually and so initiate the 'trail of suspicion'. Accordingly the moisture meter should be used at points

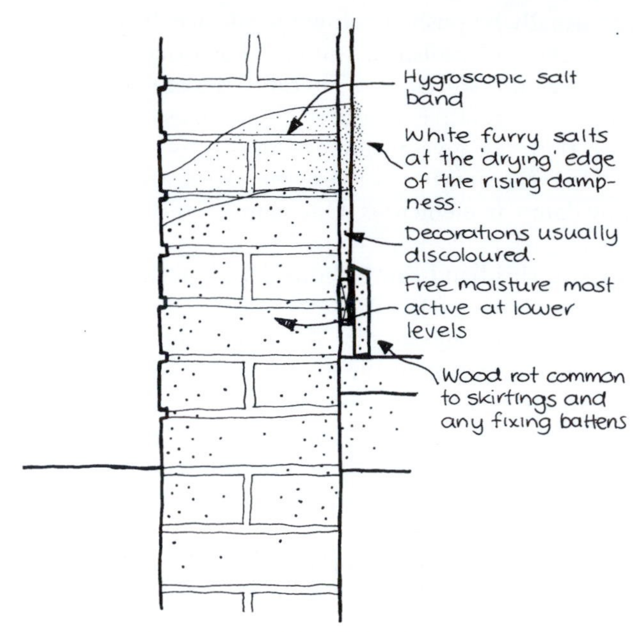

Figure 5.1 Cross section through a typical solid wall that has been affected by rising dampness and hygroscopic salts.

that are vulnerable to rising or penetrating dampness. This must always be left to the surveyor's professional judgement but typically includes:

- at the base of all ground floor walls. At each measurement position a number of readings should be taken at vertical spacings (see figure 5.2). Along walls, readings should be taken every 1.0 metre unless interrupted by furniture or fixings;
- floorboarding and solid floor surfaces where they are easily exposed by lifting the floor coverings;
- below every window sill and across its width;
- where there any visual signs of dampness (i.e. tide marks, brown watery stains, disrupted plaster and decorations, areas of mould growth, etc.);
- the reveals of every window and door opening;
- in the roof space to the party walls and chimney stacks where they are close to the underside of the roof covering.

5.2.3 Other methods of evaluating dampness

Where dampness has been detected, there is a variety of techniques and other equipment available to investigate the cause more thoroughly.

Moisture meter accessories

Many of the manufacturers of moisture meters provide a whole range of accessories that are designed to assist the surveyor evaluate other signs and symptoms associated with dampness. These include:

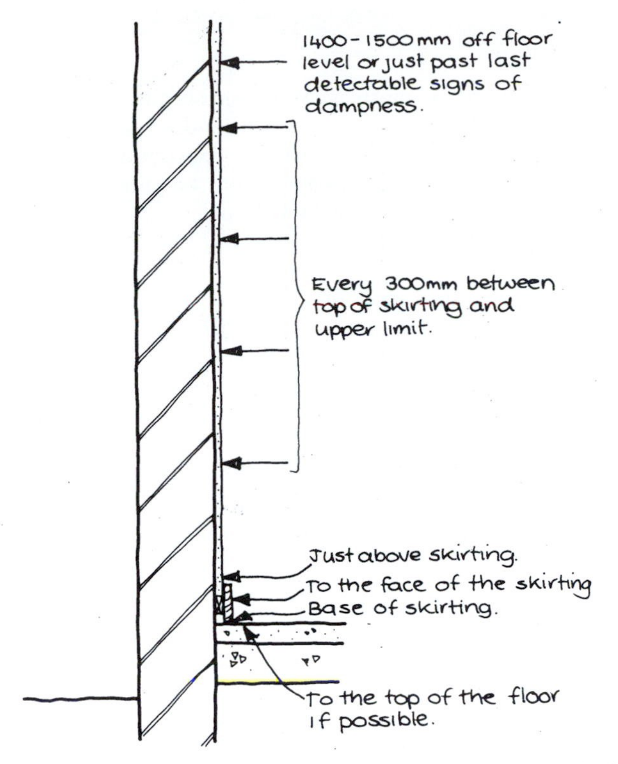

Figure 5.2 Cross section through a typical wall showing the vertical spacings at which moisture readings should be taken.

- **Deep wall probes** – these are long probes that can be connected to the meter. They look like a pair of metal knitting needles with their shanks insulated with PVCu. Two holes can be drilled into the wall and moisture readings taken with the deep probes at successive stages. The level of dampness throughout the depth of the construction can be evaluated. This is a way of distinguishing between rising dampness and surface condensation. Because it involves a destructive process (i.e. drilling) it is usually beyond the scope of most surveys.
- **Salts detector** – this is an attachment that claims to be able to detect the level of salts on a surface. The process involves recording measurements via a sensor pad before and after a small circular piece of wet 'blotting' paper has been placed on it. In certain circumstances this can be a useful aid in the diagnostic process but as with any other on-site test an acceptable level of accuracy is hard to achieve.

Destructive testing methods

Where a more precise level of diagnosis is required, specialists may use more destructive testing methods. Although these are well beyond the scope of most surveys, they may be encountered where the presence of dampness is disputed or during analysis prior to remedial work. Oliver (1997, p.262) describes the techniques in more detail but they include:

What to do with furniture?

Both the Homebuyers survey and mortgage valuation specifically exclude moving heavy furniture. Because dampness levels can be higher behind furniture than in the rest of the room, the surveyor must be very confident that it is appropriate not to move the object. There is no accurate guidance as each situation must be judged by the surveyor on its own merits. The following comments may be helpful.

Furniture should not be moved when:

- it is physically fixed to the wall or floor;
- it is so heavy that it would be beyond the capabilities of one person to move it safely (e.g. a substantial sideboard, etc.);
- it has got so many things in it or on it (e.g. ornaments, etc.) that moving it could cause damage or too much disturbance.

Where a piece of furniture could not be moved then this fact should be clearly noted in the site notes and moisture meter readings taken either side of the item. This may enable a speculative judgement of whether dampness is likely to exist behind the furniture.

Furniture should be moved when:

- it is light and easily lifted by one person (e.g. waste bin, magazine rack, standard lamp, etc.);
- the item is on castors or wheels which makes it easy to pull around (e.g. chairs, small sofas, etc.).

Whatever the circumstances, it is sensible for the surveyor to produce a simple sketch plan and note down where moisture readings were taken and what level they recorded. This was highlighted in the case of Fryer v. Bunny (1982) where a surveyor had used a moisture meter but had not detected any dampness. After the owner had moved in, dampness was discovered in the floor. The cause was a leak to the central heating pipes. The court held that more extensive use of the meter would have revealed the existence of dampness not apparent to the human eye or by placing hands against the wall. According to Murdoch and Murrells (1995, p.94), the moral of this decision '... must be for surveyors to prepare a very simple sketch of a property as it is surveyed, marking roughly where a moisture meter has been applied'.

- **Calcium carbide meter** – ('speedy' moisture meter) originally developed for testing the moisture content of more bulky materials (e.g. aggregates, grain, etc.) these kits are used where a more accurate measure is required on site. Drilled brick samples are placed in a sealed flask along with powdered calcium carbide. Any moisture reacts with this chemical to produce acetylene gas. The higher the levels of dampness, the more gas is produced. A dial on the top of the flask measures the amount of gas produced and this is expressed as a percentage moisture equivalent.
- **Gravimetric method** – this is a more precise laboratory-based method fully described by the BRE (1985b). Drilled samples are taken away and weighed both before and after drying in an oven. A standard calculation can then determine the true moisture content. Additional tests on the same samples can also determine the hygroscopic salt content of the material which can help establish the precise cause of the dampness.

5.3 Penetrating dampness

This is usually associated with rain penetration through wall and roof constructions. Roofing problems will be considered in chapter 7 and so penetration through walls will be the focus of this section.

5.3.1 Solid walls

Whether brick or stone, solid walls operate on a sponge principle. They soak up rain during wet periods holding it within the thickness of the wall until drier periods come along when the moisture evaporates again. This relies on the free movement of moisture both in and out of the material. It also depends on the wall being thick enough to prevent the moisture from reaching the internal surfaces during the wettest of weather (see figure 5.3). Problems occur when:

- the mortar pointing between the masonry units is in a poor, porous condition;
- the masonry units themselves are porous or have deteriorated; and
- the building is in a very exposed position allowing a lot of wind-driven rain to hit the wall.

Any of these deficiencies may allow moisture to travel through the wall and affect the internal surface.

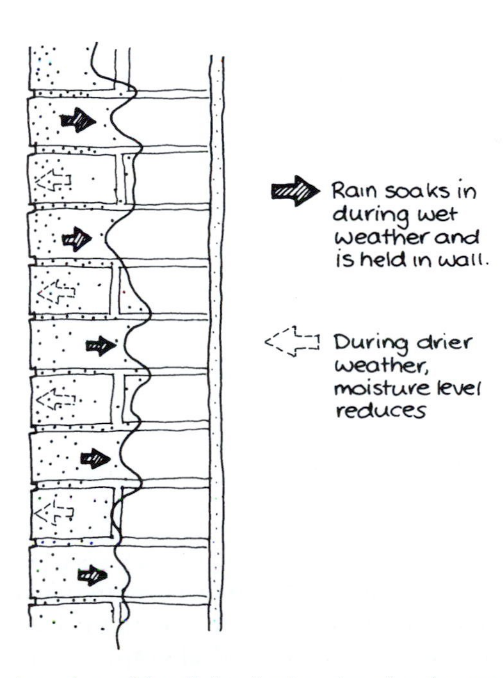

Figure 5.3 Cross section through a solid wall showing how it resists the passage of dampness.

Rendered walls

On many established houses, cement renders are often used to improve the performance of solid walls. As long as these are mixed and applied appropriately they can have a beneficial effect. If they are of a poor quality they can actually make the problem worse. Figure 5.4 shows a typical example. A strong impervious render can channel the water behind it and then prevent it from drying out. To be successful, the characteristics of the render must match those of the wall.

Other coatings

Over the last few years a wide variety of proprietary coatings have been developed and used by home owners. These include:

- **Bituminous applications** – bitumen-based paints and coatings often combined with hessian reinforcing fabric. Used for many years (sometimes called 'tunnerising') they can have short-term beneficial effects. Because bitumen soon breaks down after exposure to sunlight and temperature variations, water can easily get behind the coating and penetrate internally. The coating will also have a permanent visual effect on the building which may not be pleasing (see figure 5.5).
- **Thin-coat cementitious paint applications** – many companies have developed thin-coat render systems. These are usually a combination of paint, cement and sometimes reinforcing fibres sprayed on to the surface of the wall. These will change the visual appearance of the property and will often have the same benefits and disadvantages as normal render applications have. Where these special coatings have been used any guarantees, British Standard approvals, agreement certificates, etc. should be asked for in an effort to assess the quality of the work.
- **Transparent applications** – difficult to detect visually, the weather resistance of walls can be improved by the application of transparent applications such as silicon-based treatments. Similar to the chemical dpc systems, they rely on preventing capillary action from occurring. These applications can give improved performance in the short term but rarely have a life longer than 2–3 years especially in exposed locations.

Concentrated flow

Water penetration through solid walls often occurs where external water flow has been concentrated by other defects such as leaking gutters, rainwater pipes and running overflows. These high levels of dampness can quickly affect internal surfaces.

Window faults

The rainwater run-off from the glazed panels of windows can be very intense especially on exposed elevations. Most of this will run down over the window sill and either drip clear of the wall or flow down the face where the throating is inadequate. Wind-driven rain can penetrate around window frames and sills especially if they are poorly decorated, partly rotten or not properly sealed with mastic to the surrounding construction.

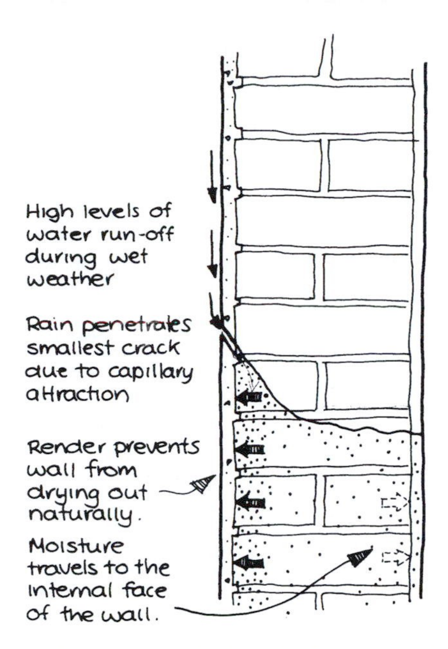

Figure 5.4 Sketch showing the effects of a cracked render finish to a solid wall.

Figure 5.5 A gable elevation of a house that has been treated with a bitumen-based compound in an attempt to improve its moisture resisting properties.

This can damage the decorations, cause the plaster to breakdown and in some cases allow wood rot to develop.

Key inspection indicators

Internal symptoms associated with penetrating dampness can include:

- damaged decorations and plaster surfaces especially below and around window openings;
- well defined and localised areas of high moisture readings that clearly match externally observed defects. These are particularly high after periods of heavy rain; and
- high levels of mould growth in the affected areas.

Externally:

- condition of the pointing and the masonry units (i.e. bricks and stones) of the affected elements with cracked and spalling rendered surfaces;
- window frames that are in a poor condition with little or no protection at the junctions with the wall construction;
- defects in the window sills such as cracks, sills out of level, inadequate projection of the sills over the face of the walling, blocked or absent throatings to the underside, etc.;
- leaking gutters and pipes;
- the orientation of the affected area. West and south-west facing elevations are particularly vulnerable in most parts of the UK. For other areas regional meteorological data should be obtained. In some cases there may have been changes in the local exposure, e.g. tree removed, adjacent building demolished, etc. This may allow a higher amount of rain to reach the building than before.

5.3.2 Cavity walls

Many of the issues affecting cavity walls are similar to those for the solid equivalent but there are some fundamental differences. Originally cavity walls were developed to combat direct rain penetration of solid walls. In some northern and western areas they have been used since early in the 1900s. On exposed locations, rain will regularly penetrate through the outer skin. This is most common at the vertical joints where the amount of mortar included between the bricks is normally insufficient. Water will then run down the internal face of the external skin within the cavity (see figure 5.6). To prevent this water from transferring to the inner skin of the wall, the method and quality of construction is critical. Problem areas include:

- poorly installed wall ties with large mortar droppings on them;
- absent or poorly installed cavity trays and dpcs above and around door and window openings;
- in modern cavity walls, poorly installed cavity wall insulation batts (BRE 1985b).

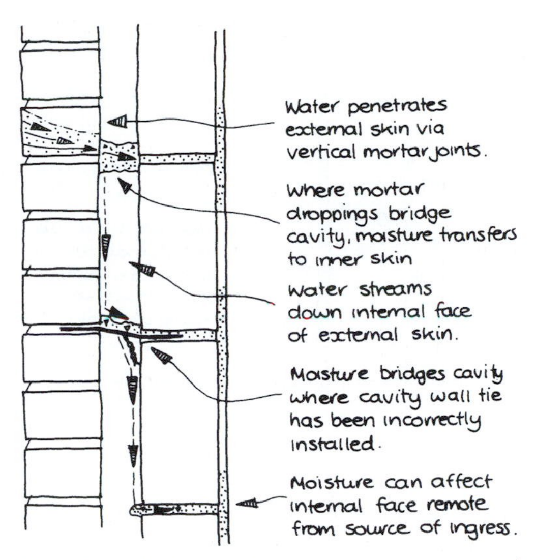

Figure 5.6 Sketch through a cavity wall showing how water can penetrate through the external skin, cross the cavity and affect the internal surfaces.

Diagnosis

The first stage is to identify whether the walls are either solid or cavity. For experienced surveyors this will seem like teaching grandparents to suck eggs but mistakes have been made. In some parts of the south (especially after the last war) some cavity walls were actually built to resemble a solid brick wall.

In many ways the signs and symptoms of dampness defects in cavity walls will be similar to those of solid walls with the following differences:

- If the fault is associated with wall tie problems, then the dampness will be more isolated.
- Where there is a dpc problem around openings the defect will be very specific to those features (BRE 1987).
- The build up of debris in the base of the cavity can give symptoms that resemble rising dampness. Telling the difference between the two is difficult. If the wall is of cavity construction then it can be assumed that it is of modern construction. Hence it is reasonable to assume that it has some form of physical dpc in place. Although this might still be defective, if classic rising dampness symptoms are observed to the base of a cavity wall then a bridging through cavity debris must at least be suspected (see figure 5.7).

5.3.3 Reporting on penetrating dampness

In their Good Repair Guide *Treating rain penetration in houses* (BRE 1997c), the BRE point out that the symptoms of these types of defects can be misleading as sometimes more than one defect is involved. Even if rain penetration to blame, the actual route it takes can be hard to pin-point especially with cavity walls. Consequently care must be exercised when reporting and advising the client:

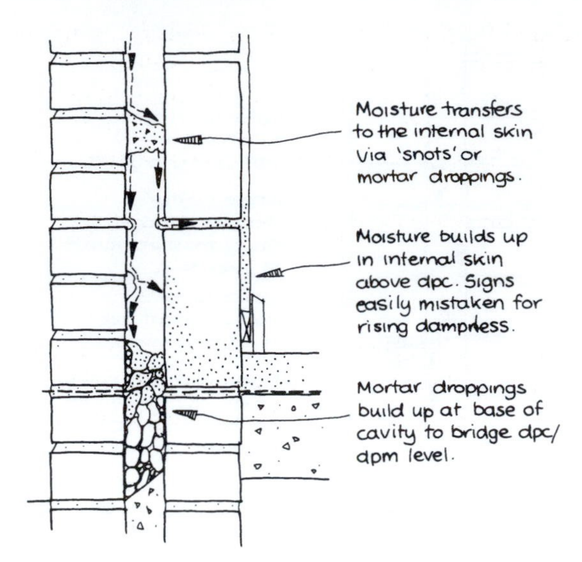

Figure 5.7 Section through the base of a cavity wall where the build up of debris and other faults can result in dampness problems internally. This can easily be mistaken for rising dampness.

- where the source of the defect is relatively obvious and easy to rectify then referral to a reputable builder for an estimate may be appropriate;
- if the cause is not so apparent then care has to be exercised.

Faults with cavity walls are probably the most problematic and clients should be advised to instruct appropriately qualified and experienced surveyors to carry out further investigations. They are more likely to have a broader range of skills and knowledge than a specialist damp-proofing contractor in this particular area who tends to be more focused on rising damp defects.

5.4 Rising dampness

5.4.1 Introduction

Rising dampness is probably the best known of all types of dampness. It is part of our cultural tradition and has even had a television sitcom named after it! It can also be one of the most damaging forms of dampness that is difficult to cure. This section will not go too deeply into the theory of how it occurs but concentrate on how to recognise and report on it. Capillary action is the main culprit allowing free water to rise from the damp subsoil into the drier wall. The height of rising dampness depends on a number of factors:

- the amount of moisture in the soil – some sites can be wetter than others. The height of the water table is very influential;
- weather conditions – rising dampness does not react immediately to changes in the amount of precipitation. Over the medium term it can become worse during wetter months;

- internal and external finishes that prevent the wall from drying out. These could include renders, tiling, wall boarding, vinyl-based wall hangings, etc. These features could force the dampness up to levels higher than could normally be expected;
- high internal humidities – if the internal air is very moist then the drying effect of the wall will be reduced. As a consequence it will rise higher and affect a larger area.

In most cases, rising dampness rarely gets higher than 1000–1500 mm.

5.4.2 The causes of rising dampness

Designers and builders over the years have been aware of the need to provide an impervious layer within the thickness of walls to prevent rising dampness. Since the Public Health Acts in 1875, physical dpcs has been common in walls and have included:

- slate
- lead/copper
- bitumen impregnated felts
- non-absorbent engineering bricks (usually blue in colour)
- specially manufactured ceramic bricks.

Depending on the age of the house, many of these original dpcs may have become defective. For example, slate dpcs can crack following movement in the walls, metals can corrode and bitumen-based products can become brittle over time. So even if there is a physical dpc present it cannot be assumed to be 100% effective.

Another major cause of rising dampness is through physical bridging of the dpc level (see figure 5.8). Many of these are associated with post-construction alterations and repairs especially those of a DIY nature. Faults in new construction can also give rise to signs that resemble rising dampness. For example the Building Research Establishment (BRE 1984) have identified a problem in new properties where the wall plaster is taken down to the surface of the floor screed that has recently been laid. The dampness

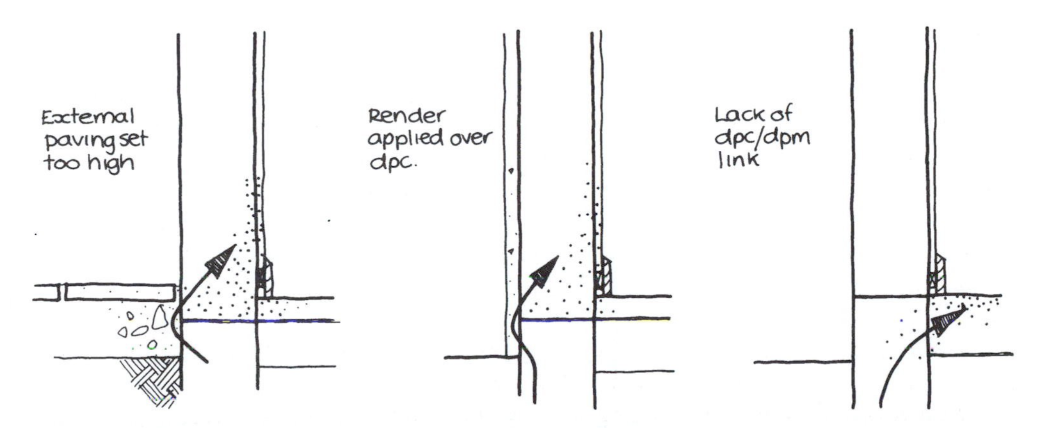

Figure 5.8 Examples of rising dampness that have been caused by the physical bridging of the damp proof course.

rising up the plaster creates a 'tide mark' so typical of rising dampness. Although this problem will recede once the floor has dried out it can still trigger a mis-diagnosis.

5.4.3 Signs and symptoms

The first step is to make sure that the dampness has not been caused by moisture from another source. The BRE (1997a) suggest that another of other sources are considered:

- Are there any leaking gutters or downpipes close by?
- Is there a problem with rainwater run-off from window sills, etc.?
- Is the house less than 5 years old? If so it could be drying-out moisture. This could also apply to houses that have been extensively refurbished.
- Are there any leaks from plumbing, wastes pipes or washing machines, etc.?
- Could it be condensation? Persistent condensation can often 'fool' the moisture meter.

Signs and symptoms typical to rising dampness include:

- persistent high level of moisture readings from the base of the skirting up to the limit of the affected area (see figure 5.9). Rising dampness rarely extends beyond 1 metre in height but can occur in bands of any height up to that level. Readings will gradually decrease higher up the wall but increase again at the boundary of the affected area. This is because hygroscopic salts are normally active at this level and will register higher readings than adjacent surfaces;

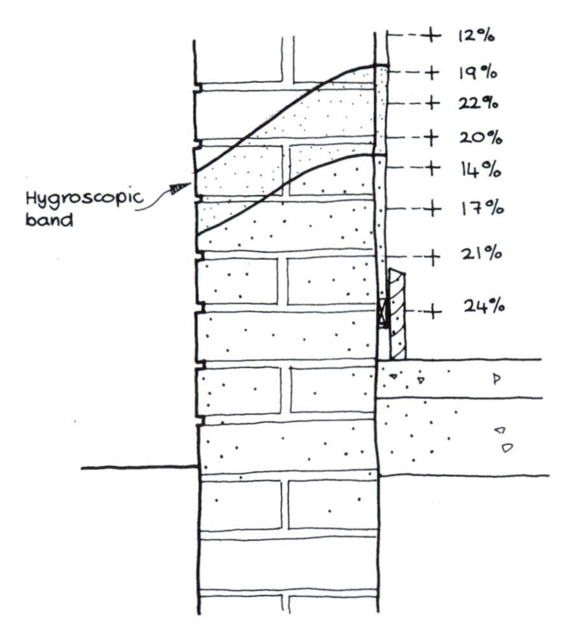

Figure 5.9 Section through a typical solid wall that has been affected by rising dampness. This shows the variation in moisture meter readings that can be expected. Note the higher values in the hygroscopic salts band.

- the decorations will be spoilt and the plaster finish may be deteriorating. There will be a characteristic 'tide' mark along the base of the wall. There may be a band of furry white crystal growth or salts at the extreme edge of the affected area. These are the hygroscopic salts drying out. Where the paint or wallpaper finish is slightly impervious, these salts may accumulate behind causing the finish to 'bubble' up. The hidden salts can be heard to 'crunch' under a slight finger pressure;
- the moisture readings are usually consistent over long periods of time. They are not as variable as penetrating dampness or condensation problems although hygroscopic salts can absorb higher amounts of moisture when it is humid in the dwelling;
- there is a theory (though not scientifically authenticated!) that mould growth will not occur where there are high levels of salts on a surface. This could be because a saline environment could be slightly antiseptic and so inhibit mould growth. This is not a clear but it could help to tell the difference between rising dampness and condensation in some cases.

The other sign to look out for is the bridging of the dpc level (see figure 5.8). High risk areas include:

- raised soil levels and paths against the external wall. The relative level of the external ground and internal floor is vital to identify;
- where boundary garden walls meet external walls. Few are isolated by vertical dpcs;
- where old timber floors have been replaced by new concrete constructions. Although the new floor may have a dpm it is rare to find that this has linked with the dpc in the wall. This will be hard to check without taking off the skirting board. A clear 'trail of suspicion' exists where a solid floor has replaced a timber one.

5.4.4 Rising dampness in pre-treated walls

Over the last 20–30 years the rise in popularity of a variety of damp-proof remedial systems means that a significant proportion of houses have had some form of treatment in the past. Even where this treatment has been originally effective, many systems may be approaching the end or even exceeding their guarantee period. Some systems may never have worked in the first place. In these situations, special care must be taken in diagnosing and reporting on the problem. The typical range of previous repairs could include:

- **Injected chemical damp-proof course** – a row of holes can be seen a few brick courses above external ground level. These may be pointed up with mortar or plugged with specially made plastic inserts.
- **Electro-osmotic damp-proof course** – these could be difficult to spot. A continuous copper or titanium wire is embedded in a horizontal mortar joint around the whole of the building. Usually this can only be seen on the wall where it joins with the 'earthing' connection. Other indicators include relatively new mortar pointing to a single continuous mortar joint at dpc level where the wire has been installed. Also where the system has an electrical trickle charge there may be a transformer connection in the region of the main electrical switchboard (figure 5.10).

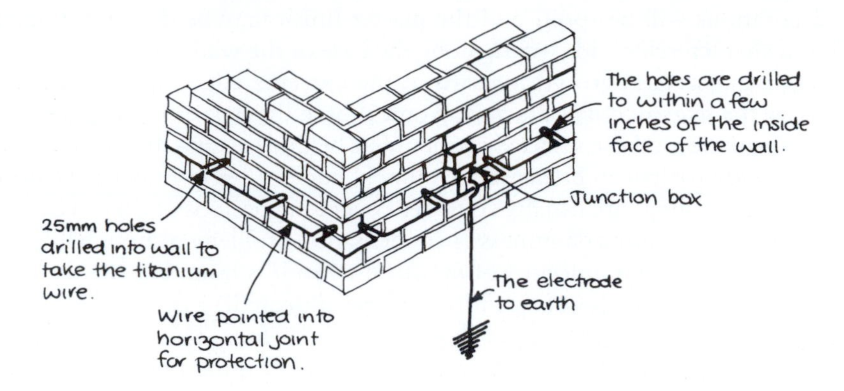

Figure 5.10 A diagrammatic sketch illustrating the normal arrangement of an electro-osmotic damp proof course.

Figure 5.11 This photograph shows ceramic 'drying tubes' that have been installed in a brick wall. The smaller filled holes close by suggests that a chemical damp proof course may also have been installed.

- **'Dalton' drying tubes** – these are large diameter (25–30 mm) ceramic tubes that are fitted into larger holes that have been 'cored' out of the wall. They are usually set at 150–250 mm centres. The open ends of these tubes are often covered by a small grid (figure 5.11);
- **new physical dpc** – depending on the method used this can usually be recognised by evidence of new brickwork and/or mortar just above ground level. Some contractors may have replaced one or two courses of brickwork when they inserted the new dpc while others may have 'cut out' just a single mortar joint in a more surgical fashion.

It is difficult for surveyors to advise clearly on these matters because there is such a wide range of opinion within the industry itself.

Chemical dpc installations, guarantees and call backs

Anyone who has been closely associated with the remedial treatment of rising dampness will know that getting a 20–30 year guarantee does not mean that there will not be any more problems! Where a property has been treated and rising dampness has re-occurred there could be a number of difficulties getting the guarantors to honour their commitments:

- The company might not be in existence anymore. If there is no insurance-backed guarantee then there is no easy remedy.
- If the company is still trading then they will wish to investigate the problem for themselves. In many cases when reputable contractors discover that the chemical dpc has indeed failed they will carry out remedial work. Sadly the industry is characterised by less professional operators that use a whole variety of excuses why the new dampness is not their fault. In the authors' experience these can include:
  - 'it's not rising dampness, it's condensation, open your windows more';
  - 'somebody used the wrong sort of plaster on this wall';
  - 'it's not the dpc, it's a leaking gutter outside';
  - 'sorry but we didn't inject this bit of the wall. Our guarantee only covers the bits we did'.

These can be appropriate explanations for why dampness has returned to a property. Yet the experience of many commentators and surveyors has resulted in a more cynical view. This is because even where a contractor accepts responsibility, the terms of the original guarantee may only cover re-injection and not any of the replastering and making good of decorations that occur as a result.

Where a contractor is prepared to carry out remedial work then there is some doubt whether it is appropriate to re-inject a wall that has already been injected. Concerns include:

- Will the new chemical be compatible with the original?
- Will any re-drilling damage the masonry causing cracking within the heart of the wall? This could lead to a loss of injection fluid which might result in a poor-quality dpc.

5.4.5 Reporting on rising dampness

Stopping moisture rising up an existing wall where the original dpc is either missing or defective is a very complicated and skilled operation. There are so many variables involved that even the BRE has acknowledged the difficulty. In *Digest 245* 'Rising dampness' (BRE 1985b), they state 'The success of these systems (chemical dpcs) in treating a case of genuine rising damp lies in reducing the moisture flow up the wall to such an extent that the dampness problem in practical terms disappears'. The phrase 'practical terms' is a clear acknowledgment that chemical injection alone is not sufficient. Most commentators and reputable contractors recognise this and state that any remedial work should be seen as a system that includes the following components:

- the chemical injection itself;
- replastering with a salt-retardent and waterproof plaster;
- building repairs and future management such as keeping the property in good order.

If all of these are not addressed properly then the treatment scheme might not be effective.

If clients are to be properly advised then surveyors need to be as proactive as possible when reporting on rising dampness. The following points may be helpful:

- Try and make a positive diagnosis of the defect. Many cases of dampness can be identified by straightforward repair of obvious defects, e.g. gutters, window sills, overflows, etc. Yet many surveyors refer clients onto 'specialists' if dampness is merely registered in a dwelling.
- If rising dampness is clearly suspected then try and give an indication where the problems were noted. This can help give an indication to the client as to the scale of the defect.
- Refer the client to a reputable damp proof specialist who:
    - is a member of the British Wood Preserving and Damp Course Association (BWPDA) and employs staff that have been on appropriate training courses;
    - issue insurance-backed guarantees that are approved by the BWPDA;
    - have been trading in their present name for a number of years (preferably ten years or more);
    - are prepared to give references from jobs recently completed.

Some commentators might be concerned that referring clients to BWPDA members might be a restraint of trade. To some extent this is true. There are many competent damp-proofing specialists who are not part of this trade association but the problem remains of how to help clients choose from a broad and unknown field. Membership of the BWPDA is not an automatic guarantee but it does represent a benchmark. Where a particularly complex problem or dispute exists, it could be better to refer the person to a building surveyor or other impartial professional who has a deep understanding of rising dampness problems.

5.5 Condensation – surface and interstitial

5.5.1 Introduction

Condensation is usually associated with older social housing but this type of defect can be found in all different housing types. One of the great myths surrounding condensation is that it is closely associated with the lifestyle of the occupants. When it is observed in a dwelling, surveyors can offer a remarkable range of advice:

- Open the windows and turn up the heat.
- Lifestyle changes including (and these are all direct quotes):
    - don't take too many baths;
    - don't let your kettle boil too long;
    - don't dry clothes indoors;
    - put lids on your saucepans; and
    - get rid of your pet fish!

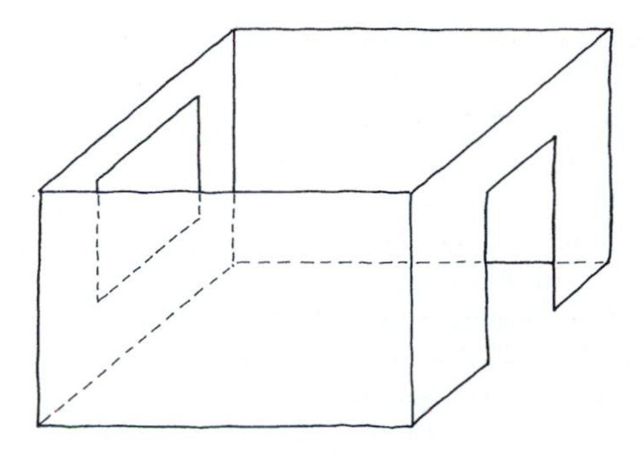

Relationship between relative humidity (RH) and temperature
Imagine an airtight room. Assume the air that it contains holds 1.6 litres of moisture in the form of water vapour (an invisible gas). The following arbitrary figures show how temperatures and humidity are inter-related.

Situation One
Temperature 15°C
Relative humidity 100%
Moisture in the air 1.6 litres
At this temperature, the air contains as much water vapours as it can carry. If the temperature drops or moisture level increases condensation will occur.

Situation Two
Temperature 20°C
Relative humidity 75%
Moisture in the air 1.6 litres
The temperature has increased, meaning that the air can 'carry' more moisture if required. Because the moisture in the air stays constant the relative humidity drops.

Situation Three
Temperature 10°C
Relative humidity 100%
Moisture in the air 1.1 litres
The drop in temperature has reduced the ability of the air to carry so much moisture. The relative humidity rises to 100% and the maximum amount of moisture the air can carry reduces to 1.1 litres. Therefore, 0.5 litres of water is deposited in the form of condensation on the colder surfaces such as windows and reveals.

Figure 5.12 Explanation of the relationship between relative humidity and temperature.

- Wash the walls down with bleach.
- Don't put furniture up against the external walls.

If this sort of advice was given following a carefully considered diagnostic process then it could help in resolving the problems. In practice it results from an ignorance of what condensation is and why it occurs. The surveying profession has a culture of blaming the occupants of a dwelling for this type of defect and needs to take a broader view.

The task facing the surveyor carrying out a pre-purchase inspection or survey is to objectively assess the true cause of the problem. Then any future vendor can be advised whether the defect:

What causes condensation?

All air contains water vapour. This is moisture in the form of an invisible gas. The warmer the air is the more water vapour it can hold. When warm air hits a colder surface it is cooled and its ability to hold water vapour is reduced. If it is cooled enough then some of that water vapour will 'condense' out forming tiny droplets of water on the cold surface. This is what is known as 'condensation'. The temperature at which this occurs is called the 'dew point'. This dew point temperature will vary depending on the amount of water vapour in the air. If there is a lot (i.e. the air is saturated, high humidities) then a slight drop in temperature will cause condensation on the coldest surfaces. If the air doesn't contain much vapour then the air temperature would have to drop by a lot before condensation begins. The measure of how much vapour the air is holding is called 'relative humidity' and occurs within a range of 0–100%. Figure 5.12 gives an explanation of how relative humidity varies with temperature.

- is indeed transient and will disappear with a change of living conditions; or
- is a result of inherent constructional inadequacies that require significant financial expenditure.

5.5.2 What is condensation?

This section will not provide an in-depth review of the fundamental science of condensation. Instead the main principles will be quickly reviewed.

One of the biggest challenges for the surveyor is to decide when condensation becomes an identifiable defect. For example after cooking a big meal, taking a bath or hosting a party the high levels of moisture produced will always result in condensation on window panes, mirrors, etc. In most cases this will disappear after an hour or two. It becomes a defect when condensation is persistent and a regular feature during the colder months. High levels of moisture can discolour decorations, cause mould to grow and in the worst cases lead to dry rot as well as affect the health of the occupants.

5.5.3 Variable nature of condensation

There are no firm rules about when condensation will and will not occur. One set of occupants can live in house that is not affected while next door a similar sized family could suffer from streaming windows, mould growth and ruined decorations. This is because the level of condensation in most homes is influenced by a complex interaction of factors. The BRE (Garrett and Nowak 1991) identified a broad range of causes. In most cases the following are the most influential:

- **The level of insulation in the dwelling** – Low levels of insulation in a dwelling can give rise to:
  - high rates of heat loss;
  - low temperatures – especially on the surface of walls and ceilings;
  - expensive bills.

Wall construction	Description	U value W/m²K
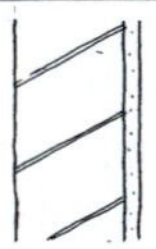	*Pre 1965 Building Regulations* 200 mm solid concrete Lightweight plaster	2.30
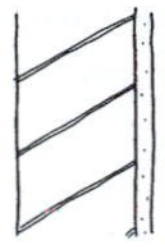	*Pre 1965 Building Regulations* 220 mm solid brick Lightweight plaster	2.00
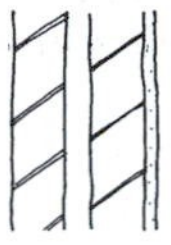	*Post 1965 Building Regulations* 105 mm brickwork 50 mm cavity 105 mm brickwork Lightweight plaster	1.40
	*Post 1976 Building Regulations* 105 mm brickwork 50 mm cavity 100 mm blockwork Lightweight plaster	0.90
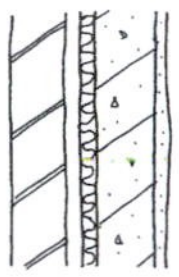	*Post 1990 Building Regulations* 105 mm brickwork 25 mm cavity 25 mm insulation 100 mm blockwork Lightweight plaster	0.45

Figure 5.13 The 'U' values of some common wall constructions.

It is essential that the dwelling should have a sufficient level of insulation to prevent this. The current Building Regulations are a reasonable measure but the vast majority of the housing stock are well below this level. Figure 5.13 compares the insulation value of typical constructions.

- **Moisture production levels of the household** – It is important to understand what are 'normal living activities'. Figure 5.14 from BS5250 (BSI 1989) helps in estimating what can be expected for a particular household. Using this table an average of 10–12 litres per day can be produced by the typical family. The dwelling must be able to cope with this. Although the way people live can affect moisture levels the following activities are the ones that have the biggest impact:

  - cooking (something all households do!);
  - drying clothes indoors (e.g. over radiators, on clothes' dryers, unvented tumble dryers, etc.);
  - liquified petroleum gas (LPG) and paraffin heating.

  Looking at other aspects of lifestyle will be largely irrelevant (e.g. the amount of baths, showers etc.) because in comparison these moisture-production activities

Cold bridging

No matter how well a dwelling is insulated there will be some parts of the structure that are colder than adjacent surfaces. These areas are known as 'cold' or 'thermal' bridges. Figure 5.15 shows some typical locations. There does not even have to be a discontinuity in the insulation; because of the greater surface area, the corner of two external walls will result in a colder area internally (see figure 5.16). The problem with cold bridges is that although the average internal environment humidities might be acceptable the conditions at the cold bridge positions allow condensation and mould growth to form. No amount of ventilation and only high levels of heating will reduce the effect. Insulation of the cold bridge area is often the only solution.

The key areas to inspect include:

- window reveals especially the jambs and heads;
- the junction of a roof and an external wall;
- the perimeter of a solid floor.

Household activity	*Moisture generation rate*
People	
• asleep	40 millilitres per hour per person
• active	55 millilitres per hour
Cooking	
• electricity	2 litres per day per household
• gas	3 litres per day
Dishwashing	400 millilitres per day per household
Bathing/washing	200 millilitres per day per person
Washing clothes	500 millilitres per day per household
Drying clothes indoor (i.e. using unvented tumble drier)	1500 millilitres per person per day

Note: These are based on average values obtained from BS 5250, Table One.

Figure 5.14 Table of typical moisture production rates in domestic dwellings (based on values shown in BS5250).

are much less significant and can be classed as 'normal'. If the occupants are producing high levels of moisture it can be dealt with:

- by putting an extractor fan over the cooker;
- by installing a vented tumble dryer or vented drying cupboard.

- **The heating system and its use** – Any dwelling needs adequate air and surface temperatures to avoid condensation. The BRE (Garrett and Nowak 1991) suggest that most dwellings should stay relatively free of condensation if the internal temperatures do not drop below an average of 14°C. Therefore, the property should have a whole-house heating system that is able of achieving this average economically. For example, the most expensive heating types are open coal fires, LPG heaters and on-peak electric heaters. These are followed by properly installed gas fires, enclosed coal heaters and electric storage heaters (off-peak). Gas central heating can be the most economical especially if highly efficient condensing boilers are used.

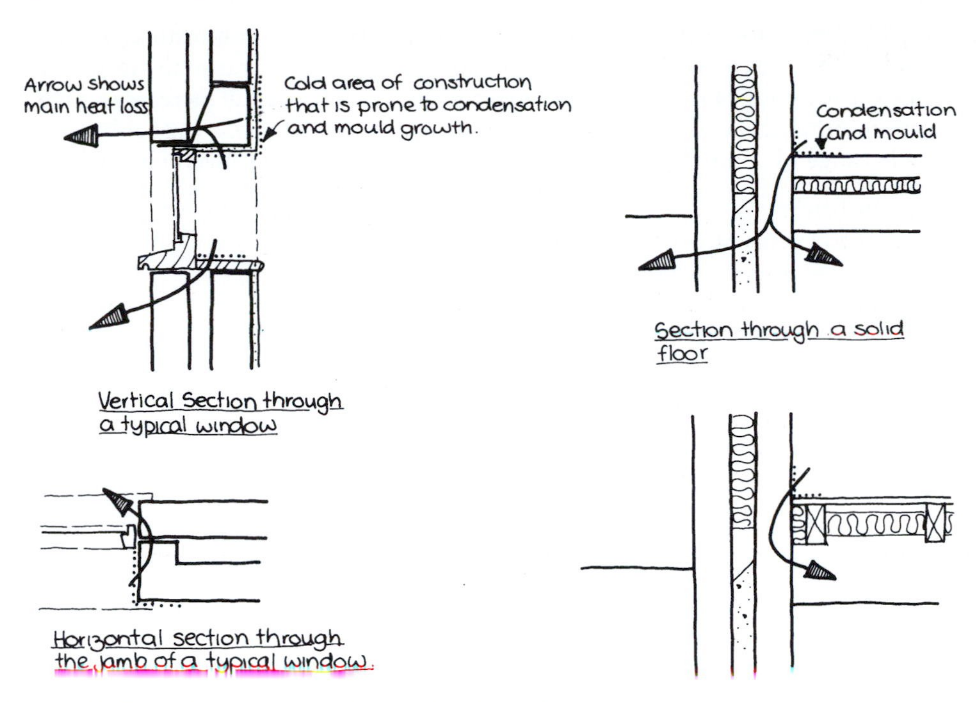

Figure 5.15 Typical locations of 'cold bridges' in domestic construction.

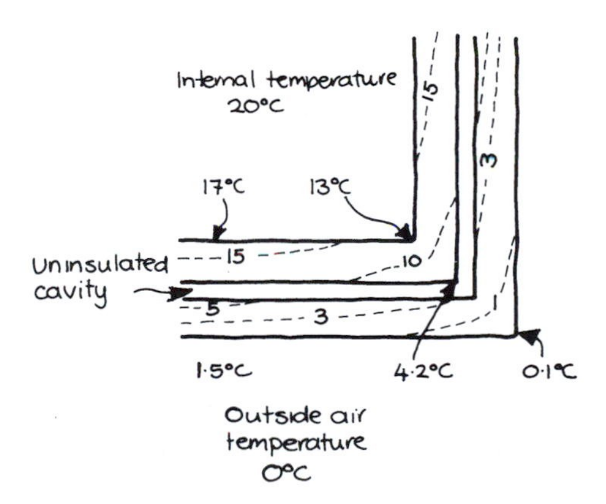

Figure 5.16 The cold bridging effect of an external corner. Because of the greater external surface area at this point, temperatures of the internal wall surface at the corner can be several degrees cooler than the adjacent areas. In some critical cases, this can allow condensation and mould growth to occur.

If a household cannot afford to keep their property warm then it will be vulnerable to condensation.

- **The ventilation system and its use** – Ventilation in dwellings is essential to remove smells, indoor pollutants, etc.; to provide ventilation for combustion appliances and remove excess moisture. It is measured by 'air changes per hour' (ach). Older houses are more 'leaky' than modern constructions. The problem is establishing the ideal rate. Too little ventilation will result in a build-up of moisture. Too much will result in excessive heat loss as the house loses all the warm air. In the latter case the lowering of internal temperatures would itself lead to an increase in condensation. Opinions vary but somewhere between 0.5–1.00 ach is probably ideal. A wide range of ventilation methods and products are used in dwellings including:

- air bricks;
- trickle vents in window frames;
- hit and miss spinner vents;
- manually and 'humidistat' controlled extractor fans;
- whole house ventilation input fan systems;
- heat exchange fans;
- passive stack ventilation, etc.

Some of these are more appropriate than others in helping to achieve an appropriate balance.

5.5.4 Diagnosis of condensation

Like all forms of dampness, condensation is easy to confuse with other causes. The following signs and symptoms may help in the process (BRE 1997b):

- There will be considerable moisture on the windows especially those that are single glazed. There may be evidence of water pooling on window sills. There may be damp staining at the base of the window jambs. In the worst cases, this staining may be present below the window sill level where the condensation has run off.
- Moisture meter readings are concentrated near to the surface of the wall. The deeper the probes go the drier it becomes.
- Moisture meter readings will often vary across even mould-affected areas. This is because the level of condensation will change throughout the day. The amount will depend on the recent moisture production activities. This can be contrasted with penetrating and rising dampness where more consistent moisture meter readings could be expected.
- Problems are worse in the winter. Even where dwellings suffer from serious condensation problems, the symptoms can virtually disappear during the summer months and are very hard to spot.
- Mould growth occurs on the coldest parts of the building first. This includes the external walls and especially the cold bridges.
- Mould growth will often be worse where air movement is restricted including behind furniture and in wardrobes and other cupboards.
- Mould growth has no defined edge and will gradually peter out across the affected area.
- There are no hygroscopic salts.

5.5.5 Reporting on condensation problems

Because this type of defect is closely associated with the way people live in their dwelling it is very difficult to give clear advice. Three different situations might be observed and

Category	*Description of condition*	*Advice to client*
Good (satisfactory repair)	There is little evidence of damage from condensation. There may be some isolated mould growth and paint damage to bathroom and kitchen windows. There is no visual evidence of mould growth in typical cold bridge locations.	There is a small amount of condensation damage that has resulted in superficial damage. If care is taken to keep the dwelling up to adequate temperature standards, keep moisture production down to a minimum and keep the dwelling reasonably well ventilated these problems should not give any cause for concern.
Fair (minor maintenance)	There is evidence of damage in a few locations. This could include mould growth to window frames in a number of rooms, limited mould growth to wall and ceiling finishes in cold bridge locations and behind some furniture.	Although the problems are not serious, some work is required to bring the internal environmental conditions back into a balance and avoid future problems. Remedial work could include some additional ventilation work, making good of damaged surfaces, improving the heat input into the dwelling and possibly some limited insulation work to the coldest surfaces, etc. To make sure that the remedial work is properly targeted and cost effective, advice from a building surveyor or other professional experienced in this sort of work may be appropriate.
Poor (significant or urgent repair needed)	There is significant damage in a number of areas. Windows and adjacent reveals and sills are badly affected by mould and condensation run off; wall and ceiling surfaces in cold bridge locations are densely covered with mould. There is considerable mould growth behind the furniture which is beginning to affect the adjacent wall surfaces.	The dwelling is significantly affected by condensation related defects. One or more of the key influences on condensation formation (i.e. heating, insulation, ventilation, moisture production) are inadequate and out of balance with the rest. It is likely significant remedial work will be required to bring the internal environmental conditions back into balance and so avoid future damage and unpleasant living conditions. Advice of a building surveyor or other professional experienced in this sort of work will be required. The scale of work could include improved insulation, heating, ventilation as well as ensuring moisture production rates are sensibly managed. The cost of this work could be high.

Figure 5.17 Table linking condensation defects to client advice.

the report could reflect this. These are summarised in figure 5.17. This approach assumes that the cause of the dampness has been correctly attributed to condensation.

This type of advice will have to be adapted to suit each particular situation. The detail will depend on the surveyor's knowledge and experience but this framework could form a useful basis for positive advice about a defect usually ignored by surveyors in the past.

5.6 Built-in moisture

5.6.1 The drying-out problem

Using traditional methods of construction, for a three-bedroom house over 6,000 litres of water can be used during the construction phase (Addleson and Rice 1994). Most of this is used in the concrete, cement mortar and plaster finishes. Because contractors are very keen to speed up the sale process and make a return on their investment, many houses are far from dry when the first occupants take up residence. This is true especially during the winter. The drying-out time for a floor screed has been estimated as taking 1 month for a 25 mm thickness. Therefore, a concrete floor slab with screed topping could take up to 6 months to fully dry. Combined with the normal moisture-production activities of the first owner, moisture levels can soon rise.

5.6.2 Key inspection indicators

- High levels of condensation on the window panes and frames;
- white 'furry' salts appearing out of the plaster;
- mould growth may occur where air movement is particularly restricted, e.g. beneath or behind cupboards;
- decorations 'bubbling' up because of entrapped moisture, tiles becoming detached from walls, etc.

These effects can be unrelated to weather conditions or the lifestyle of the occupants. The key is the recentness of the construction process. It could take at least a year for some houses to properly dry out and so condensation-type defects during this period may be closely associated with construction moisture. After this period, moisture from another source must be suspected.

Not only new houses can be affected by this problem but also recently refurbished dwellings. A major modernisation can use 1,000–2,000 litres of water and have similar effects.

5.6.3 Giving advice

If this type of problem is suspected then the owner should be advised that additional ventilation and heat should be provided to assist the drying process. Forced drying methods (e.g. industrial heaters and dehumidifiers) are not recommended as the building should be able to reach an equilibrium naturally. Forced drying out can cause cracking and warping of timbers, etc. The client should be advised that if the symptoms persist for more than a year, further investigations need to be carried out as there might be another source of moisture.

5.7 Traumatic dampness

This is a general term given to dampness defects that are caused by water leaks or bursts from such things as water and waste pipes, tanks, cisterns, radiators, sinks, baths, basins, etc. These can be truly traumatic and associated with large volumes of water, collapsed ceilings, etc. In other cases the leaks can be small and gradual and the symptoms mistaken for other forms of dampness. Here are some examples from real cases:

- A new bungalow was showing high moisture meter readings at the base of the internal and external walls. The paint to the wall surface began to flake off and a characteristic tide mark developed. For many months this was mistaken for rising dampness associated with a dpc/dpm fault. Further investigations revealed that the central heating pipes had been buried in the floor screed. A joint in one of the pipes had developed a leak allowing water to soak into the screed. The screed had become so saturated that the moisture had soaked up the wall causing the same symptoms as rising dampness.
- A watery brown stain had developed on the ceiling of a bedroom to an upstairs flat. Because there was an old slate roof above a roofer was instructed to overhaul the roof slopes. The stain continued to develop and a loft space inspection revealed a water tank directly above. Even then the cause was not that simple. An inspection of the tank failed to identify any plumbing leaks but did show a high level of condensation run-off from the water tank. It was this that was gradually seeping through and causing the stain.
- A cold water storage tank developed a large leak that led to 50 gallons of water flooding through the house. Ceilings collapsed and carpets and possessions were ruined, etc.

These cases illustrate how difficult it can be to make a precise diagnosis especially when it has to be based on a single visit. Despite this, typical characteristics can be associated with these types of defects:

- the dampness is usually in isolated areas and the level of dampness can be very high;
- moisture meter readings show a sharp change from wet to dry;
- not usually associated with the weather or time of year;
- there are often drainage, plumbing, heating or sanitary fittings nearby;
- hygroscopic salts are not usually present.

Giving advice on this type of defect is usually straightforward as long as the diagnosis is accurately made.

5.8 Thermal insulation and energy conservation

5.8.1 Introduction

Over the last few years energy issues have appeared on social and political agendas. VAT on fuel and the possible future ‘carbon taxes’ (where energy inefficiencies will be penalised financially) have made consumers more aware of the costs of heating their homes. Money is not the only issue. People want to be warm and comfortable in their home environment as the nation is no longer content to live in damp and draughty houses. A significant section of the population is also motivated by environmental concerns: wanting to help conserve non-renewable resources and contribute to the reduction of greenhouse gases. The domestic energy sector accounts for nearly 30% of the nation's consumption so significant improvements can be made.

5.8.2 Scope of surveys and inspections

Despite this interest and concern, mortgage valuations make no reference to energy use and Homebuyers reports includes only a short paragraph entitled 'Thermal Insulation'. Although this gives an opportunity to comment on the property's overall energy efficiency few surveyors go beyond hot water tanks, loft insulation and double glazing. Yet some lending institutions already include significant energy advice with mortgage valuations as a standard part of their service. This section aims to provide a framework that can help surveyors meet the demands of energy-conscious clients.

5.8.3 Energy efficiency – the building regulations

Since the mid-1960s the building regulations have stipulated a minimum level of insulation that is to be included in the walls and roof of a dwelling. As awareness about energy issues has increased, so have the building regulations. Part L of the 1995 Regulations has been summarised in figure 5.18.

Energy rating procedures

One of the latest developments in energy efficiency has been to look at dwellings as a whole. Energy efficiency is more than just the amount of insulation included in the walls and roof, it depends on:

- the climate – including temperature, wind and sun;
- physical characteristics of the dwelling and heating system;
- the way in which the house is used;
- internal temperatures and the use of domestic appliances.

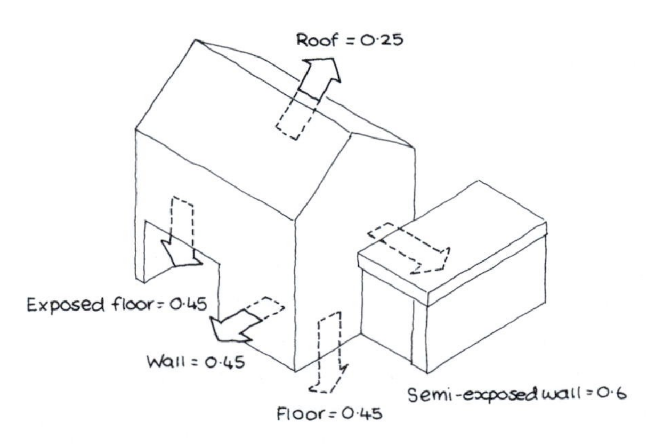

Figure 5.18 Main requirements of Part L of the Building Regulations (Crown copyright. Reproduced with the permission of the controller of Her Majesty's Stationery Office).

To account for some of these issues, the Government originally encouraged the development of energy 'labelling' products. These assessed the energy efficiency of dwellings as a whole then assigned a simple 'label' or value that could easily be understood by the population. Two of the earliest products were the National Home Energy Rating Scheme produced by National Energy Services and MLM Starpoint.

To make sure these different systems could be related to each other, the Government introduced a third method called the Standard Assessment Procedure (or SAP for short). This is now the Government's preferred energy rating. Using a manually completed 'worksheet' or a computer program any dwelling can be given a SAP rating of between 0 and 100. A new house built to current building regulations would have a SAP rating of about 70–80. A poorly insulated end-of-terraced property would have a rating in the 40s. All energy rating schemes now have to produce an accurate SAP rating in addition to their own label.

5.8.4 SAP levels and existing dwellings

Although the Standard Assessment Procedure has its critics, it provides a basis for surveyors to begin to get a 'feel' for the overall energy efficiency of a dwelling. The National Home Energy Rating Scheme is just one of the energy labelling systems on the market and has arguably one of the best developed products. A considerable amount of research has gone into their method of energy rating. In the instruction manual for their 'Home Rater' Program (NES 1997) they have outlined a few rule-of-thumb methods that are useful in giving surveyors an insight into SAP ratings of different dwelling types.

5.8.5 New dwellings

For newer dwellings, likely SAP levels are easier to predict. NES (1997, pp.8–9) have illustrated this by identifying the impact of variations in energy efficiency on a 'model' house. For example, consider a 100 m^2 detached house that has:

- wall and floor U-values of 0.45
- roof U-value of 0.25
- timber double glazing
- standard wall-mounted balanced-flue gas boiler
- boiler controlled by room thermostat, programmer and thermostatic radiator valves (TRVs)
- water heating – from the hot water cylinder fitted with thermostat and insulated.

This dwelling would have an SAP value of around 73. Figure 5.19 shows the impact of various constructional and appliance changes on the SAP. Studying this table it can be seen that changes in the heating system have the greatest impact. Increases in the level of insulation (especially of well-insulated dwellings) have only a marginal impact on energy efficiency.

5.8.6 Assessing the energy efficiency of an existing dwelling

The situation is not that straightforward for an existing dwelling. Because the existing housing stock can vary so dramatically, it is important to try and categorise the particular characteristics of individual dwellings. After carrying out a survey the first stage will be to categorise the insulation level and heating systems of the dwelling in accordance with the descriptions in figures 5.20 and 5.21. This will be broad-brush and not too precise. These qualitative descriptions can the be transposed to the categories in figure 5.22, which gives indicative SAP 'factors' for typical dwellings. These figures do not represent real SAP values but are relative indicators. They should not be used to give clients an energy rating, but instead give a rough indication of overall energy efficiency of the property. The 'SAP' factor can then be used to indicate what type of client advice might be appropriate by choosing the appropriate category in figure 5.23.

Important note: this process does not replace the careful and considered energy assessment of a licensed operator. Great care must be exercised when giving advice based on this approach.

Changes to model dwelling	*SAP change*	*New SAP*
Change from detached to semi-detached	+6	79
Change from semi-detached to mid-terraced	+6	85
Increase size from 100m^2 to 150m^2	+5	78
Decrease size from 100m^2 to 75m^2	–8	65
Change standard boiler to condensing type	+7 → +10	80 → 83
Mains gas to LPG	–15 → –20	53 → 58
Mains gas to Oil	+15→ +20	88 → 93
Mains gas to storage heaters	–10 → –15	58 → 63
Wall U-value from 0.45 to 0.3	+4	77
Roof U-value from 0.25 to 0.15	+1	74
Floor insulation thickness from 37mm to 50mm	+1	74
Standard double glazing to low emmissivity double glazing	+1	74

Note: The SAP for the original model house is assumed to be 73.

Figure 5.19 Table showing the impact of constructional changes on the SAP level of a model house (reproduced courtesy of National Energy Services).

Category description	*Description of typical dwelling*
Very bad	Solid brick walls less than 50 mm insulation in loft; no floor insulation; single glazing; no draught proofing.
Poor	Solid brick walls or unfilled cavity (prior to 1965); 100 mm loft insulation; not much draught proofing; single glazing and no floor insulation.
Moderate	Cavity walls post-1972 or filled cavity; 100 mm in loft; no floor insulation; well draught proofed or double glazed.
Good	Filled cavity or timber framed; 150 mm in loft; double glazing and draught proofed throughout; floor insulation (although this is rare).

Figure 5.20 Descriptors that categorise the thermal insulation standards of existing dwellings (reproduced courtesy of National Energy Services).

Category description	*Description of typical dwelling*
Very bad	On peak electric room heaters or open fires.
Poor	Poorly controlled old central heating systems (e.g. coal, oil or floor mounted gas boiler without room thermostat); old fashioned storage radiators; back up heating provided by on peak heaters.
Moderate	Standard gas boilers with controls that turn the boiler off when no heat required; modern storage radiators with thermostats and time clocks; modern coal boilers; modern gas room heaters with appliance thermostats.
Good	Gas condensing boilers or electric boiler with independent temperature control in zone 2 (either TRVs or zone control).

Figure 5.21 Descriptors that categorise the heating efficiency standards of existing dwellings (reproduced courtesy of National Energy Services).

	Heating system			
Insulation	*Very bad*	*Poor*	*Moderate*	*Good*
Very bad	0 –20	30	40	50
Poor	25	45	50	60
Moderate	30	50	60	75
Good	50	65	80	90

Figure 5.22 Table linking the thermal insulation and heating efficiency descriptors with an appropriate 'SAP' factor.

Advice category	*Typical indicative SAP 'factor'*	*Advice to client*
Poor (significant or urgent repair needed)	0–45	Where a property is clearly energy **inefficient** in fundamental and basic ways. The surveyor should refer the client to an energy assessment specialist as the potential savings may be considerable. In addition the surveyor should identify those works that will bring greatest benefit (such as loft insulation, hot water tank insulation etc.) so the client can improve the basic level of energy efficiency without spending more money on fees.
Fair (minor maintenance)	46–65	Where a property falls between good and poor, the client should be advised to seek further specialist advice. This is because although there may still be potential for considerable savings, more targeted advice is required to accurately identify priorities.
Good (satisfactory repair)	66–100	Where a property is clearly very energy **efficient** then a recommendation to a specialist energy consultant may not be appropriate as the potential savings to the client will not justify the fee.

Figure 5.23 Table linking 'SAP' factor with advice to the client.

5.8.7 Giving advice

One of the advantages of carrying out a survey on a property is that you get a good insight into its construction, heating appliances and condition. This gives the opportunity not only to advise on broad energy matters but also link guidance on what physical repairs are required. For example, most energy assessments of existing houses will show that the payback time for replacing single-glazed windows with double-glazed alternatives will often be over 200 years! For energy reasons such a course of action would not be justified. If, on the other hand, the existing single-glazed windows were in poor condition and require replacement anyway, the formula becomes very different. This will give the client a broader view than an energy assessment specialist who may look at energy performance only. The main conclusion that can be drawn is that it is difficult to get good ratings unless the property has a modern and efficient heating system.

References

Addleson, L., Rice, C. (1994). *Performance of Materials in Buildings.* Butterworth-Heinemann, Oxford.

BRE (1984). 'Floors: cement-based screeds – specification'. *Defect Action Sheet 51.* Building Research Establishment, Garston, Watford.

BRE (1985a). 'External masonry walls insulated with mineral fibre cavity-width batts: resisting rain penetration'. *Defect Action Sheet 17.* Building Research Establishment, Garston, Watford.

BRE (1985b). 'Rising dampness in walls: diagnosis and treatment'. *Digest 245.* Building Research Establishment, Garston, Watford.

BRE (1987). 'Windows: resisting rain penetration at perimeter joints'. *Defect Action Sheet 98.* Building Research Establishment, Garston, Watford.

BRE (1997a). 'Diagnosing the cause of dampness'. *Good Repair Number 5.* Building Research Establishment, Garston, Watford.

BRE (1997b). 'Treating condensation in houses'. *Good Repair Number 7.* Building Research Establishment, Garston, Watford.

BRE (1997c). 'Treating rain penetration in houses'. *Good Repair Number 8.* Building Research Establishment, Garston, Watford.

BSI (1989). *Code of Practice for the Control of Condensation in Buildings.* British Standards Institution, HMSO, London.

Fryer v. *Bunney* (1982). 263 EG 158.

Garrett, J., Nowak, F. (1991). *Tackling Condensation. A Guide to the Causes of, and Remedies for, Surface Condensation and Mould Growth in Traditional Housing.* Building Research Establishment, Garston, Watford.

Lacy, R.E. (1977). *Climate and Building in Britain.* Building Research Establishment, HMSO, London.

Murdoch, J., Murrells, P. (1995). *Law of surveys and valuations.* Estate Gazette, London.

NES (1997). *National Home Energy Rating 'Home Rater' Program Manual.* National Energy Services, Milton Keynes.

Oliver, A. (1997). *Dampness in Buildings.* Blackwell Science, Oxford.

Ormnady, D., Burridge, R. (1988). *Environmental Health Standards in Housing.* Sweet & Maxwell, London.

Oxley, T.A., Gobert, E.G. (1994). *Dampness in Buildings*, 2nd edn. Butterworth Heinemann, London.

TRADA (1991). 'Moisture meters for wood'. *TRADA Wood Information Sheet*, section 4, sheet 18. Timber Research and Development Association, High Wycombe.

Chapter 6

Wood rot, wood-boring insects, pests and troublesome plants

6.1 Introduction

These topics have a few features in common:

- they are all living things;
- they are loosely related to dampness defects in buildings;
- they can quickly cause problems for a building owner. This can in turn affect the value and the saleability of the property.

Therefore, they have been included in the same chapter.

6.2 Wood rot

6.2.1 Introduction

Wood rotting fungi is the chief cause of wood decay in this country. Essentially the fungus grows on the timber extracting the sugars out of the cellulose causing the material breakdown. There are two main types:

- dry rot (sepula lacrymans)
- wet rot (coniophora puteana).

6.2.2 Dry rot

This is the most destructive of all the different types of wood rot. If left undetected, it can seriously weaken timber structural components and in the worse cases lead to collapse. The eradication of an outbreak involves a complex procedure of repair and treatment that could cost thousands of pounds. To properly detect an actual or potential outbreak, understanding a few fundamental principles is useful (Singh 1994, p.36):

- To properly grow, dry rot needs a source of moisture, oxygen and cellulose (usually from the timber).
- The ideal conditions for growth are:
  - moisture content of timber between 20–40%;
  - adequate temperatures, best 18–23°C;
  - poorly ventilated stagnant conditions.

Dry rot is a very delicate plant. It does not like varying conditions and ironically it can very easily be killed if these ambient conditions are altered. When conditions are favourable or the fungus is threatened, an outbreak can produce 'fruiting bodies' that usually appear in more obvious parts of the building. These widely distribute brick-coloured orange spores.

Key inspection indicators

The main indications of an outbreak of dry rot include:

- There is a silky cotton wool-like mass or growth that may be of considerable size. The fungus can be very 'fresh' and 'fleshy' and white in colour at the extreme edge of the growth. Behind this long thin vein-like hyphae develop that connect the growing front with the source of food at the original outbreak. These hyphae can be several millimetres in diameter. When dry they can be snapped.
- The affected timber is gradually broken down and shrinks in size. It becomes darker in colour and will often crack both across and along the grain. This produces 'cuboidal' cracking that is so typical of dry rot. Small cubes of very light timber can be easily broken away.
- The surface of the timber will become uneven. The rot can develop behind a paint film and producing a 'wavy' effect on the surface. When prodded with a blunt probe (a car key is the ideal instrument!) the probe will quickly penetrate deep into the timber.
- When conditions are favourable, fruiting bodies will appear in the vicinity of the main outbreak. These look like fleshy growths or 'brackets' that appear relatively quickly (a matter of weeks) and are soon covered with tiny, bright brick-red particles or spores. These spores can spread across a wide area and if undisturbed can literally cover adjacent surfaces (see figure 6.1).

Location is the key for identifying dry rot. Because it requires stagnant and stable environments, the main growths are usually below suspended timber floors, in under-stair cupboards, within the thickness partitions, etc. Only the fruiting bodies tend to be noticed when they appear in the more open areas of the dwelling. Inspection of basements and sub-floor voids are important in this respect. Richardson (1991, p.84) points out that dry and wet rot can often affect the same piece of timber; wet rot infecting the wetter area with dry rot in the areas of lower moisture content.

6.2.3 Wet rot

Wet rot occurs where timber is excessively wet for most of the time. The moisture content would be in the region of 28–40%. The rot has a preference for the wetter end of this scale. Wet rot turns the timber very dark brown and causes longitudinal fibrous-type cracking. In the final stages of attack the timber becomes very brittle and is readily powdered. Wet rot can affect the heart of timber sections. When it becomes obvious on the surface the middle of the timber is often completely destroyed. Where the timber is painted, a surface waviness is a typical sign. Wet and dry rot in the early stages can be hard to tell apart. In many ways it does not matter that much in relation to mortgage

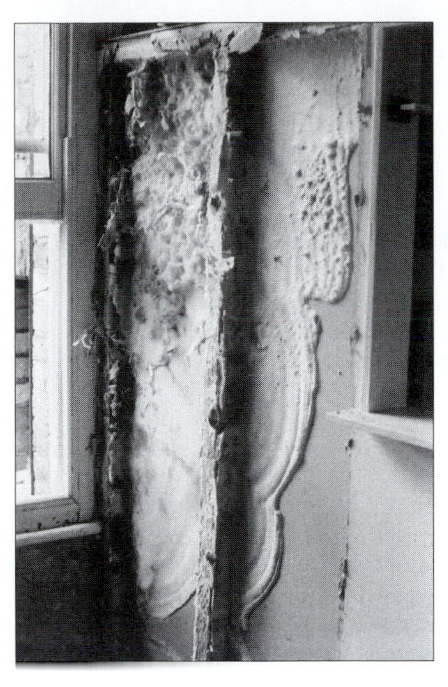

Figure 6.1 Photograph of dry rot within a timber partition. The plasterboard to the wall has been removed to reveal a luxurious dry rot growth. The outbreak began in the vertical timber stud that is fixed to the inside face of the external wall (to the LHS of the picture). The source of moisture and the cause of the outbreak was a leaking gutter that saturated the wall.

valuations and Homebuyers reports. If any type of rot is discovered then a referral to a specialist would be required. Despite this, location is often the distinction between the two. Wet rot can be found in wetter location:

- fully exposed window and door frames especially at the joints;
- joist ends where they are built into very damp walls;
- dry rot is not common in roof spaces but wet rot can affect the ends of rafters, wall plates, chimney trimmers, etc.;
- timber that is dampened by regular leaking from piped water services.

6.2.4 Inspecting for rot

The main controlling conditions are the presence of moisture and timber products. Consequently a moisture meter is vital if potential outbreaks of dry rot are to be identified. Berry (1994) also recommends other items of equipment that may be useful:

- various 'opening-up' tools to take up floorboards, drill inspection holes, etc.;
- a surveyor's mirror on a telescopic handle (a mirror-on-a-stick); and
- a boroscope with a light source and flexible light guide.

Typical locations that should be inspected include:

Inspecting and reporting on a suspended timber floor

An older suspended timber floor can be very vulnerable to both wet and dry rots. Poor ventilation, lack of isolating dpcs and poor workmanship generally can result in damp, rot-infected timbers. It was only recently that the surface of sub-floor voids had to be waterproofed so many older dwellings have bare earth beneath the floorboards. This can lead to high levels of moisture in the timbers.

To properly assess the condition of a suspended timber floor a full inspection would be required. This would include getting access to the floor void either by lifting a number of floorboards around the room or literally crawling beneath the floor itself. Apart from being beyond the scope of most standard surveys, the presence of fixed floor coverings and well-nailed floorboards would prevent this approach. A typical service would include:

- Lift the corner of any loose floor coverings to identify the nature of the flooring itself.
- If this is possible then the surface of any timber components should be tested with a damp meter.
- If an access hatch or loose floorboards are conveniently positioned (i.e. carpets or heavy furniture do not have to be moved) then these should be lifted and an inverted 'head and shoulders' inspection carried out with the aid of a strong torch.
- Where a 'trail of suspicion' presents itself (e.g. high damp readings to floor joists, signs of wood-boring insects or wood rot, unusual features, etc.) and the sub-floor void is easily accessible (say 750 mm clear space and above) it might be 'reasonable' for the sub-floor area to be inspected further. There cannot be any hard and fast rules for this – it will always be a matter of professional judgment.

If there is no hatch and the floorcoverings are fixed it is still possible to make an assessment of the likely condition of the floor. Assuming that it is a typical suspended floor construction that would be found in a three-bedroom inter-war semi-detached house, the following indicators should be noted:
Internally:

- Is the floor springy? The informal 'standard' test is to jump up and down on the floor. Excessive rattling could suggest either undersized or deteriorated timbers (or an overweight surveyor!).
- Is there evidence of dampness to the walls, skirtings and floorboards around the edge of the room?
- If the odd corner of carpet can be lifted, are there any signs of wood-boring insects?

Externally:

- Is there enough ventilation to the sub-floor void from the outside? On average there should be one 225 × 150 mm air brick every 1.5 m run of external wall. These should be clear of all debris and extend right through the wall (the standard test is to poke through a piece of rigid wire, e.g. an unwound wire coat hanger).
- Even if there are sufficient air bricks, are they well distributed around the floor? The classic situation would be the mid-terrace house where the rear extension floor is solid while that under the main house is suspended (see figure 6.2). In these cases, large areas of the underfloor area may not have any through ventilation and would remain stagnant. Rot could easily develop.

- Is there a dpc in the wall and is it at least 150 mm above the external ground level? If not then joist ends could be damp and possibly rotten.
- Is the internal floor level well above the external ground level? If not then it might be very difficult to achieve an adequately ventilated floor void.
- Have external paths or flower borders been built up against the wall possibly bridging the dpc?

If there are no adverse signs then further investigations may not be necessary. If there are any indications of potential problems then further investigations will be warranted. This is the classic case of the 'trail of suspicion'.

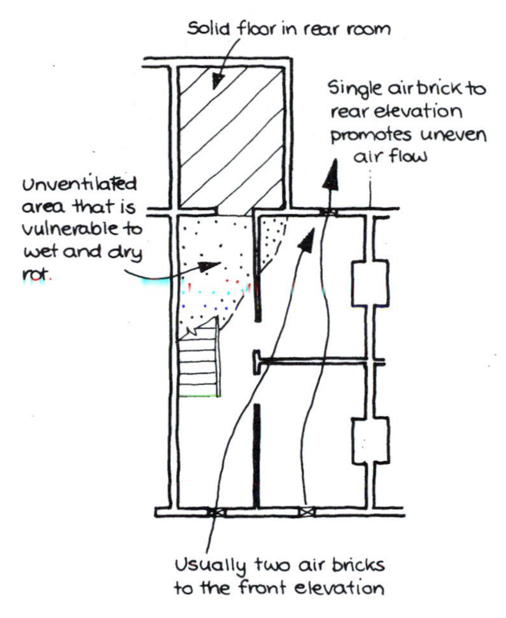

Figure 6.2 Plan of a typical suspended timber floor showing inadequate ventilation because of a lack of through flow of air.

- skirting boards – these can act as a 'wick' for dry rot and are often the first visible element affected by rot. The moisture readings of the timber and adjacent wall are very important;
- the base of door frames and linings where they are adjacent to a solid floor;
- woodblock flooring or other timber floor finishes especially where a dpm is missing or defective;
- the rear faces of built-in cupboards or heavy furniture that has not been moved for some time;
- timber ground-floor joists where they have been built into adjacent brick walls or have not been isolated from damp masonry by a dpc;
- timber joists of upper floors, fixing grounds or wall plates that are often built into the thickness of older walls. High levels of rain can penetrate the wall on exposed elevations leading to rot in the exposed timbers;

- timber close to leaking water services. Examples could include rising water mains, radiators, bath wastes, toilet connections, etc.;
- timber sections in roofs that are constantly damp such as valley beams of 'butterfly' roofs, timber trimming pieces around chimney stacks especially the secret gutters at the rear.

6.2.5 Reporting on rot

Outbreaks of dry rot are very serious. Where actual outbreaks are detected then further investigations are required so:

- the impact on the value of the property can be assessed;
- the client can be properly advised.

Although many surveyors have the skills and knowledge to organise the necessary investigation and remedial work, guaranteed specialised treatment methods are required by most lending institutions. Therefore referral to the specialist contractors will be appropriate. Where no rot was discovered but conditions likely to lead to dry rot exist, then further investigations still might be required. Typical advice could include:

- Evidence of dry rot was observed during the survey/inspection in the following locations (specify). Because this is such a potentially damaging fungus it will be necessary to appoint a timber treatment specialist. They should carry out a full investigation and provide a report with an estimate for the necessary remedial work. This should be done before you enter into any commitment to purchase.
- Although no evidence of dry rot was noted during the inspection, a number of timber components were either damp or close to other parts of the building that are damp (specify location and extent). Under these conditions, there is a possibility that wood-rotting fungi could develop in the foreseeable future. To ensure this problem is rectified a damp-proofing and timber specialist should be asked to inspect the property and provide a report together with any appropriate estimates.

When referring clients to timber specialists, precautions similar to those outlined in section 5.5.4 should be taken.

6.2.6 Inspecting basements and cellars

Where part or all of the ground floor has a cellar beneath then the whole floor structure can often be fully inspected. In many other cases, these will be 'underdrawn' or have an old lath and plaster ceiling beneath. Not only does this limit the extent of the survey but it can restrict free ventilation. In these cases the client should be advised that the underdrawing needs to be removed to protect the timber. Despite the assertions of many owners that their cellars are 'dry', any older basement walls that are in full contact with the adjacent ground will invariably be damp. Special care should be taken to test joist ends as they are often built into damp brickwork.

Other trails of suspicion could include:

- signs of standing or running water. This could be caused by excessive ground water or leaking drains. This can be seasonal so evidence of 'tide marks' or water staining should be noted;
- large coal chutes or other openings that could allow water to enter the space;
- large amounts of stored timber that could allow rot to develop.

6.3 Insect attack

6.3.1 Introduction

According to Bowyer (1977) '... infestation of timbers is probably, to the layman (sic), the prime reason for a ... survey'. He goes on to explain that this public interest is probably due to '... the extensive publicity given by commercial bodies interested in the treatment and eradication of the scourge'. This identifies the two key problems in assessing properties for insect attack:

- owners are very concerned when 'woodworm' is discovered in their home; and
- many remedial treatment firms often carry out blanket treatments of whole dwellings when the presence of insects has been noted.

Both reactions are often out of proportion to the scale of the original infestation.

Before this is discussed any further it will be useful to review the nature of wood-boring insects. Many that use wood as a food source prefer the damp timber of standing trees, freshly felled logs or decaying matter. Few will attack dry timber in buildings. Yet if the conditions are right, wood-boring insects can badly damage timber to the extent that structural integrity can be threatened and the value of the dwelling reduced. Their life cycle begins when the eggs of the beetles are laid on the surface or in cracks in the timber. Small grubs or larvae hatch out and bore into the wood feeding as they go and forming a network of tunnels. As the larva moves along it excretes bore dust or frass. After a number of years of feeding within the timber, the larva pupates close to the surface. When the adult emerges, it bites through leaving the 'flight hole' that is such a typical symptom. The adults do not cause any further damage. After mating, the females will lay their eggs in suitable timber and so continue the infestation.

6.3.2 Main types of wood-boring insects

There are a large number of different species of wood-boring insects. Although the BRE (1992) includes a more complete review, most surveyors will only normally encounter the following five types (BRE 1998):

Common Furniture beetle (*Anobium punctatum*)
Shape and size of flight hole: circular 1–2 mm
Bore dust (under a ×10 lens): cream, granular, lemon shaped pellets
This type is probably the insect that most people have come across and is usually wrongly called 'woodworm'. Characteristics:

- It is found in the sapwood of all softwood and European hardwood timbers.

- It can infest timbers with moisture contents that are typically found in ventilated roofs and suspended floors. In such an environment, the extent of the damage will be moderate and the activity of the insect low. A heavy infestation on the other hand is a sure sign that a dampness problem will almost certainly exist as well.
- Where it affects joinery, staircases and intermediate floors in centrally heated environments the beetle tends quickly to die out.
- It rarely causes structural weakening of timber except in small areas that may be damp as well.

House Longhorn beetle (*Hylotrupes bajulus*)
Shape and size of flight hole: oval 6–10 mm
Bore dust (under a ×10 lens): cream powder, chips and cylindrical pellets
Generally limited to the south-east of England and measures have been included in the building regulations to try and control the insect. Characteristics:

- Only found in parts of Surrey.
- Found mainly in roof timbers and because of size of the larvae it can cause structural weakening.
- Not encouraged by damp conditions.

Death Watch beetle (*Xestobium rufovillosum*)
Shape and size of flight hole: circular 2–3 mm
Bore dust (under a ×10 lens): brown disc-shaped pellets
Probably the most famous wood-boring insect about which a great deal of mythology has developed. Characteristics:

- Distribution – common in southern half of the UK; less frequent in the north of England and virtually unknown in Scotland.
- Only found in buildings more than 100 years old; mainly affecting hardwoods.
- Requires damp conditions and often associated with fungal attack. Therefore, found in wall plates, truss ends, bonding timbers, and panelling, etc.
- May cause severe damage so the implications have to be considered carefully.

Lyctus Powderpost beetle (*Lyctus brunneus*)
Shape and size of flight hole: circular 1–2 mm
Bore dust (under a ×10 lens): cream, talc-like.
This beetle is found all over the world and can be very destructive. It can reduce the sapwood to a powdered mass within a few years. Characteristics:

- Attacks the sapwood of hardwood timbers.
- Does not attack softwoods so most modern dwellings are safe.
- Usually infests timber before it is delivered to site.

Wood-boring weevil
Shape and size of flight hole: ragged 1 mm
Bore dust (under a ×10 lens): fine brown and angular

Very closely associated with very damp timber and easily controlled by reducing the moisture content. Characteristics:

- Attacks only damp and decaying timber.
- It does not cause timber deterioration but it can speed it up.
- Found in rot-affected joist ends, backs of skirtings on damp walls, etc.

6.3.3 Locating infected timbers

The BRE (1998) identify a number of key locations in a dwelling where insect attack is most common especially if the timber is damp. These are illustrated in figure 6.3. Exposed timbers should be inspected with strong torchlight and look for evidence of the flight exit holes. Indicators of activity include:

- Freshly cut holes, which tend to have sharp and well-defined edges. Where the inside walls of the tunnel can be seen they tend to be light in colour when compared with those holes that have been open for some time. Where holes are observed, these should be measured so the type of beetle can be identified.
- On adjacent surfaces beneath the affected timber, small piles of freshly ejected bore dust or frass can often be seen. Again the more recent the dust the lighter the colour.
- In the worst cases, sometimes the actual larvae can be found by probing the damaged timber with a bradawl or penknife, etc. The BRE point out this can be very rare in practice.

6.3.4 Insect attack and standard surveys

Although there are a number of key locations where insect attack can be expected, in occupied premises few of these areas will be fully exposed. Fixed floor coverings, bath panels, internal linings and inaccessible roof and floor voids may all prevent access. When exposed timber surfaces can be seen, every opportunity should be taken to look for evidence of beetle attack. The flight holes of all the common types can easily be seen by the naked eye and torchlight and then measured. A magnifying glass is not usually part of the standard surveying kit but can be useful addition. A sharp implement (such as a bradawl or a stout knife) can be a useful to assess how bad the attack could be. In a dwelling where a lot of timbers are found to be damp then the likelihood of infestation is higher and so special attention should be paid. This is a typical 'trail of suspicion' that should be either followed or at least emphasised in the report.

6.3.5 Assessing the extent of the damage and reporting

Where insect attack in timber is found, it is beyond the scope of most surveys to precisely identify the type of attack and recommend treatment. The BRE (1998) identify several categories of damage in timbers in an attempt to limit unnecessary work. Where Common Furniture, House Longhorn, Lyctus Powderpost and Death Watch beetles are found to be active, further remedial treatment is usually needed. Consequently referral to a specialist timber treatment contractor is justified who then should identify

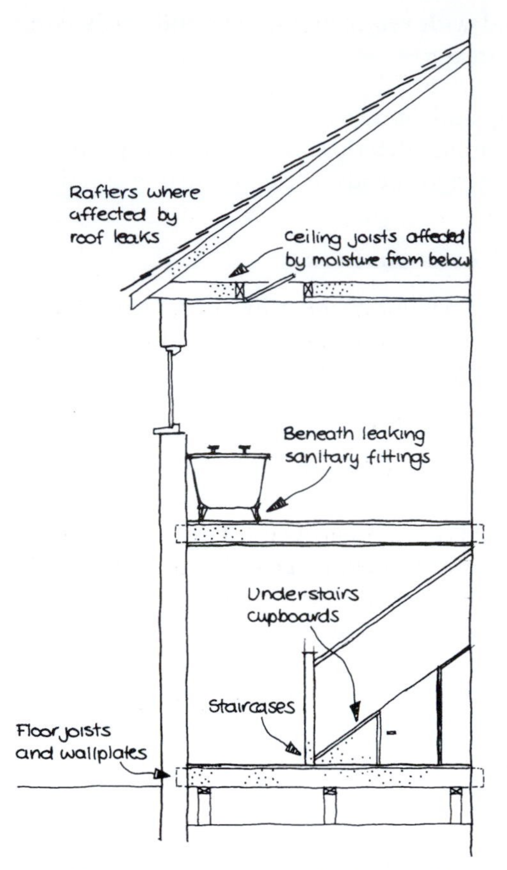

Figure 6.3 Common locations of insect attack in a domestic dwelling. The risk of insect attack is higher where the moisture levels in the timber are greater.

the type, extent and treatment required. In addition to this referral, additional advice can usefully be given to the client to indicate the seriousness of the infestation. For example:

- If the timber has been attacked by House Longhorn, Death Watch or Lyctus then the damage could be considerable. As well as remedial treatment, the replacement of structural timbers are often required. The BRE (1998) recommend that a full building survey should be arranged where Death Watch and House Longhorn are found.
- Common Furniture beetle on the other hand is usually of little structural significance although remedial treatment will often be necessary with a small amount of associated repair work.
- Wood-boring weevil is so closely associated with decayed damp timber that remedial work to rectify the dampness problem will eradicate this type of insect.

It is worth noting that the close association between rotting timber and insect attack is the reason why joint referrals to specialists are normally required. Further information on the relative effectiveness of the various remedial treatments is contained in BRE Digest 327 (BRE 1992a). The problems posed by excessive use of chemicals in treating outbreaks are included in BRE Digest 371 (BRE 1992b).

6.4 Other pests

There are a number of other living things that share living space with human beings. Not all of these will affect the condition of the dwelling but some might be very expensive to eradicate and can threaten the health of the occupants. If they affect the dwelling most clients will want to know. Bateman (1990, p.294) described the range of pests that normally affect buildings. The main types are described below.

6.4.1 Pigeons and other birds

Birds can indirectly damage buildings:

- Nests can block drains, gutters and flues. This can lead to water damage and sometimes dangerous heating appliances when the proper operation of flues is prevented.
- Droppings can disfigure surfaces of buildings and encourage other biological growths including lichens and algae. Limestones and sandstones can be badly affected by the high acid content released from the droppings.
- Dead birds can affect the operation of appliances and fittings especially where they are roosting in the roof space.
- Birds can introduce parasitic insects that can cause nuisance, damage and health risks especially where food is stored or prepared.

Pigeons are the most common culprits but sparrows and starlings can also be a problem in large enough numbers.

Identifying bird nuisance

In many cases the problem can be self-evident. Sometimes dozens of birds can be seen roosting on ledges, roofs, telephone wires, guttering, etc. Where they have gained access to roof spaces they can be seen on purlins, struts, etc. There are few surveying experiences to compare with unexpectedly disturbing a dozen flapping and scared pigeons in a dark confined roof space of an empty property! Where the birds are not evident accumulations of droppings is a key indicator. These may be fouling the ground some distance below. Internally, nests and dead bodies as well as droppings may be evident.

Giving advice

As with all other pests, the Environmental Health Department of the local authority may be able to offer help, advice and in some cases appropriate treatments. There are also a range of pest control contractors. In general the approach will include:

- removal of all droppings, dead bodies, nests, etc.;

- before the areas are inhabited, access must be denied. This can include:
  - blocking up any gaps or holes in the fabric;
  - protective measures to roosting points including:
    - steel-sprung wires; metal or plastic vertical spikes;
    - netting across roosting sites;
    - various gels and resins that make the birds feel insecure because of their stickiness.

Removing the source of food can also discourage roosting in a particular area. Depending on the extent of the attack, some of these measures can prove expensive.

6.4.2 Rats, mice and squirrels

Although rats are the most troublesome, all rodents can damage the fabric and fittings of a building. To wear down their continually growing teeth they regularly gnaw hard materials and will shred softer items to form nests. Rats in particular can:

- gnaw through plastic pipes and electrical cables;
- chew through copper pipes and cause water damage; and
- caused localised subsidence through extensive tunnelling.

The main concern about rats is the spread of disease. Salmonella and Weil's Disease can be deadly. Rats can also have a deep psychological impact on the occupiers. The rodent has never got over the public relations disaster of the Black Death and will always be associated with unhealthy conditions. The irony is that in many urban areas rats either pass by or through most dwellings.

Key inspection indicators

To the untrained eye, spotting rodent infestations is not easy.

- Look for small trays or pots of poisoned bait (usually a grain that has been dyed blue) installed by pest control agencies. These can usually be found in roof spaces, semi-concealed ducts, behind heavy furniture, etc.;
- neat circular entrance holes to network of rat burrows in external areas. These can go down to 750 mm and extend horizontally for great distances;
- gnawed and damaged building components and finishes;
- nests of neatly piled shredded material in underfloor areas, lofts, etc.;
- with squirrels, trees that have branches within 3 metres of the building and gaps between components especially at eaves and fascia board junctions;
- large amounts of pellet-shaped dung (see figure 6.4 for more details).

Giving advice

Where these indicators are noted the matter should be reported to the client. The Environmental Health Department or independent pest control contractors should be

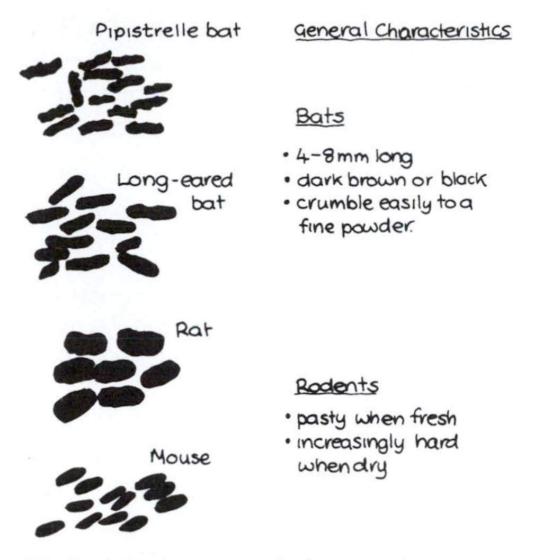

Figure 6.4 Comparison of bat droppings to those of other animals (reproduced courtesy of English Nature 1995).

asked to identify the true nature and extent of the infestation. The remedial work will depend on the circumstances but could include putting down poison baits, removal of dead rodents and possibly the blocking of holes, blocking off disused drainage connections, etc. and other disruptive ancillary building work. Further advice on how to reduce the risk of pest infestation in buildings is outlined in BRE Digest 415 (BRE 1996a).

6.4.3 Bats

These mammals are different to the other living things described in this section. Although some people may see them as pests they are protected by the law. The Wildlife and Countryside Act 1981 and the Conservation Regulations 1994 make it an offence to:

- kill, injure, catch or keep bats;
- damage, destroy or obstruct bat roosts;
- disturb bats by entering known roosts or hibernation sites;
- sell, barter or exchange bats, alive or dead.

It is unlikely that surveyors will intentionally be involved in harming bats directly. In normal circumstances it is the contractor rather than the surveyor who will do the damage. Surveyors can get involved indirectly by being part of a repair or remedial work 'chain of events'. For example, the surveyor may notice a serious outbreak of wood-boring beetle in a loft and recommend that it should be treated. A specialist contractor might then follow and spray the affected area, inadvertently killing the bat colony.

Should the law be contravened in this way, the surveyor will be judged by applying the test of 'reasonableness'. Should a reasonably competent surveyor have noticed the bats during the inspection or survey? As with building defects, if there was not an obvious 'trail of suspicion' then the surveyor cannot be expected to have spotted these very small mammals.

Precautions before the survey is undertaken

It may be an offence under the Acts to enter known roosts or the hibernation sites of bats. This is because if they are disturbed their chances of successfully breeding or surviving the winter may be affected. This poses two problems for the surveyor inspecting a dwelling:

- If the owner knows that there is a bat colony in the property and informs the surveyor before the survey a special licence may have to be obtained from English Nature (1995) before the inspection can go ahead.
- If the colony is undiscovered but detected by the surveyor during the inspection, (s)he may have to withdraw from the roost and obtain a licence as described above.

In both cases it is likely that the survey will not be completed. One way to avoid this occurring is to ask the owners if they have any bats in their property when the inspection appointment is being arranged.

Recognising bats in buildings

There are two main issues to consider:

- how to recognise the presence of bats; and
- once they are known about, what advice should be given?

Spotting the signs

Most bats are seasonal visitors forming small colonies during May or June and leaving for more sheltered roosts in August. They do not migrate but stay in this country to hibernate through the winter. The following summary may help to identify bat roosts:

- Bats have adapted to live in all kinds of buildings including houses, churches, farms, ancient monuments and all sorts of industrial buildings. They can be found in urban, suburban and rural environments.
- Bats choose their roosts carefully:
  - During the summer they prefer sites that are warmed by the sun and are most often found on the south and west sides of a building.
  - Most types prefer small roosting spaces. The pipistrelle prefers places outside such as in the eaves, behind tile and timber claddings, etc. and has been known to colonise recently built houses. The brown long-eared bat on the other hand is usually found just below the ridge in the loft spaces of older buildings.

- Although it may be possible to see the bats themselves, bat droppings are usually the clearest indicator of any roosts. They are roughly the same size as mouse droppings (see figure 6.4) but crumble to a powder when dry. They are generally found stuck to walls near to the roost exit or in small piles beneath where the bats hang. As colonies can contain between 50–1,000 bats some of this evidence might be significant!

Giving advice

The Countryside and Wildlife Act 1981 states that any building owner must contact English Nature before anything is done that is likely to affect bats or their roosts. This could include:

- alteration and maintenance work including thermal insulation;
- re-roofing works;
- getting rid of unwanted bat colonies;
- re-wiring and plumbing works in roofs;
- timber treatment for beetle and rot infestations;
- treatments for wasps, bees, cluster flies and rodents.

Many of these activities may be recommended by a surveyor following an inspection. Therefore, where bats are found in a building clients should be advised that:

- A possible bat roosting or hibernation site has been identified in the property and give the precise location.
- If the discovery of the bats has limited the inspection, clearly identify the restrictions and explain the possible consequences of not carrying out a full appraisal.
- Outline the implications of having bats in a property. This could include:
  - the need to contact English Nature to gain approval for any work to be carried out;
  - possible restrictions on how any planned work might be carried out. For example:
    - work might be restricted to certain times of the year;
    - contractors may be allowed to only use specific types of chemicals;
    - bats might have to be carefully moved before work can start.

  These restrictions could add to the cost of the work.

On a more positive note, the client could be told that sharing their home with bats will present very few practical problems and be no threat to their family's health. Bats are considered by many as remarkable animals with some amazing features. Home owners have a crucial role to play in the conservation of bats in this country and this can be enhanced by clear, objective advice from the surveying profession.

6.4.4 *Masonry bees*

'Masonry bees' have occasionally hit the headlines when they have been discovered burrowing through the brickwork of a house. These tabloid headlines tend to conjure an image of a spreading swarm of house-eating bees. In reality *Osmia rufa* (or masonry bees) usually inhabit earth banks and soft rocks (BRE 1996b). On the odd occasion the bees will choose the soft mortar of older buildings as a suitable nesting site. The damage is usually done by the female bees who burrow into the masonry to form galleries and tunnels to house the pupal cells of the next generation. Although they act in a solitary capacity, a number of different bees can inhabit the same area if the conditions are right. Sometimes the bees may overwinter within the masonry and enlarge the tunnels. This can result in extensive damage over a number of years. In a few cases this has led to isolated rebuilding of the wall.

Key inspection indicators

- Numbers of large holes and linked tunnels in the mortar joints of older and softer brickwork and masonry;
- evidence of frequent bee flights during the summer months.

Giving advice

The Environmental Health Department or a pest control contractor can give advice. Remedial work will usually involve:

- possible injection of suitable pesticide into the holes and galleries;
- raking out and repointing the affected areas with a mortar that is not too hard for the bricks but too hard for the bees;
- where soft masonry has been affected then a render coat may have to be considered.

The BRE (1998) point out that the use of pesticides and other injection treatments are unlikely to be effective on their own.

6.5 Troublesome plants

Trees and shrubs have been discussed in section 4.6 and can undermine the very foundations of the dwelling. There are a number of other plants and organic growths that although they will not lead to structural failure can adversely affect the fabric of the building. The main ones are listed below.

6.5.1 *Creepers and climbing plants*

These can include a whole variety of plants that are encouraged to climb up the dwellings by the owners. Few of the fashionable gardening programmes on TV or radio point out the less desirable effects!

- Walls and roofs may be disturbed by the plant's roots.

- Gutters and downpipes can be blocked leading to water damage.
- The wall is kept damp for most of the time and is unable to be properly inspected for defects. This may lead to neglect.
- It is difficult to repair or paint the wall satisfactorily.
- Root suckers can disfigure the wall surface.

The main types of plants are:

- Ivy and other creepers:
  - Ivy has small roots that grow in search of moisture and darkness. They can gain a foothold in open or cracked joints forcing them apart even more. As the growth thickens the damage can become quite serious. On balance, unless the building is in very good condition and can be kept under strict observation then ivy is best removed.
  - Creepers – these tend to attach themselves to the building via small surface suckers. Although these are not as damaging as ivy roots they can secrete small amounts of acid that can pit susceptible stonework.
- Other climbing plants – plants such as roses, jasmines and honeysuckle all need to be supported by a framework of trellis or wires attached to the walls. These frameworks should be appropriately fastened to the wall. Ideally the whole plant and supports should be able to be bent forward to allow maintenance and repair of the wall beneath.
- Plants on or near buildings – the BRE point out a number of other situations where plants can cause damage:
  - ground-cover plants should be kept clear of ventilation bricks;
  - woody plants that have root systems that can penetrate masonry should be cut down and the stumps poisoned;
  - seeds from trees can lead to plants growing almost anywhere on a building. Flat roofs, secret gutters, valley gutters, etc. are all at risk. Where this growth occurs all plant growth should be removed.

6.5.2 Giving advice

At the very least the climbing plants should be kept well trimmed back from the eaves and gutters and clear of door and window frames. On balance it is better to recommend the complete removal of ivy plants. This is normally done by cutting the main stem and digging up or poisoning the stump to kill it off. The remainder of the branches and leaves should be left in place until it completely dies back and dries out. Only at this time should the plant be removed and the roots carefully dug out. Repointing is usually required.

References

Bateman, P.L.G. (1990). 'Pests. In buildings and health'. *The Rosehaugh Guide to the Design, Construction, Use and Management of Buildings*, (pp.234–315). RIBA Publications, London.

Berry, R.W. (1994). *Remedial Treatment of Wood Rot and Insect Attack in Buildings*. Building Research Establishment, Garston, Watford.

Bowyer, J. (1977). *Guide to Domestic Building Surveys*. The Architectural Press, London.

BRE (1992). 'Identifying damage by wood boring insects'. *Digest 307*. Building Research Establishment, Garston, Watford.

BRE (1992a). 'Insecticidal treatments against wood boring insects'. *Digest 327*. Building Research Establishment, Garston, Watford.

BRE (1992b). 'Remedial wood preservatives: use them wisely'. *Digest 371*. Building Research Establishment, Garston, Watford.

BRE (1996a). 'Reducing the risk of pest infestation'. *Digest 415*. Building Research Establishment, Garston, Watford.

BRE (1996b). 'Bird, bee and plant damage to buildings'. *Digest 418*. Building Research Establishment, Garston, Watford.

English Nature (1995). *Bats in Roofs – a Guide for Surveyors*. Nature Conservancy Council for England, Peterborough.

Richardson, B.A. (1991). *Defects and Deterioration in Buildings*. E&FN Spon, London.

Singh, J. (1994). *Building Mycology: Management of Decay and Health in Buildings*. E&FN Spon, London.

Chapter 7

Roofs

7.1 Introduction

In their book *Structural Surveys of Dwelling Houses*, Melville and Gordon (1992, p.45) state that a dwelling needs a 'sound head and feet'. There is nothing more evocative of a dilapidated building than a roof leak dripping water into a metal bucket! Murdoch and Murrells (1995, p.97) identified that problems associated with roofs account for 13% of the claims of negligence against surveyors, second only to damp and timber defects. This underlines the importance of the roof inspection.

7.2 Extent of inspections

The first question is what is expected of the surveyor when inspecting a roof? One of the best sources of information is the standard guidance from the professional institutions. This varies with the different types of inspections and surveys and will also be modified by the specific requirements of any lending organisation. The main elements are summarised below.

7.2.1 Mortgage valuations

Subject to reasonable accessibility, the roof space is inspected only to the extent visible from the access hatch without entering it. The first problem is the access hatch. If there is not one this should be clearly stated as it limits the judgements that can be made. Care must be taken looking for one. It will normally be on a landing above the stairs but sometimes it could be tucked away in a small cupboard or above a tall wardrobe. An access hatch still might be considered inaccessible if:

- the ladder cannot be erected safely and easily without moving heavy furniture;
- the hatch is fixed or painted shut so it could not be opened without significant force and potential damage to the finishes.

If the hatch is secured with cups and screws that are easy to unscrew with a standard screwdriver then it would be reasonable to remove them. This judgement will always be up to the surveyor. The RICS are just as vague. In explaining the provisions of the new Homebuyers survey, Holden (1998, p.25) states that the surveyor should unscrew a loft hatch '... if you think it sensible and safe to do so'.

If it is openable, the ladder should be assembled and the loft hatch opened. A brief torchlight inspection should be carried out from that position. If an obvious defect is observed should the surveyor venture further into the roof space? Does the surveyor always have to follow the trail of suspicion? This is difficult to define and will depend on the weight of the visual evidence. In Smith v. Bush (1990) the issue of loft space inspections was discussed. The case famously involved a surveyor not noticing that a chimney breast was unsupported in the loft. This later collapsed into the bedroom below. The fact that the chimney stacks were observable externally and the chimney breast below was absent created a 'trail of suspicion'. This should have led the surveyor to look in the loft even if it was not standard practice at the time. Applying this principle a surveyor could be expected to enter the space if the following defects are clearly observable by torchlight:

- significant structural problems with the party walls, chimney breasts, roof timbers, etc.;
- the ingress of water is clearly evident, e.g. staining, daylight showing through the covering, etc.;
- visible wood rot or insect damage to the roof timbers;
- external signs that suggest potential problems, e.g. sagging roof slopes, bowing ceilings, extensive damp staining to the ceiling below, etc.

This is not an exhaustive list but a few examples that may help make the necessary judgement in marginal cases.

7.2.2 Homebuyers surveys

The surveyor will have to get off the ladder and have a general walk around the roof space. The extent of this 'walk around' will depend on health and safety issues (such as balancing on the tops on ceiling joists) and the amount of stored items. The inspection should be confined to details of design and basic construction. Individual timbers are not specifically examined for defects although where defects have been observed as part of the general examination such details are noted in the report. The surveyor should not move occupant possessions or lift insulation unless there is a reason to and it can easily be done. The principle of 'follow the trail of suspicion' would apply here. Damp meter readings should be taken close to where walls and chimneys penetrate the roof coverings. Roof timbers in this area should be thoroughly checked as they could be vulnerable to damage. The roof space inspection should be more proactive during a Homebuyers survey. Depending on the amount of stored possessions and general safety issues, the surveyor should spend a significant amount of time in the loft. Assuming a 2 to 2½ hour total inspection time, 20–30 minutes could easily be spent in the loft space.

7.2.3 Building surveys

The interior of accessible roof voids should be inspected in detail and accessible connecting voids from the nearest vantage points to the extent practicable with the equipment available. All timbers should be checked for damage and it is important to record the moisture content of the timbers, size and design of the roof structure and particularly the pitch of the roof which will help in reporting on the suitability of the covering. During a building survey the surveyor should be prepared to spend a considerable amount of time inspecting all timbers, crawling down to the eaves, moving occupant possessions and generally getting very dirty!

Although it is important to define the limits of any inspection, no surveyor has ever been sued for going beyond the parameters of the professional guidance. In fact courts will expect this where a trail of suspicion exists.

7.3 Different types of pitched roof structure

A surveyor can expect to encounter a wide variety of roof structure types including those described below.

7.3.1 Common traditional roof types

Some common types are listed below and two are illustrated in figure 7.1.

- **Single lean-to roof** – spans of up to 2.5 m, usually used above back additions of older terraced property.
- **Double lean-to roof** – common on Georgian town houses because it allowed a free façade. Consists of central valley beam supporting rafters bearing on to the party walls.
- **Collar roof** – the introduction of a collar between rafters did allow savings on brickwork as a room was formed partly in the roof space. The collar was usually so far up the rafters it was not very effective at tying rafters together. This was associated with cheap work.

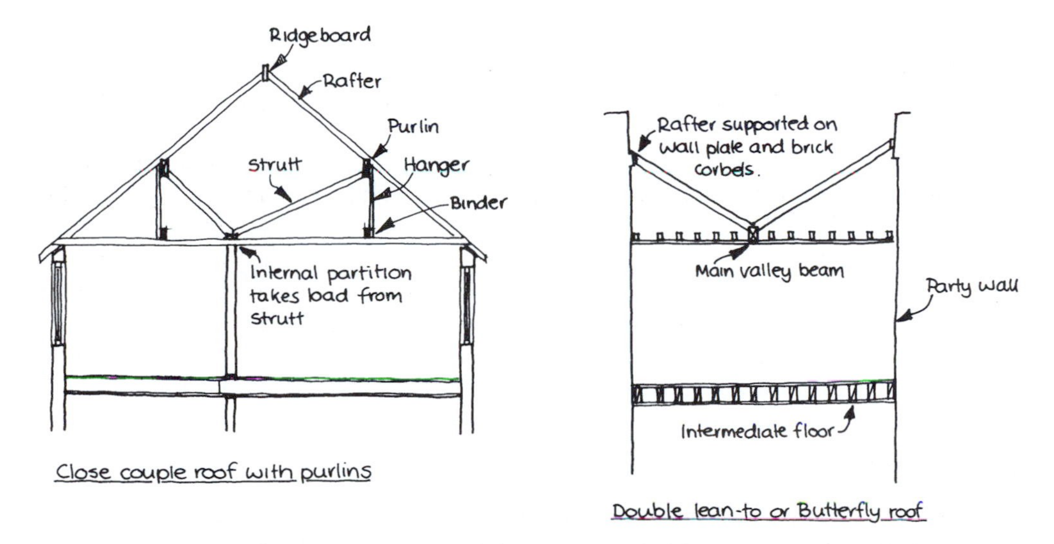

Figure 7.1 Two common types of traditional pitch roof structures.

- **Couple roof** – very simple roof consisting of rafters on wall plate with no provision for ceiling tie.
- **Close-couple roof** – ceiling tie introduced between rafter feet and as long as it has been constructed correctly can be very stable. Limited to spans of 4 m.
- **Close-couple roof with purlin** – to get wider spans, purlins were introduced to keep rafter size to reasonable dimensions. It is usual to have one or two purlins per roof slope parallel to elevation. Struts should support the purlins and must be taken down to internal load-bearing walls. In many narrow-fronted houses purlins can span party wall to party wall. Collars are often found at purlin positions. A variation occurs when a room in the roof is provided. The floor level to the second storey is different to that of the rafter feet so triangulation problems can occur.

There are many variations often based on regional custom and practice. A more complete review is contained in Melville and Gordon (1992, p.57).

7.3.2 Specialised roof structures

There are many different types of historic and other specialised roof structures that can be encountered. It is beyond the scope of this publication to look at any of these in detail; reference to more specialist publications will be necessary. These types could include:

- king and queen post trusses
- crown post trust
- cruck construction
- thatched roofs.

Specialist knowledge and experience are required to assess these types of structures. Any surveyor should carefully consider whether they have the right background to make an assessment. Ignorance of a particular roofing type will be no defence in court.

7.3.3 New roof structures

The most common method of constructing modern roofs is by using prefabricated trussed rafters. These are discussed in more detail in section 7.4.7.

7.4 Defects associated with pitched roof structures

7.4.1 Roof spread

There is a wide range of defects that can affect roof structures. One of the most common is associated with weakness of the structure itself. This can lead to the settlement or 'spread' of the roof frame. Typical signs and symptoms of this defect include (see figure 7.2):

- a bowing or dipping of the ridge line;

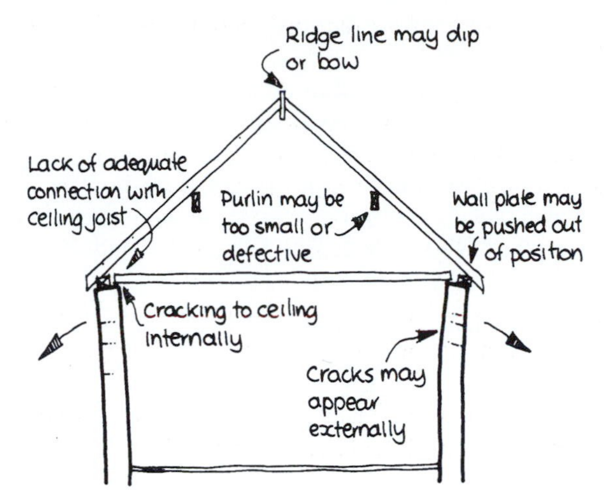

Figure 7.2 Sketch showing the typical signs of roof spread.

- where the roof covering extends over the party wall, distortion in the structure can cause unevenness or 'humping up' of the covering over the party wall positions;
- the feet of the rafters pushing outwards can cause the top of the supporting walls to distort, in some cases displacing the wall plates;
- horizontal cracking to mortar course just below eaves level where the brickwork is pushed outwards by the spread of the rafters;
- observed physical distortion of the roofing members internally. In extreme cases this could lead to splitting of the timbers themselves.

Roof spread is often caused by structural deficiencies described in the following section.

7.4.2 Adequacy of timber sections

The size of the timber sections in many traditional roofs may have been inadequate even for their original purpose. The assumed level of loading on roof structures due to wind and snow has previously been underestimated and has been increased in recent codes of practice. Therefore it is common to find that older roof structures do not match up to current standards. There are two quick checks that can be used to see if existing roof components are approaching an adequate size:

- by using the 'Tables of sizes of timber' in Appendix A of Approved Document A of the Building Regulations 1991 (HMSO 1991). These give the required sizes for timber components of given spans and loadings;
- traditional 'rules of thumb'. Melville and Gordon (1979) suggest the following for rafters 'span in feet (305 mm) divided by 2 equals the depth of 2 inch (50 mm) wide rafter at 15 inch (381mm) centres'. So for a rafter spanning 2.44 metres (8 feet) between supports would require a 50 mm by 100 mm section size. This would also meet the building regulations.

Assessing adequacy of timber sizes

The above are two approaches of how roof timbers could be assessed for adequacy. It must be emphasised that this is not a full structural appraisal. If other aspects of structural integrity are not reviewed (e.g. lateral stability, triangulation, etc.) then serious deficiencies might well be overlooked. Despite this it is possible to develop a 'feel' for what should be within the margins of acceptability. Figure 7.3 gives some guidance on how to assess the size of rafters. The main evidence is the performance of the roof. If the roof is free from major defects and distortion then the technical adequacy of the roof timbers are of secondary importance. This is because most components of traditional roofs will be undersized when compared with modern standards. To avoid numerous referrals to roofing specialists or structural engineers, a sense of balance must be retained. If the sizes of the existing timbers are only marginally below the recommended sizes (say no more than 20%) and

- there is no evidence of distortion, sagging or deflection; and
- the structure appears well triangulated;

then no further work or referrals may be required.

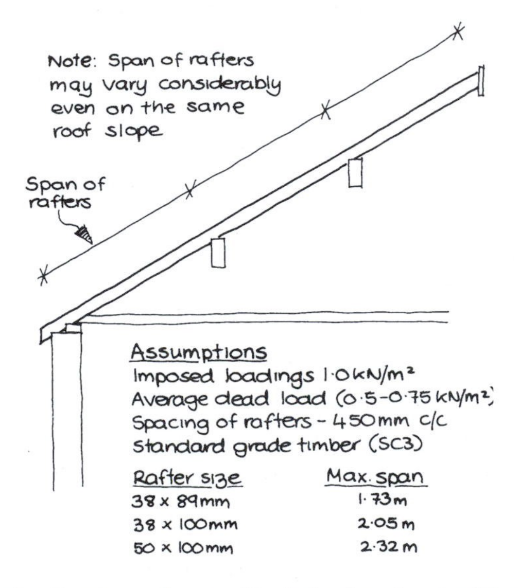

Figure 7.3 Guide to assessing the size and span of typical rafters in a traditionally constructed roof.

7.4.3 Purlin problems

The purlin is usually the main supporting component to many roof structures and can often suffer from a variety of problems:

- The purlins can be sized incorrectly. Figure 7.4 gives a broad indication of the sizes that should be expected.

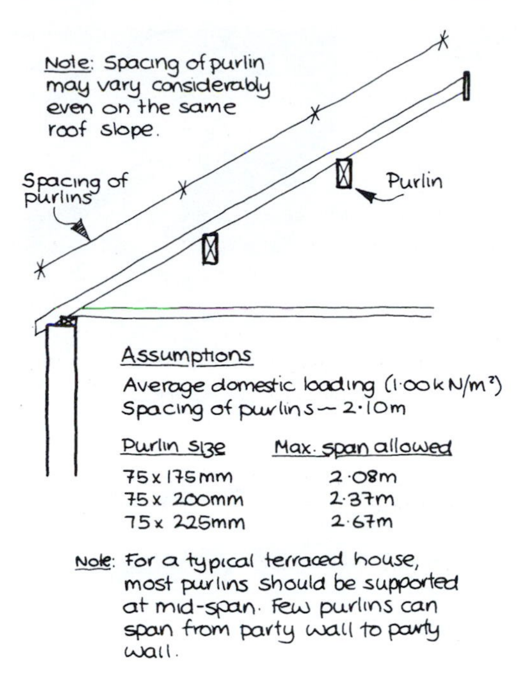

Figure 7.4 Guide to assessing the size and span of typical purlins in a traditionally constructed roof.

- Support to purlin ends. Typical faults can include:
    - Where the purlins are built into exposed gable walls the ends can be vulnerable to wood rot where they are not properly protected. This can weaken the end bearing of the purlin.
    - Some purlins can be supported on brick corbels that protrude out of the supporting wall. These can break under load unless properly constructed.
    - Other purlins may be poorly supported by the wall they are bearing on (see figure 7.5).
- Purlins should be in one length between the main supports. Where they are jointed they should be properly 'scarfed' equally each side of the support. If not then the purlin can rotate around this joint.
- Where purlins are supported at mid-span by a vertical or near-vertical strut, this must bear on to an appropriately supported element (i.e. an internal spine wall or a partition around a staircase). In some cases this strut can be carrying up to a sixth of the weight of the roof slope. If it is not properly supported, deflection of the roof slope could result. Check the following features:
    - that the struts bear directly on to the supporting wall below and not merely on to a timber plate positioned on the ceiling structure (see figure 7.6);
    - that the internal wall taking the strut matches up with walls on all the floors below all the way down to the foundations. Special checks should be made where 'through' rooms extend the full width of the house especially when these have been as a result of recent DIY alterations.

Figure 7.5 Inadequate support to purlins at the end bearing. These purlins are poorly supported by this partial party wall.

Figure 7.6 Inadequate strut support to a purlin. This traditionally constructed roof had been recovered with concrete tiles and had begun to deflect under the excessive weight. The purlin was visibly deflecting and in an effort to limit this problem a contractor installed a small strut. This was poorly supported by a timber plate at it's base. The timber plate was fixed to the back of the ceiling joists at mid span causing considerable deflection to the ceiling below.

- Ring purlins – purlins to hipped roofs – often have no visible support. This is because they act as an informal 'ring beam' and rely on the whole structure acting together as one. Consequently a roof of this type should have the following characteristics:
  - The purlins should be of substantial size (i.e. 75 × 225 mm minimum).
  - All the timber sections should be cut closely together and properly spiked or nailed into position especially at the junctions with the hip rafters and purlins.
  - The rafters should be properly fixed to the purlins.
  - There should be adequate restraint to the feet of the rafters by either the ceiling joists, collars or other forms of restraint.
  - There should always be a tie (sometimes called a 'dragon tie') reinforcing the connection between the wall plates at the corner of the building below the hip rafter. Without this, the wall plates will move outwards allowing the hip rafter to deflect (see figures 7.7 and 7.8).

If a hipped roof has most of these characteristics and there is no evidence of distortion then no further action would be necessary.

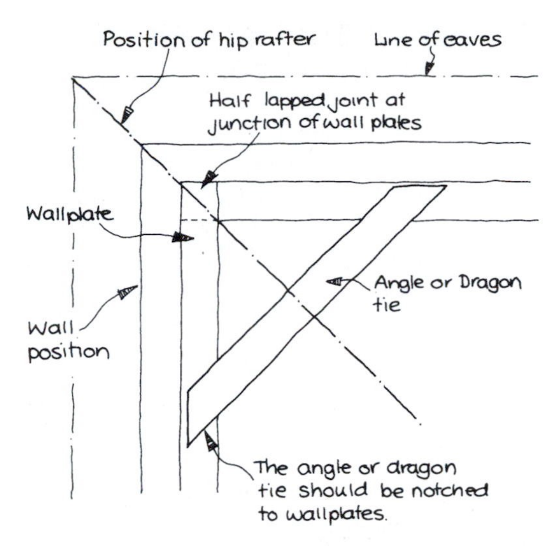

Figure 7.7 Sketch showing the typical arrangement for a 'dragon tie' that joins the adjacent wall plates to the gable slope of a traditional hip roof.

Figure 7.8 The hip rafter to this roof was deflecting because the wall plates were being spread apart by the thrust of the roof slopes. There were no dragon ties to restrain them. A builder had inserted a crude vertical strut to take the weight of the hip rafter. This was supported off the top of the ceiling joists below. A clearly inadequate solution.

7.4.4 Ceiling ties and joists

The ceiling structure has a number of functions:

- to support the ceiling finish;
- to support an appropriate amount of occupant storage;
- to help triangulate the roof structure by restraining the feet of the rafters.

Ceiling joists are often 'stiffened' by ceiling binders (see figure 7.9).
Some of the most common problems include:

- The ceiling joists can be either too small or lack support at mid-span. If the ceiling joists are not big enough to span between their main supports then a binder with hangers should be provided. Figure 7.9 gives an indication of this arrangement and indicative sizes.
- The ceiling joists are not at the same height as rafter feet and so do not offer any resistance to rafter spread. In this case look for typical signs and symptoms of roof spread (see section 7.4.1).
- Poor nailing at junctions with other roof members.
- Where ceiling joists do not restrain the rafters, check for timber collars. To be fully effective these have to be in the lower third 'zone' of rafters to ensure restraint. They are more effective if the collars are dovetailed to the rafters rather than just nailed on to the side.
- Where the roof structure has a hip slope the feet of the hip rafters are often not tied to the ceiling joists as the latter often run parallel to the gable wall. Additional ties may be needed at this point to help prevent rafter spread.

7.4.5 Room in the roof

In some areas of older housing, a room in the roof was provided at the time of construction. Characterised by a steep winding staircase up to a small room served by a dormer or skylight, these spaces can break the continuity of roof structure. Ties for rafters can be difficult to provide as the floor structure and the wallplate are often at different levels. In many cases where there are access hatches to the front and rear roof spaces the true nature of the roof structure can be assessed. If these hatches are absent then care must be taken. Experience has shown that few of these types of roof structures are adequately triangulated (see figure 7.10).

7.4.6 Other defects

There are a number of other defects associated with the roof structure can be regularly encountered and include the following.

Over-cutting of joints

Where one timber section meets another they are often 'birds-mouthed' together. This is where one is cut to fit tightly against the other. The depth of this birds-mouth should be no greater than one-third the depth of the timber itself.

Weakening of structure through woodworm and rot

Dry rot is not common in roof spaces because the environmental conditions change so quickly. Wet rot can affect timbers especially where they are built into or in contact with damp masonry. Typical locations would include trimming rafters around chimney

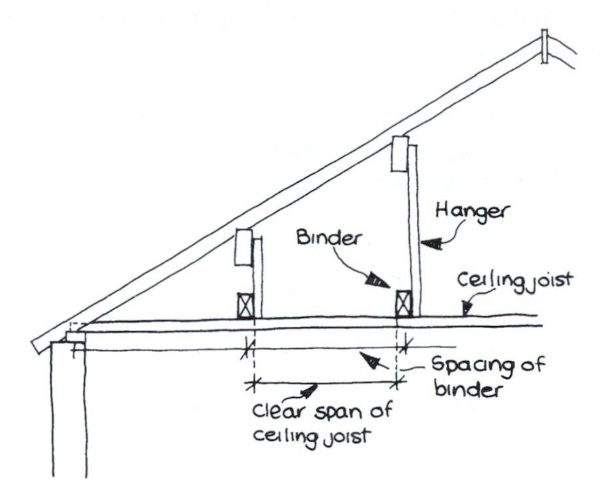

<u>Assumptions</u>
Average domestic loading
spacing of binder 2·10m
spacing of ceiling joist 450mm

Binder Size	Max. clear span of binder
75 x 150mm	2·05 m
75 x 175mm	2·42m
75 x 200mm	2·79m

Ceiling joist size	Max. clear span of ceiling joist
47 x 97mm	1·81m
47 x 147mm	3·05m
47 x 195mm	4·28m

Figure 7.9 Guide to assessing the size and span of typical ceiling joists and binders to a traditionally constructed roof.

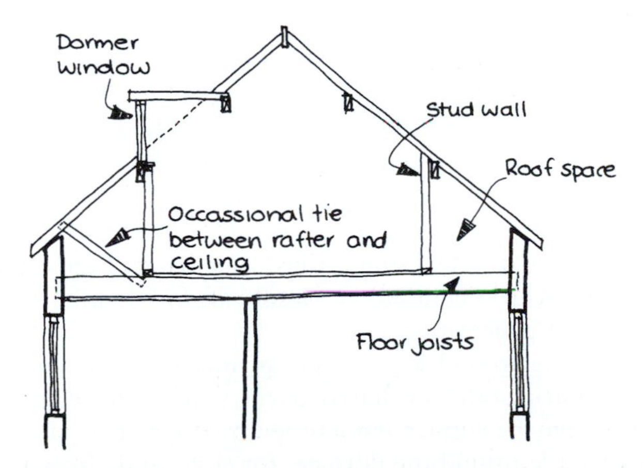

Figure 7.10 Room-in-the-roof structure. The lack of connection between the floor/ceiling joists and the rafter feet can often result in roof spread.

breast; rafter feet and wall plates adjacent to leaking gutters, valley beams and gutters where they are built into the wall.

Woodworm can be found in most older roof spaces but usually there are only low levels of infestation that pose no serious threat to structural stability. Where the infestations are particularly heavy especially with Death Watch and Longhorn Beetles, the structural timbers can be considerably weakened.

DIY alterations

Home owners who undertake their own unadvised alterations can create a range of structural problems in a roof space. A few examples from real cases are listed below:

- removal of struts to increase storage space;
- cutting out the chords of prefabricated trussed rafters so an 'Alpine' model railway could be installed;
- informal loft room conversion (see section 7.6.6 for further discussion);
- removal of partitions in rooms below leaving struts and purlins unsupported;
- fitting of new water storage tanks at mid span of the ceiling joists;
- installation of new roof light/dormer windows that resulted in the main purlin being cut (see figures 7.11a and 7.11b).

In each of these cases remedial work was required to ensure the stability of the roof structure. If any similar alterations to the roof structure are noted, the owner should be asked whether the appropriate building regulation approval had been obtained and if a competent contractor had carried out the work. If not, then further investigations may be required.

Dormer fronts

Many traditional houses have large dormer windows where the front wall of the dormer is formed by the continuation of the external wall of the house above the roof line. Unless this wall is well restrained then it can become unstable. A classic sign is the wall leaning backwards out of vertical putting pressure on the adjacent roof structure. Internal distortion of roof timbers and ceiling can often be seen.

Dormer windows in modern houses are often affected by a number of defects. These features need to be detailed very carefully if rainwater penetration is to be avoided (HAPM 1991, section 4.3). Typical problems include:

- Surface and interstitial condensation – the dormer cheeks and ceiling detailing can make it difficult to achieve continuity of insulation and ventilation. Check for signs of mould and water penetration.
- Inadequate trimming around the dormer opening – the top and bottom trimmers and the side supporting rafters will have to carry the truncated portion of the main roof that results from the dormer. If not properly designed then the timbers members could deflect. Check around the dormers for signs of deflection and cracking to the finishes.

Figure 7.11a The roof to this dwelling has been affected by roof spread. Based on the external visual clues, the likely causes are heavier replacement concrete tiles and the installation of the roof windows.

Figure 7.11b An internal view of one of the new roof windows. The purlin has been cut through so the window could be fitted. The vertical posts supporting the ends of the purlins are supported off the existing floorboards and floor structure. Little surprise the roof covering is uneven!

- Rainwater penetration – there are a number of vulnerable points around a dormer including the junction of the window frame and the dormer structure; the junction of the dormer and main roof covering and the apron flashing at the base of the dormer usually under the window sill. Internal surfaces adjacent to these features should be checked for signs of water penetration.
- Inadequate durability of the window in the dormer – because this window is usually very exposed and is less accessible for maintenance than lower windows it is often in poor condition. It is often better to have a low maintenance PVCu or metal window even if the rest are timber.
- Flat roof problems – where the dormer has a flat roof the covering is very exposed to wind, rain and sunlight. Although it will be almost impossible to inspect, the underside shown be checked carefully for signs of water leakage.

7.4.7 Trussed rafter roofs

The vast majority of houses constructed since the war have trussed rafter roofs. These prefabricated sections have been designed in accordance with modern design codes (although they have changed over the years) and tend to be of a more reliable quality (Noy 1995, p.130). Because they are so well engineered their reserves of strength can be less than traditional roofs. They also are more reliant on the proper interaction with other parts of the building structure. As a consequence, they are more vulnerable to damage caused by inappropriate installation and alterations. Typical problems include:

- To ensure that trussed rafter roofs have sufficient strength to resist lateral wind forces they need to be adequately braced (see figure 7.12). Often these bracings are omitted or are sometimes cut through by operatives who are fixing terminal sections to flues, soil and vent pipes, etc.
- Trussed rafters are not usually placed at greater spacings than 600 mm. Any more than this could result in excessive deflection.
- The final three trussed rafters should be fixed to the gable/flank wall with galvanised restraint straps to make sure that the walls and roof structures interact. It is important that the correct number and type of fixings are present. More information is given in figure 4.13.
- A trussed rafter roof is very sensitive to alterations and so no part of it should be cut or removed without a properly engineered solution.
- Any water tanks need to be adequately supported with the weight spread across a number of separate trussed rafters (see figure 7.13).
- Trussed rafters can often be damaged and split where they are 'skew' nailed to the wall plate. The two should be connected by metal truss clips.
- Check the metal nail plate timber connectors for signs of corrosion. This is important if high humidity levels are suspected.
- Earlier trussed rafter roofs will often have less ventilation than the current regulations require. Extra insulation added over the years might have reduced this even further. Evidence could include mould growth and damp staining to the surface of the timber members.

7.5 Pitched roof coverings and their defects

7.5.1 Introduction

There are many different types of roof covering to pitched roofs. It is important to inspect them from above and below if possible. The standard guidance suggests the following:

- **Mortgage valuations** – the roof coverings, chimneys, parapets, gutters, etc. should be inspected from ground level within the boundaries of the property and from adjacent public and communal areas. This is usually done through binoculars (8–10 × magnification). These features should also be inspected through any windows whilst standing at the various floor levels within the property. There is no specific mention of using ladders for roof inspection purposes.

Figure 7.12 Typical lateral bracing to a trussed rafter roof (reproduced courtesy of BRE).

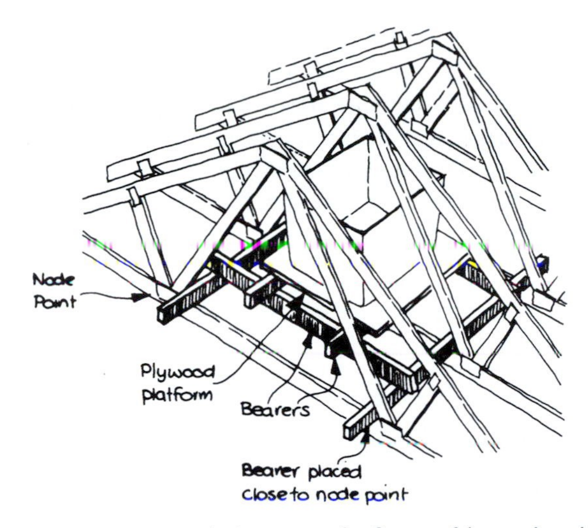

Figure 7.13 Support to water storage tanks in a trussed rafter roof (reproduced courtesy of BRE).

- **Homebuyers survey** – similar in nature but usually involves the use of a small portable ladder. This is taken to mean a 3-metre 3–4 piece sectional ladder that is usually adequate to allow inspection of a flat roof of a single-storey building. There is no requirement to get off the ladder and actually get on to the roof area unless there is a 'trail of suspicion'.

7.5.2 Different types of roof covering

This section identifies the main types of roof covering and briefly outlines the most common defects.

Natural slates

This is a very durable covering that has been used for hundreds of years. There is a great variety in both type and characteristics. Some are longer lasting than others. Main problems include:

- Sulfurous acids in rain can cause breakdown of carbonates in the slates. Capillary action can allow the rainwater to travel between slates and so the corrosion can begin from the underside. The nail hole can be enlarged allowing the slate to slip out of position. This will especially affect slates that only have one centre nail fixing that is sometimes found in cheaper work. Excessive tingles or lead clips that secure former slipped slates are evidence of problems.
- Incorrect size in relation to pitch – the lower the pitch the bigger the slate needs to be so that increased capillary action does not extend up to the next joint. Examples of overlap include:
    - steeper than 45 degrees: 63 mm
    - 30–45 degrees: 75 mm
    - less than 30 degrees: 88–100 mm, up to 150 mm in very exposed areas.

Fibre-based slates

These were originally asbestos cement slates with surface finish so they resembled natural slates. These have now been replaced by non-asbestos fibres. Early types often warped, lost the surface finish and supported high levels of mould and moss growth. Several different types are made by different manufacturers. It is important to become familiar with different systems. Regularly updated manufacturers literature will be a useful source of information.

Key inspection indicators

- Fixing methods – especially the correct provision of copper discs or rivets that hold down the lower end of the tiles.
- Evenness of the tiles – newer tiles have to be laid on a reasonably true and even supporting structure. If they are not properly supported from beneath they can easily crack when walked over during normal maintenance work.
- Accessories – i.e. eaves and verge details, ridge and hip fittings. They should be from the same manufacturer's range as the tiles themselves.

Clay tiles

These have been used for centuries, hand and machine made. The main problems are:

- if too porous then can break down with frost;
- check the nailing – will vary according to exposure:
    - normal – every fourth or fifth course;
    - exposed location – this can be every course;
    - the nailing can be checked from the loft space by gently pushing up the tiles but only as long as there is no roofing felt.

In some older properties tiles may be held in by wooden pegs. Considerable doubt over durability of this type of fixing should be indicated to the client.

This type of tiling also includes pantiles, Italian and Spanish tiling. Some machine-made varieties can have variable durability. There can also be a problem of matching profiles when they break and need replacing.

Concrete tiles

There has been large-scale use of concrete tiles since the 1970s. Many concrete tiles can be of poor quality. The main problems include:

- they can be much heavier than original roof coverings they replace and so cause sagging of the roof structure;
- the surface finish often weathers away and can contribute to the blocking of gutters and drains;
- difficult to form properly weathered details especially at valleys, hips and abutments. This can often lead to extemporisation on site (i.e. another word for a bodge job!);
- the fixings especially at the verges and eaves need to be in accordance with the manufacturers recommendations to prevent uplift. This usually takes the form of metal clips.

Stone slates

Can be sandstone in Yorkshire or limestone in the Cotswolds. Because they are very heavy the roof structure needs to be assessed with special care. Many older roofs may have become distorted over time because of this loading. If the building is listed or in a conservation area, new or re-used stone slates will have to be used in any recovering or maintenance work. As this requires specialist knowledge and skills the client should be warned about the possible cost consequences.

Problems with the sarking felt

Newer or recovered roofs can often suffer from leaks due to faults in the sarking felt. For example, if the felt has not been properly supported where it is dressed over the eaves into the gutter or over the edge of any barge board, water may build up and leak internally. The same effect can occur where the felt has not been fitted closely around soil and vent pipes that pass through the covering. Typical signs can include:

- evidence of dampness and water staining to the ceilings, inner leaf of the wall, loft insulation and eaves and barge board construction;
- poorly fitted, loose or partly missing sarking felt to the underside of the covering.

Where the roofing felt is dressed into the gutters if it is the standard type of felt (reinforced bitumen) it can degrade on exposure to sunlight. To prevent this, the BRE (1985a) recommend that a supplementary strip of dpc quality material should be provided at eaves and the verge.

The sarking felt should not be:

- stretched too tightly over the rafters because it can result in tears occurring when it is nailed and water running down the felt being trapped behind the batten;
- too limp. In this case there would be a danger of water not draining away properly or ponding in any folds, etc. (HAPM 1991, section 4.1).

This can be assessed during a roof space inspection.

7.5.3 Assessing the condition of roof coverings

Slate and tile roof coverings can have a long life. Many natural slate roofs are still giving good service 80–100 years after they were first installed. Even where some slates have slipped or are broken, a competent roofing contractor can extend the roof's effective life even further. The problems begin where a high proportion of the slates or tiles are defective or have been repaired before. Roofs in this condition can be ravaged by high winds and storms. Not only would this be expensive to repair but it could also cause considerable damage to the rest of the dwelling and the occupants' possessions as well.

A survey or inspection report will have to look ahead and advise the client of the implications there might be in the medium to long term. The guidance in figure 7.14 may prove useful. This is broad advice that must always be applied on a case-by-case basis. It aims to identify the range of advice that surveyors should be giving clients.

7.5.4 Temporary repairs to roof coverings

Many building owners will carry out temporary repairs to roof coverings to save money. Rather than going to the expense of recovering the roof, cheaper proprietary repairs are often used. These can include:

- **Bitumen coatings to the roof slope** – sometimes called 'tunnerising' this involves the application of liquid bitumen products to the whole or part of the roof area. Occasionally this is combined with hessian or glass fibre sheeting to reinforce the repair. Although this can keep the water out in the short term the bitumen can soon break down allowing water to penetrate again.
- **Rigid foam coatings to the rear face of the covering** – adverts for these products are often spotted in the classified sections of the Sunday papers. The installers claim that the treatment extends the life of the roof, stops leaks and increases thermal insulation. Concerns about this method include:
  - with some systems, battens and other timber components are encapsulated by the foam. This could lead to rot;
  - ventilation to the roof space could be reduced to very low levels.

Some of these methods might be partially effective but none are covered in the British Standard Code of Practice for Slating and Tiling, BS5534 (BSI 1997) and few have fully relevant agreement certificates. Therefore, they should not be recommended to a client. If the coverings are in place, the client should be made aware of the implications and advised to consider replacement in the near future.

Category	*Description of condition of roof covering*	*Advice to client*
Good (satisfactory repair)	The roof slopes are true and even and there is little evidence of replaced or slipped slates or tiles. The flashings, ridge and hip tiles appear in good condition and there is no evidence of leaks internally. The internal roof space inspection gives no cause for concern and there is adequate roof ventilation.	The roof is in a sound condition and is in an acceptable state of repair taking account of the age of the property. Any further investigations will be of little benefit.
Fair (minor maintenance)	The roof slopes are slightly uneven and there are a number of slipped or missing slates or tiles. The covering shows evidence of previous repair work and the flashings, ridge and hip tiles are at least serviceable but require some repair. There may be evidence of a few minor roof leaks internally. There could be a few deficiencies with the roof structure and the ventilation may have to be improved.	Although the roof covering still has a number of years of serviceable life, some repair works normal for a property of this age and character are required in the short term. This will extend the life of the roof covering. Due to its age it may require major expenditure in the medium term (say 5–10 years). It is advised that the roof is regularly maintained. As some repair work will be required, referral to a competent roofing specialist will be appropriate.
Poor (significant or urgent repair needed)	The roof slopes are in very poor condition. They are very uneven and have got a high number of slipped, missing or damaged slates or tiles. Many areas have been repaired before and the flashings, hip and ridge tiles are in a poor condition. There are many leaks internally and the internal loft space inspection revealed a number of concerns about the structure of the roof.	The roof is at the very end of its life and it may need complete replacement in the short term. The defects are considered to be an actual or developing threat to the fabric of the building or to personal safety. Because this might involve obtaining building regulation approval, the client should be referred to a building surveyor or structural engineer or other specialist that can provide a full and comprehensive service.

Figure 7.14 Table linking the condition of the roof covering to client advice.

7.5.5 Recovering of roofs – statutory control and implications

Since 1985 the recovering of a roof must have formal approval from the local Building Control Authority. Section 3 of Part A of the Regulations (HMSO 1991a) stipulates that this should:

- compare the loading of new and old roof coverings;
- the existing roof structure must be inspected by a building control officer to ensure that it can take any increased loading. Where the new roof covering will be lighter, vertical restraint might be needed to prevent uplift;
- adequate strengthening may be required and could include:
    - replacement of defective members, fixings and restraints;

- provision of additional structural members to enable new load to be supported;
- provision of additional restraint to prevent uplift.

In most cases the Building Control Authority will require calculations to justify the new proposals. Therefore, it is important to identify if the roof covering has recently been replaced. There are a few key features to look for:

- Does the covering appear relatively new and unweathered?
- Is it different from neighbouring properties?
- Is there evidence of recent work from within the roof space such as structural repairs, new roofing felt, etc.?

If the signs suggest renewal, the vendor should be asked to provide evidence that the appropriate approvals have been obtained. If not it could be an indication of sub-standard work. The client should be clearly advised that the roof may require some remedial work. There are two options:

- Ask for retrospective approval to be obtained from the Building Control authority for which a fee will be payable and remedial work will almost certainly result (see section 13.2).
- Appoint a building surveyor or structural engineer to assess the adequacy of the roof and recommend any remedial work. Once this has been completed issue a certificate of structural adequacy or its equivalent.

To give an insight into the possible effect of a change in covering the relative weights are outlined in figure 7.15. Any loading calculation not only has to take account of the self-weight of the structure and coverings, and wind and snow loads but also the water absorption of the covering itself. This can be as much as 10% of the original weight for some concrete and clay tiles. The BRE *Digest 351* (BRE 1990) outlines the procedure for recovering roofs and is very useful.

7.6 Other features associated with pitched roofs

There are a number of other features associated with pitched roofs that may cause problems. These are outlined below.

7.6.1 Party walls

Party walls between terraced and semi-detached properties can either terminate below the roof line or extend above it. Each has its own problems:

- Walls through roof line – these are vulnerable at the junction of the roof covering and party wall masonry. The flashings and soakers may be defective or totally absent. In some cases the junction is waterproofed by a simple cement fillet which is never very effective. The copings also present a weakness allowing water to bypass the covering. Any one of these can lead to dampness penetration into the loft space or even the rooms below. The rafters close to the wall may become very vulnerable to rot.

Covering type	Dry weight (kg/m²)	% of absorption	Total weight (kg/m²)
Eternit Duracem slates	21	say 3%	28
Natural slates	25–78	0.3%	26–81
Redland 49s (concrete)	47	10%	52
Plain clay tiles	65	10.5%	72

Figure 7.15 Table showing the relative weights of common pitched roof coverings (reproduced courtesy of the BRE).

- Walls terminating below roof line – these have a separating function in the case of fire and a security function. Many dwellings in the same row of terraced housing have been burgled from 'above' using the loft hatch as access. The walls can be completely missing or stop short of the underside of the roof covering. These partial party walls can offer inadequate support to purlins. In all of these cases, appropriate remedial work or further investigations should be recommended.

7.6.2 Gable walls

These can be very exposed to the weather and structurally unstable:

- because the walls of older buildings can reduce in thickness at this height making them very slender and vulnerable to lateral instability and bulging and bowing. This can be made worse by relatively heavy loadings from the roof structure and lack of adequate lateral restraint between the structure and the gable wall itself;
- because of the reduced thickness, wind-driven rain can penetrate the mortar joints and cause damage internally.

Because of these issues, where gable walls exist they should form a key part of the inspection process.

7.6.3 Chimneys

These can be the source of many problems. Some of the main defects are listed below.

- dampness penetration due to (see figure 7.16):
    - defective flashings, soakers and secret gutters at the rear of the chimney. The flashings to many newer properties may never have been properly wedged into the mortar joint. Instead it is held into position by mortar only. This can become easily dislodged in high winds;
    - defective flaunching, chimney pots and flue terminals;
    - poor pointing to stack allowing water to by-pass the flashings.
- Stability problems due to:
    - slenderness ratio, the chimney being too high for its width. The height of the chimney should be no more than 4.5 times its width (see figure 7.17). Chimneys to rear offshots are specially vulnerable as they were often taller so the combustion products avoided the main house;

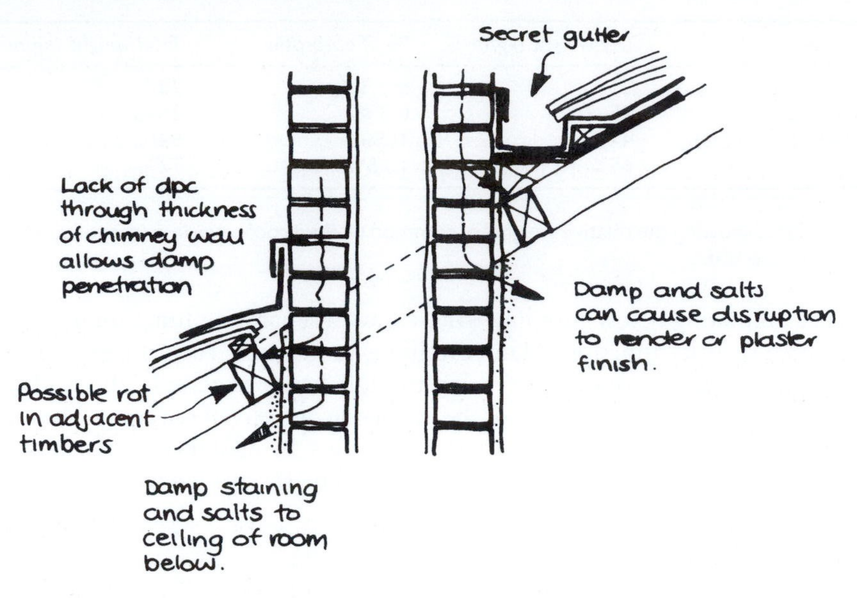

Figure 7.16 Possible routes for penetrating dampness around a chimney.

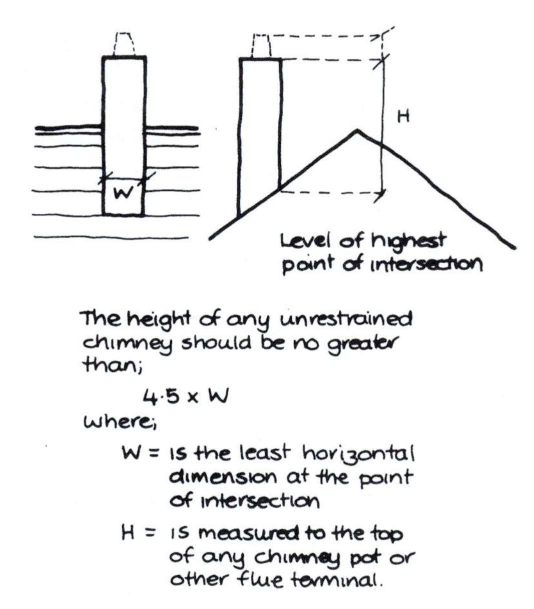

Figure 7.17 Maximum heights for free-standing chimneys above pitched roofs (Crown copyright. Reproduced with the permission of the Controller of Her Majesty's Stationery Office).

- sulfate attack may cause the mortar joints to expand (see section 4.7.3). Maximum expansion normally occurs on the wet side of an exposed chimney, causing the whole stack to bend over;
- where the chimney breasts have been removed internally rendering the whole stack unstable;

- the lack of bond between chimney and the adjacent wall can result in the separation of the two. In the worse cases this can lead to the escape of flue gases into the roof space.

Because of this variety of problems, chimneys require special attention during any survey. The BRE have produced a useful guide on the repair and rebuilding of traditional chimneys (BRE 1990b).

7.6.4 Thermal insulation of pitched roofs

Part L of the Building Regulations (HMSO 1991b) now links the insulation values of the different elements of a building to that building's overall energy efficiency. This is measured on a scale of 0–100 and is known as the SAP Rating (Standard Assessment Procedure). Section 5.8 looks at this in more detail. For roofs, the Regulation states that:

- for a dwelling with an SAP of 60 or less – the roof U-value should be equal to or better than 0.2 W/m^2K;
- for a dwelling with an SAP of over 60 the figure can rise to 0.25 W/m^2K.

This equates to about 150–200 mm thickness of normal mineral fibre insulation. For the sloping parts of a room-in-the-roof, a U-value of 0.35 W/m^2K will be acceptable (approximately 100 mm of mineral fibre insulation). This is to make it easier to achieve adequate ventilation of any confined roof spaces (see section 7.6.5 below). If the insulation thickness is less than this then it will beneficial to increase the insulation.

7.6.5 Roof ventilation

Part F of the Building Regulations covers ventilation of roofs (HMSO 1991b). Without it moisture levels in roofs spaces can quickly build up and result in mould growth on the roof timbers and in the worse cases even wood rot (BRE 1985b). The general requirements are:

- pitched roofs (15° or more) – cross ventilation between each side equal to a 10 mm continuous strip;
- lean-to roofs – 10 mm wide strip at eaves with ventilation at high level equal to at least 5 mm wide strip along the whole length of the junction with the wall. This can also be provided by a series of individual ventilating slates or tiles fitted evenly along the roof slope.

Ventilation of roof spaces is made more difficult where the ceiling follows underside of the rafters. Two main features are required:

- ventilation at eaves equal to 25 mm wide strip *and* ridge level equal to 5mm wide strip to promote flow of air above insulation and;
- minimum air gap of 50 mm between the top of the insulation and the underside of the roof covering.

If there is no roofing felt then this requirement may be waived because the roof space would be so draughty anyway! For recovered roofs where a continuous strip at the eaves is not possible, proprietary roof tile vents can be used. For example, if 'Glide-vale' tile ventilators were used the requirements would be:

- for standard loft space ventilation where there is no habitable space – one ventilating tile every 1.0 m;
- where a 'room-in-the-roof' exists with a sloping ceiling to the underside of the rafters;
    - eaves ventilation – one ventilating tile every 400 mm centre (i.e. every rafter spacing);
    - ridge ventilation – one ventilating tile every 2.0 m centre.

For an average terraced property this could result in a considerable number of ventilating tiles being fitted along the lower part of the roof. If there are only one or two (or none at all!) then this suggests that the work might not have building regulation approval. Appropriate advice should be given to the client (see figure 7.18).

Condensation can also be made worse if there are lots of routes into the roof space for water vapour from the house to pass through. These can include:

- gaps around pipes, wiring and other services where they pass through the ceiling;
- loose fitting loft access hatch, etc.

Where a ceiling appears to be 'leaky', the client should be advised to have the gaps filled and the loft access hatch draught stripped (BRE 1985c).

Figure 7.18 Ventilating tiles to a pitched roof. Where continuous ventilation is not provided at fascia level, a number of special ventilating tiles should be present near to the eaves.

7.6.6 Loft conversions and rooms-in-the-roof

A significant number of dwellings have had their lofts converted into habitable rooms. This can have both positive and negative effects:

- It can enhance the value of the property.
- If not carried out properly it can lead to structural instability in the dwelling.
- Usability and amenity – where the new space is poorly designed, difficult access and low head heights can make the space uncomfortable to use.
- Safety – a poorly designed conversion can compromise safety especially in relation to escape in the case of fire.

It is essential that surveyors are aware of the issues which give rise to these effects so clients can be properly advised. Some of the key issues are outlined below.

Older loft conversions and original room-in-the-roof spaces

In many houses, a room-in-the-roof may have always existed. It is helpful to use the current standards as a benchmark to assess the room's suitability but it would be unreasonable to call for upgrading to meet the full scope of the current building regulations. Three questions can help in evaluating existing 'rooms-in-the-roof':

- Is the room accessible enough to be considered a habitable room? Are the access stairs so small and winding that the room should really be considered (and valued!) as a store room?
- Does the space offer enough protection in the event of fire or would the safety of any occupant be seriously compromised?
- Has the structural integrity of the roof been undermined by the room? Is there evidence of roof spread? Is the structure well triangulated?

These factors can help the appraisal and ensure that balanced advice is given to the client.

Loft conversions since 1985

Loft conversions have been specifically included in the Building Regulations since 1985. If the conversion has not got the necessary approvals, a Regularisation Certificate can be obtained from the Local Building Control Authority (see section 13.2).

During a standard survey it will be impossible to assess standards of the conversion on inspection without an in-depth knowledge of the regulations and time-consuming exploratory and opening-up works. Therefore, you should look for key features or triggers that will bring a series of recommendations into play. These are outlined below:

- Has the property got a loft or attic room? This may sound very obvious but they have been missed in the past!
- If yes, has it been created since 1985 and therefore should be covered by the building regulations? Key indicators might include:

- Information from vendor.
- Are the rooms newly fitted out and decorated?
- Has the roof-covering recently been replaced or altered especially around any new roof lights or windows?

- If the room has been constructed recently then look for key features that might indicate it has been converted to appropriate standards. These might include:
  - Have the doors to staircase been fitted with self-closing devices down to the exit doors from the house?
  - Is the partitioning of robust standard that can resist the passage of fire (i.e. plasterboard and not hardboard)?
  - Do the rooflights or windows conform to escape requirements? The opening areas must be at least 500 × 850 mm minimum and be positioned a maximum of 1.5 m away from the eaves of the roof.

Advice to the client

If the loft conversion has been completed since 1985 then the client should be advised that planning permission and building regulation approvals should have been obtained. Even if this has been granted make sure that the local authority has issued a 'final completion certificate' to say that the work has been carried out in accordance with the original approval. Often home owners will get the plan approval but not carry out the work properly (see section 13.2 for more discussion on this aspect). If these conditions are not met, then there are several options:

- Ask for building regulation and planning permissions to be obtained in retrospect. See section 13.2 for more discussion on regularisation.
- Ask for full inspection by a building surveyor or a structural engineer that assesses the loft room in relation to current standards and itemises any remedial work to be carried out.

The client's solicitor may need to be specifically instructed on what to ask for.

7.6.7 Pests in roof spaces

Living things can often be found in loft spaces. Not only can these be a nuisance but they can also be a threat to health. Most have these have been described in section 6.4 but are listed here for convenience:

- Pigeons, swallows and swifts – these can gain access through missing slates or tiles or gaps at eaves and verges. Pigeons can pose a particular threat to health.
- Bats – these are a protected species. Their presence has to be reported which may delay alterations and timber-treatment regimes.
- Squirrels – these have become a particular nuisance especially if trees are close by. They can cause considerable damage to timbers and cabled services through their gnawing.

- Rats – these have been known to climb up boxing around soil and vent pipes and can cause similar damage to squirrels.
- Wasps and bees – dangerous (and painful!) if the nest is disturbed.

7.6.8 Services in loft spaces

Services in domestic dwellings are described in some detail in chapter 10 and this will not be repeated here. Because loft space inspections can give such a good insight into the condition of these elements a brief reminder has been included at this point. The two main elements are:

- cold water storage and feed and expansion tanks;
- electric cabling.

For a full checklist see chapter 10.

7.7 Flat roofs

7.7.1 Introduction

Inspection of flat roofs is different to pitched ones as it is rare to have the opportunity to view the structure. Therefore any report must assess not only the condition of the roof covering but also speculate about the state of the structure hidden beneath. This must be done by judging the condition of '... what cannot be seen by that which can'.

On domestic properties, flat roofs can be found in a number of different locations:

- over the whole dwelling. Flat roofs were very fashionable during the 1960s and early 1970s;
- over rear additions or 'offshots', originally built at the rear of the house;
- new extensions added since the house was constructed;
- flat roofs over front and rear bay windows, balconies, etc;
- linking front and rear roof slopes of the main roof construction.

When carrying out a survey two main elements have to be considered:

- the roof structure
- the roof covering.

7.7.2 The roof structure

There are two different types of roof structure:

- **Suspended timber construction** – usually consisting of roof joists, timber decking and covering. This may include cement screed over the decking.
- **Solid construction** – solid reinforced concrete as well as cast concrete between filler joists, hollow clay pots between concrete beams, etc.

The most effective method of telling the difference is to knock on the surface; timber will give a hollow sound. Beware of the usual surveying method of jumping up and down on the roof because if the structure has been seriously weakened by rot injury could result! Because timber roofs are more vulnerable to defects they need greater consideration during the survey so this initial distinction is important.

During the inspection, the key factors to consider are as follows:

- Does the roof surface look flat and even?
- Are there any signs of ponding or pooling on the roof? Even it is dry weather there should still be some staining around the edges of former puddles.
- Does the roof have any ventilation? This is usually provided by:
  - ventilation strips around all sides of the roof under the eaves behind the fascia board. This should be equal to a 25 mm wide continuous strip to two opposite sides of the roof;
  - ventilation fittings set into the roof structure itself that extend through the covering.
- The roof space ventilation must also be well distributed around the roof. If there is no cross-ventilation then parts of the flat roof void may become stagnant. This could lead to wood rot developing even though the overall level of ventilation may be acceptable (see figure 7.19).
- Are there any significant faults with the roof covering?
- Is there any evidence of leaks to the rooms and/or spaces below? If yes, hidden timbers may have been adversely affected.
- Is the flat roof being used appropriately? Some occupants may use them as breakfast balconies or roof gardens even though they have been designed to support only minimal loading. Indications of this mis-use would include informal access (i.e.

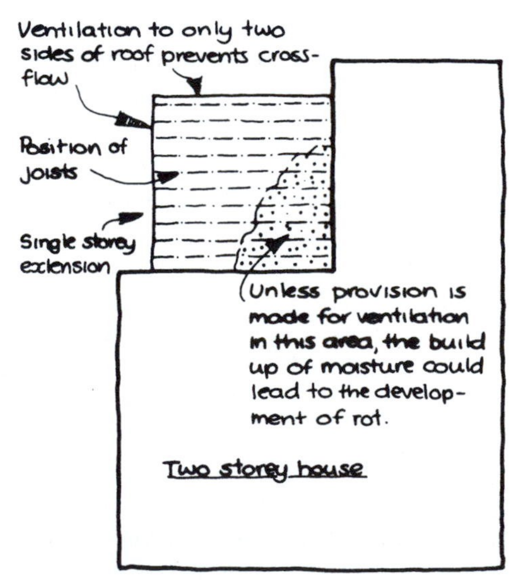

Figure 7.19 Diagrammatic sketch showing inadequate ventilation to flat roof because of lack of through ventilation.

through a window rather than a proper door), poor quality and inadequate detailing and workmanship, evidence of excessive defection (i.e. cracking, bulging, etc.) to roof surface or ceiling below.

If a significant number of these adverse signs are noted, then making a recommendation for further inspection and investigation would be appropriate.

7.7.3 Roof coverings

The main types of flat roof covering include:

- **Asphalt** – this is a combination of limestone aggregate and bitumens. It is laid to thickness of 20 mm and can have a life of 25–30 years. It is usually black when newly laid but will change to light grey after exposure to sunlight. It is important to protect asphalt from the sun by either mineral chippings or solar reflective paint (usually white or silver).
- **Metal roof coverings** – these include lead, zinc and copper as well as newer alloys. These will have particular features such as rolls and cappings. The installation and repair of these types of covering is a specialised operation.
- **Bitumen-based roofing felts** – fabric impregnated felts that are laid in wide sheets or strips bedded in hot bitumen. Originally based on asbestos or jute fabrics, modern versions are more durable relying on polyesters and glass fibres. Although traditionally protected with mineral chippings many felts are now self-finished with coloured mineral chippings or even reflective metal foils.

Leaks to flat roofs

Flat roof coverings are more vulnerable than pitched roof equivalents because:

- rainwater stays on the surface for much longer because of the shallow slope or fall of the roof. Any small leak can result in a considerable amount of water entering the building;
- a flat roof can also be very exposed to damaging winds, a wide range of temperature changes and solar degradation as well as occupant misuse.

It is not surprising that flat roofs have a poor reputation. The life expectancy of many flat roofs is considerably less than for pitched roofs and so must be viewed with considerable suspicion during a survey. The BRE have produced a *Good Repair Guide* on how to assess flat roof coverings (BRE 1998) and some of the main points are summarised below.

Key inspection indicators

During the inspection look for the following signs and features.

MAIN ROOF AREAS

- Any splits or cracks in the covering;
- any bubbles, blisters or bumps in the surface;
- evidence of ponding or deflected areas to the main roof area;
- areas of the roof that have no solar protection, e.g.
    - lack of stone chippings;
    - lack of reflective paint or other finish.

VERGES AND ABUTMENTS DETAILS

Many leaks from flat roofs result from inadequately designed or installed 'edge detailing' where the flat roof meets other parts of the building. These include:

- upstands to party, parapet walls and rooflights. The vertical parts of the roof covering can slip down, tear and split;
- junctions with the main pitched roof slopes. Sometimes the flat roof covering does not go up behind the pitched roof covering far enough;
- the eaves and verges of the roof. This detail can often split or crack;
- drainage fittings, outlets and any internal guttering are often poorly made;
- any pipes, flues or handrail supports that penetrate the roof covering and allow water to penetrate;
- any paved areas on fire escape routes, maintenance access, etc. where the slabs can damage the covering.

If any faults are noted then an internal check should be made to see if there is evidence of roof leaks. Even if the defects have not yet resulted in roof leaks the client should still be made aware of the faults and future implications.

TEMPORARY ROOF REPAIRS

Because flat roofs are so vulnerable to faults, it is highly likely that older ones will show evidence of temporary repairs. These may include:

- waterproof adhesive tapes and proprietary products applied over splits and cracks;
- black bitumen poured over the surface sometimes with reinforcing matting beneath;
- patch repairs where areas of felts and asphalt have been cut out and replaced.

Temporary roof repairs are evidence that the existing covering is at the end of its useful life. Where there are a number of these repairs then a recommendation of replacement should be made or at least a warning that this could be required in the near future.

7.7.4 Roofs of bays and porches

The roofs of porches and bays are often covered to a lesser standard than a roof to the main dwelling would be. The BRE (1990) point out that roofs over these features often include simple zinc on timber, bitumen on primitive concrete roofs and small-tiled pitched roofs that have multiple hips, valleys and abutments. The junction with the main roof is often poorly waterproofed. These defects become even more critical because in many older dwellings there is often a large timber 'bressumer' or beam spanning the width of the bay opening. These can be only just covered by the roof construction. Water leakage can lead to rot and consequent structural failure. Rainwater disposal may be blocked, inadequate or absent allowing rainwater to discharge over the bay or porch below causing increased deterioration of timber components. Where the roofs of these features cannot be inspected from ground level they can often be seen from the windows of adjacent upstairs rooms. If this is not possible and the bay or porch is single storey then the inspection should take place with the aid of ladders.

Key inspection indicators

- Is the junction of the roof and main structure watertight and properly formed?
- Are the flat/pitched roofs properly detailed?
- Does the feature have appropriate methods of shedding, collecting and disposing of rainwater?
- Is there evidence of water staining to any ceiling or soffits below?

7.7.5 Balconies and canopies

The BRE (1990) defined these features as being horizontal projections from the main structure of the building. They may range from small timber canopies that are fixed to the wall with gallows brackets to a substantial extension of an intermediate concrete floor that has been specially engineered to act as a cantilever. These features can be evaluated by identifying a number by characteristics as follows.

Safety

Safety is a major consideration for balconies in relation to those using the balcony and those walking beneath it.

Key inspection indicators

- Are the balustrades strong enough and securely fixed (figure 7.20)?
- Is the balustrade high enough with no unacceptable gaps that will allow children to fall through? The normal rule is whether a 100 mm ball can pass between any component.
- Is there a kerb to stop objects rolling off the balcony?
- Are pedestrians walking below likely to hit their head or bump into the supports?
- Does the balcony act as a fire escape? Some flats were designed with these as a vital part of the escape route. Does it still function in this way? Has it been informally barred?

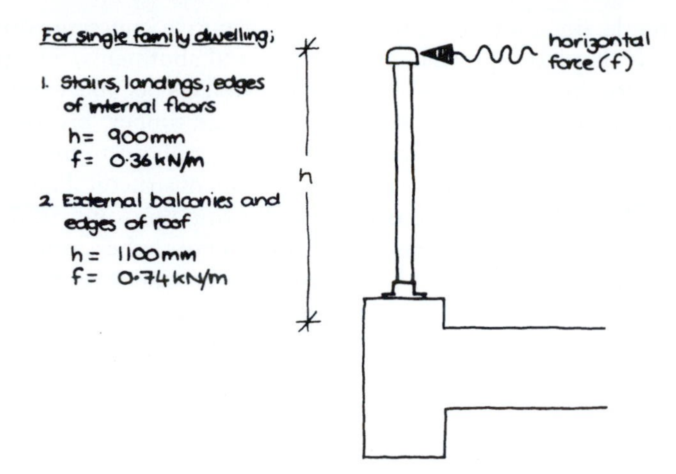

Figure 7.20 Balustrades to a balcony. The Building Regulations (Part K) require all balustrades and handrails to resist a certain level of horizontal force. This is higher for external balustrades.

Condition

- With concrete balconies:
  - Are there any signs to suggest that the structure is under stress, e.g. cracking or local crushing around the bearings in the main structure, cracking to the top surface near to the junction with the wall suggesting lack of adequate reinforcement, etc. In these cases a detailed engineering appraisal might be needed.
  - Is there evidence of spalling or cracked concrete to the underside or exposed edges of the balcony? This could be caused by corrosion of reinforcement too close to the surface or carbonation of the concrete itself that has allowed further deterioration of the material. All loose material may need to be hacked back, the concrete reformed and the surface sealed.
  - Check the junction of the main balcony with the junction of the external wall construction for signs of damp penetration. Indicators can include:
    - Are there any signs of dampness internally? Look for staining below the balcony itself and within the adjacent rooms and spaces.
    - Obvious deficiencies with the waterproofing layer to the balcony including lack of a proper upstand, splits or slumps in the waterproofing layer, etc.
    - Is there any evidence of a link with a cavity tray in the wall? This should include an upstand and open vertical mortar joints above that drain a cavity tray (see figure 7.21).
- With timber or metal structures:
  - Is there evidence of rot or corrosion to the decking that might suggest future problems?
  - Carefully check the support to the balcony especially for:
    - columns that are bent, twisted or out of plumb. Columns often have inadequate foundations and readily suffer from settlement;

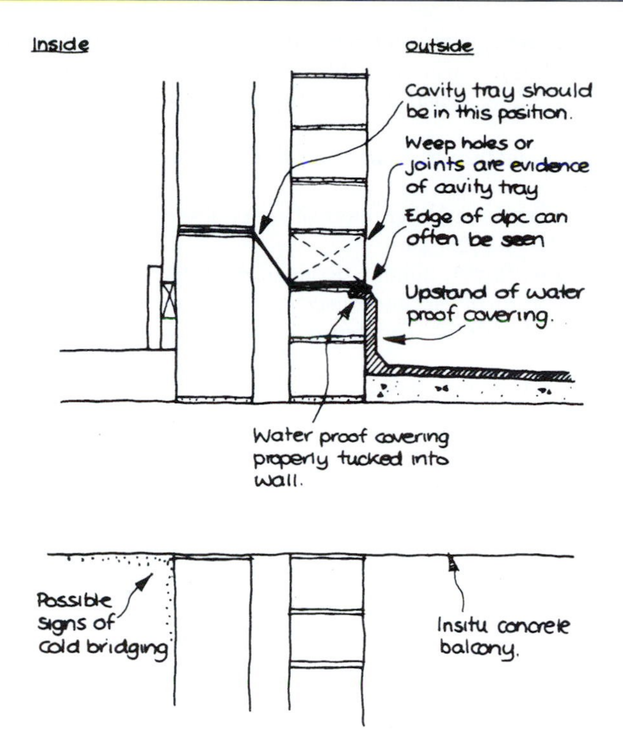

Figure 7.21 Link between a concrete balcony and a cavity wall. It is essential that the balcony has a waterproof covering and upstand that links in with the dpc and cavity tray of the wall.

- gallows brackets that have been insecurely fixed to the wall resulting in them 'pulling' out the top fixing with buckling or crushing at the bottom.

- Are the waterproofing materials in a sound condition?

The replacement of these features can be very expensive and so have an impact on value.

References

BRE (1985a). 'Pitched roofs: sarking felt underlay – drainage from the roof'. *Defect Action Sheet 9*. Building Research Establishment, Garston, Watford.

BRE (1985b). 'Slated or tiled pitched roofs: ventilation to outside air'. *Defect Action Sheet 1*. Building Research Establishment, Garston, Watford.

BRE (1985c). 'Slated or tiled pitched roofs: restricting the entry of water vapour from the house'. *Defect Action Sheet 3*. Building Research Establishment, Garston, Watford.

BRE (1990). 'Recovering old timber roofs'. *Digest 351*. Building Research Establishment, Garston, Watford.

BRE (1990a). *Assessing Traditional Housing for Rehabilitation*, Building Research Establishment, Garston, Watford.

BRE (1990b). 'Repairing or rebuilding masonry chimneys'. *Good Building Guide 4*. Building Research Establishment, Garston, Watford.

BRE (1998). 'Flat roofs: assessing bitumen felt and mastic asphalt roofs for repair'. *Good Repair Guide 16, Part 1*. Building Research Establishment, Garston, Watford.

BSI (1997). *BS 5534. Code of Practice for Slating and Tiling*. British Standards Institution. HMSO, London.

HAPM (1991). *Defects Avoidance Manual – New Build*. Housing Association Property Mutual, London.

Holden, G. (1998). 'The new homebuyer: more of your questions answered'. *Chartered Surveyor Monthly*, 7(8), May 1998, p.25–26.

HMSO (1991a). *Structure, Part A*, Building Regulations 1991. HMSO, London.

HMSO (1991b). *Conservation of Fuel and Power. Part L*, Building Regulations. HMSO, London.

Melville, I.A., Gordon, I.A. (1979). *The Repair and Maintenance of Houses*. Estates Gazette, London.

Melville, I.A., Gordon, I.A. (1992). *Structural Surveys of Dwelling Houses*. Estates Gazette, London.

Murdoch, J., Murrells, P. (1995). *Law of Surveys and Valuations*. Estates Gazette, London.

Noy, E., A., 1995. *Building Surveys and Reports*. Blackwell Science, Oxford.

Smith v. *Bush* (1990). 1 AC 831.

Chapter 8

External joinery and decorations

8.1 Introduction

Most dwellings have some external timber components which can include windows, doors, external cladding, fascia and barge boards, etc. Depending on how well these have been constructed and maintained there will inevitably be some minor shortcomings. As Melville and Gordon (1992, p.177) point out, a few casements that are difficult to shut may be tolerated by most building owners, whereas when timber windows are rotten to the point of replacement then not only will the client be facing a large bill but also they will be very angry with their surveyor if he or she does not point out the problem! Consequently external joinery needs very close and careful attention.

8.2 Windows and doors

8.2.1 General defects to all types of windows and doors

A number of defects that are common to all types of windows and doors are described below.

Timber decay

According to the BRE (1997) windows manufactured between 1945 and 1970 were often of poor quality and decayed prematurely. Much of this has been reduced by pre-treatment of the frames and components. Because this has only been general practice since 1970 there are still a lot of components that are vulnerable to deterioration. The BRE point out that the more effective use of preservative should not be seen as a substitute for good design (BRE 1985a). Several locations on windows and doors are particularly vulnerable. Where these are accessible these should be probed to see if wet rot has developed. In many cases, the paint film may be intact but the timber below completely rotten. Windows and doors on the most exposed elevations (normally south and west facings) are particularly at risk and so should be inspected closely. The precise locations include (see figure 8.1):

- the bottom rails and sections of any door and window frame. Most of the rain water run-off will flow over these components;
- the junction of the side mullions or jamb sections and the timber sills;

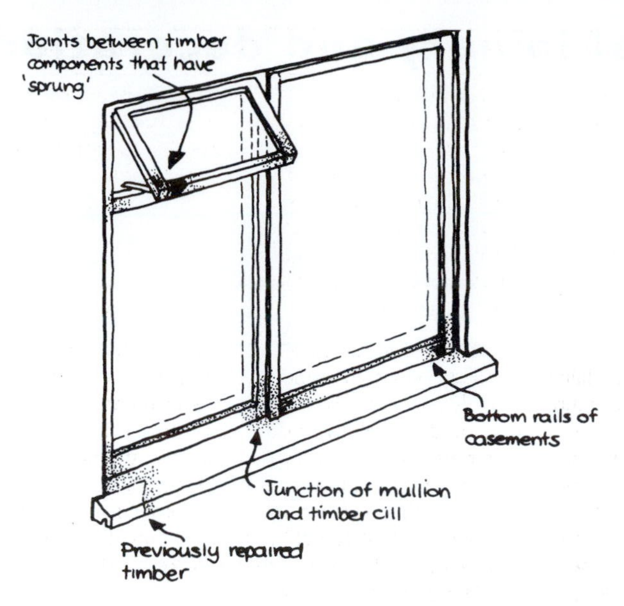

Figure 8.1 Vulnerable parts of a typical casement window.

- around any rebates where the glazing putty has cracked or dropped away. Thin, dry bedded glazing beads can also present a problem;
- the bottom edge of sills especially where they are bedded directly onto stone or tile sub-sills;
- any open or gaping joints between timber components;
- any obviously rotten areas of timber or parts that have been spliced, repaired or made good with a proprietary filler.

If the amount and extent of rot are limited then the decay can sometimes be arrested by cutting out and splicing in new timber. The insertion of borax rods or special resins can also slow down the rate of timber decay. If the problems are more extensive, replacement of the whole window may be the only option. Figure 8.2 attempts to provide some simple criteria against which standard timber window frames can be assessed and advice given to clients.

Rain penetration

Rain penetration can occur around the jambs, sills and heads of window and door frames if the dpcs have not been installed properly or the sealants between the frame and the jambs are ineffective. The exposed elevations of dwellings are particularly vulnerable. Key indicators include:

- damp staining to the internal reveals of window and door openings. In the worse cases this could lead to a breakdown in the plaster and decorative finish;
- a lack of sealants at the junction of the frame and the jamb or defective sealants that are cracked, split or sagging (BRE 1985b);

Category	*Condition of the windows*	*Advice to the client*
Good (satisfactory repair)	The majority of the windows are sound. There are few areas (if any) of rotten timber and the glazing putties are in good condition. Most of the casements operate easily and there is little evidence of condensation internally. The decorations are in good condition and appear to have been renewed regularly in the past.	The windows are in sound condition and should give many years good service as long as they are regularly maintained.
Fair (minor maintenance required)	Although a number of windows are in a fair/good condition a significant percentage (say 40%) have developed defects. These could typically include small areas of rotten timber, cracked or missing glazing putties, a number of open joints, deteriorating decorations, a number of casements that are difficult to open. Evidence of condensation internally, poor or missing sealants at the junction with the wall. Some evidence of previous timber repairs.	The windows are in a fair condition only. Although total replacement is not required at the present time a number of repairs need to be carried out to ensure adequate future performance. Once completed the windows should be redecorated and regularly maintained (say every 4–5 years) in the future.
Poor (significant or urgent repair needed)	Most of the windows are in a poor condition with a number (say 55–60%) showing significant defects. These might include several rotten areas of timber, cracked and missing glazing putties, many sprung and open joints, a number of casements that are fixed shut or difficult to open, lack of sealants to frame/wall junctions. Some evidence of rainwater penetration internally as well as serious condensation problems. A number of windows may be insecure and there is evidence of previous timber repairs to the casements and frames.	The windows are in a poor condition and are affected by a number of serious defects. They will require a significant amount of repair work to bring them back into a serviceable condition and then will have a limited life. It may be economical to consider complete replacement with equivalent double-glazed alternatives.

Figure 8.2 Table linking condition of windows with advice to clients.

- lack of an overhanging sill with a throating that can shed water run-off from the face of the wall below.

Remedial work may range from re-application of an appropriate sealant through to the removal of the frame and the re-forming of the dpc where there has been a fundamental failure. This is discussed at greater length in section 5.3.3.

Condensation

All dwellings will suffer from condensation from time to time. In some dwellings, condensation can be such a regular occurrence that it affects the durability of elements of the building (see section 5.5). Single-glazed and metal-framed windows are most commonly affected. Properties that are badly affected can show the following signs:

- evidence of mould growth to the inside of the window frames and sometimes the reveals of the walls. Even where occupants regularly clean down surfaces, traces of mould growth will still be left;
- adjacent plaster surfaces may be stained by condensate run-off especially below the internal window sill;
- in the worst cases timber components may show signs of wet rot especially at the junction of the mullion and rail sections;
- with double-glazed units, check for condensation especially between the panes of glass. This could indicate that the seal between the panes has failed.

Condensation signs can be easily confused with general deterioration so care must be taken when making a diagnosis.

Functional performance

Even if the windows and doors are not showing signs of deterioration they should still operate effectively. Carry out the following checks:

- The windows and doors should be opened and closed. Comments should be made on whether they are easy and convenient to use. If keys are available, any security locks should be released and re-secured. If the keys are not available then this should be pointed out in the report.
- Are the doors and windows reasonably draught free? This may be difficult to assess during the summer without a wind. It can be estimated by seeing if there is any draught proofing and how close the casements and doors fit to the frame.
- Do the windows let in enough daylight? Does the room seem dark? In some cases replacement PVC windows with large mullions may have reduced light levels.
- If secondary glazing has been fitted is it to the right quality? Some fittings are not be very robust and distort when being used. Is there much condensation between the two sets of glazing?
- Security is always a concern for clients and any good survey will give advice where appropriate. The BRE (1994) suggest that secure windows will have the following characteristics:
    - Glazed areas should be as large as possible so the noise of breaking glass can deter a burglar.
    - Look for small opening lights that may allow access to locks or catches on adjacent windows.
    - Louvre windows are very insecure. Trickle vents allow ventilation without having to open windows.
    - Externally fixed glazing beads can easily be levered off and internally fixed beads can be 'kicked in' if they are too small.
    - Window locks should need a key to open them. Because of the need to escape during a fire, a key should be kept close by the window (but concealed).

Sash windows

Double hung sash windows are probably one of the most common of all older windows in this country. They consist of top and bottom glazed 'sashes' that are hung on sash cords, pulleys and weights within a boxed frame on either side (see figure 8.3). In good condition they can perform well but if poorly maintained they can result in a number of problems. Key inspection indicators include:

- Do the sashes open? Are they painted shut? If they do open, are they easy to operate?
- Are the sashes well balanced? When unlocked and operated do the sashes stay in one position or do they drift up or down on their own?
- Are the sash cords frayed or broken?
- Are the sashes well fitted or loose and rattling?
- Are there signs of wet rot? Look carefully at the junction of the timber and stone sub-sill.
- Have the windows been fitted with secure and robust window locks? Sash windows can be very easy to open from the outside unless properly secured.

Many older sash windows may add considerable character to the elevation of older dwellings. If replacement is recommended, then advise the client that the new windows should at least complement the design of sash windows. In some cases if the building is listed or is in a conservation area, then replacement windows that exactly match the originals will be required.

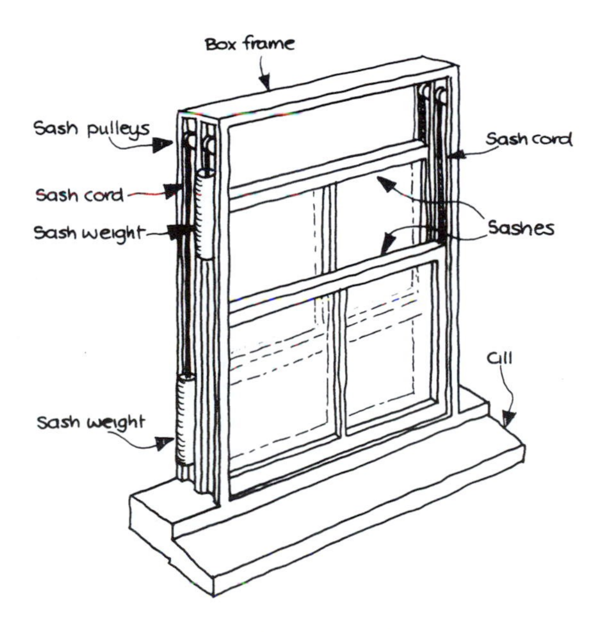

Figure 8.3 Sketch showing the construction of a typical sash window.

8.2.2 Metal-framed windows

Between the wars, single-glazed metal-framed windows were used extensively in housing schemes. Many of these 'Crittal windows' are still in serviceable condition but typical faults can include:

- corrosion or rusting of the metal sections especially the horizontal components towards the bottom of the window. This can be caused by rain from the outside and condensation internally;
- because the actual frame itself can occupy up to 20–25% of the total window opening area, excessive condensation can be common fault. Look for evidence of mould growth on the frame and signs of pooling of condensation on the window sill;
- the metal frames can warp out of shape resulting in excessive draughts and making the windows difficult to open and close. This distortion is usually caused by corrosion of the frame but can also be induced by movements of the wall around the opening.

Many metal window frames were set in timber sub-frames. The main windows may be in good condition but replacement might still be required if the sub-frame is rotten. Special attention should be paid to the timber sill section as these are usually the first to go.

Modern metal-frame windows can include:

- single- and double-glazed aluminum framed windows;
- single- or double-glazed powder-coated steel windows.

Both types can include sliding or hinged casements and can be set directly against the openings or in timber (often hardwood) sub-frames. Some types can have 'thermally broken' frames. This is where the inner and outer faces of the frame are isolated from one another by a solid resin or other material. It is claimed that frames of this type are more thermally efficient. The criteria for evaluating these are similar to those identified for replacements in section 8.2.3 below. In addition, look out for the following:

- Are the windows stable or do some parts flex and distort when operated?
- Do the frames suffer from surface condensation? Investigations by the authors has cast doubt over the real thermal value of 'thermally broken' frames.
- Sills to some windows are provided with holes so excessive condensation run-off can be drained to the outside. These can often become blocked allowing condensation to overflow onto the internal sill board and wall finish below.

8.2.3 Installation of new windows and doors

Over the last 20 years or so windows and doors of many older properties have been replaced as the 'fashion' for replacement windows has grown. This type of improvement has been seen by many owners as the best way of adding value to their property and improving thermal performance (despite this not being the case – see section 5.8). Consequently many perfectly sound and serviceable windows have been expensively replaced before their time.

In most cases, competent contractors would have installed acceptable quality components. But occasionally replacement windows result in more problems than they solve. Therefore even if the windows are relatively new they still have to be assessed carefully. The following key questions may help in this process:

- Has the owner got any details of the installation work available? Was the contractor a member of an appropriate trade association? Did they install components that were constructed to British standards?
- The openings may have been adjusted during the installation of the windows, e.g. sills and jambs built up, sub-sills removed, etc. Look for new or altered brickwork around the opening. Has this been done properly? Is it properly sealed? Is there any evidence of water penetration around this new work?
- Older window frames often had an informal structural function especially in bay windows. Once removed, the weight of the supported bay structure is transferred to the new frames that will not have been designed for that purpose. Typical signs could include distortion of the construction above the window including cracking and sloping of upper floors, etc. The new window frames may also have distorted making it difficult to open casements.
- Are the replacement windows secure? Can the external beading be easily removed allowing the glazing to be removed? Carefully check large lounge/patio doors. Because of possible deflection of the framing members, locks and catches can easily be 'sprung' through the application of force to the element.
- Check for misting between the double glazing. This is evidence that the edge seals have failed.

PVCu replacement windows – standards

For further advice on the general requirements for PVCu windows the BRE Digest 404 (1995) contains a good review of the key issues in the design, manufacture and installation of this type of window. Relevant British Standards include:

- BS 7412:1991 PVC-U extruded sections;
- BS 7413: 1991 PVC-U extruded sections with welded joints (material type A);
- BS 7414: 1991 PVC-U extruded sections with welded joints (material type B).

BS 7722: 1994 updates some of these provisions. There is also a BSI Quality Assurance Kitemark scheme to BS 7412 that is being awarded to an increasing number of firms and is rapidly becoming an industry standard. The installation of windows is covered by the following publications:

- BS 8213: Part 4: 1990 Code of Practice for the installation of replacement windows and doors;
- British Plastics Federation Windows Group/Glass and Glazing Federation. Code of Practice for the survey and installation of PVC-U windows and doorsets. March 1995.

Replacement windows may also threaten the safety of the occupants:

- A large panel of glass at a lower level can be dangerous if someone should fall against it (see chapter 13 on the building regulations). These should conform to appropriate safety standards.
- If the opening lights are small and/or positioned high up in the window then escape in the event of fire may be impeded if not prevented altogether. Ideally opening casements should be at least 500 × 850 mm so people can climb out or be easily rescued.

8.2.4 Doors and door frames

Assessing the condition of doors and their frames is a similar process to that of windows. One of the main differences is that doorways are the main ways in and out of the dwelling. Therefore they have to be robust enough to withstand normal usage and prevent illegal entry. The following checklists may assist with the assessment.

Condition

Doors to domestic dwellings vary but most are made of timber. Panelled and partially glazed doors are particularly vulnerable to deterioration. Key inspection indicators include:

- Have the glazing putties or beads deteriorated?
- Is there any rot at the junction of mullions, rails and weatherboards especially where there are open joints?
- Check the door frame where it meets the sill and make sure the frame is securely fixed back to the wall.

Security

According to the BRE (1994), as many as 30–40% of residents in some inner city areas live in fear of burglary. Yet the actual risks are much lower than this. In the worst areas, actual burglaries can rise to 10%. In most suburban locations the figure will be around 1–3% per year. Any survey report can give new owners useful advice and the entrance doors are one of the key components. Secure doors should have most of the following features.

MAIN ENTRANCE DOORS

- They should be robust and be able to resist being kicked or charged. Usually they should be at least 44 mm thick.
- The frame should be securely fixed within the opening although this will be difficult to check during the survey.
- The door should be hung on three 100 mm steel butt hinges.
- There should not be any glazed panels that could allow easy access to locks if the glass was smashed.

- The door should be fitted with the following (see figure 8.4):
  - a mortice deadlock operable from either side with a key (minimum 5 lever);
  - a rim automatic deadlock;
  - a door chain or limiter;
  - a door viewer;
  - a letter box that is at least 400 mm away from any locks.

  In flats a quick release of the mortice lock in the case of fire is required.

OTHER EXTERNAL DOORS

Other external doors do not normally have to perform as escape doors and can generally be secured from the inside before the occupants leave the house or go to bed. They should be fitted with the following:

- mortice sash lock;
- two or more surface mounted or mortised bolts, preferably the type that are key operated.

These doors should conform to the general standards laid down for front entrance doors. This is because they tend to be in places that are not so observable, so they may be subject to prolonged and violent attacks, e.g. with hammers, crow bars, etc. Therefore, special reinforcing plates and other precautions are advisable. Inter-connecting doors between attached spaces such as a garage, etc. are particularly vulnerable.

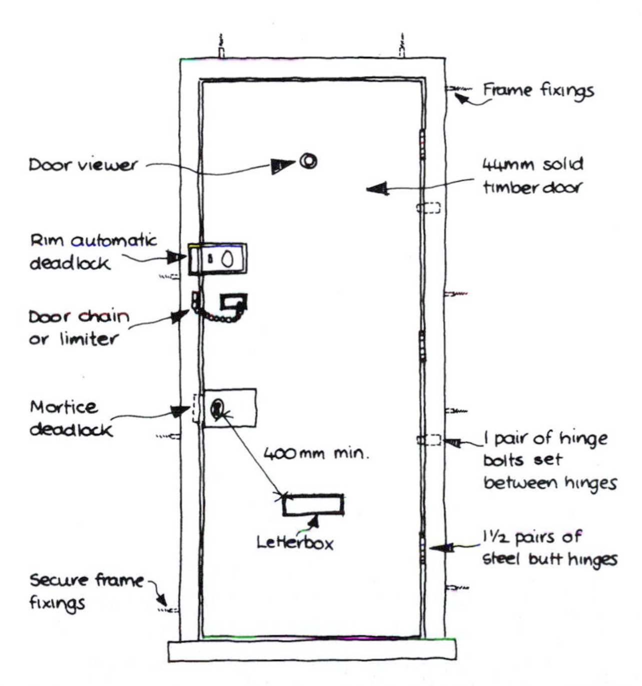

Figure 8.4 Security fittings to a main entrance door (reproduced courtesy of BRE).

External door thresholds

Where external entrance doors are exposed to the elements, wind-driven rain can often penetrate across the threshold and around the frame (BRE 1985c). Typical indicators will include:

- staining to the timber threshold internally;
- discolouration and high dampness readings to the floor finish just inside the entrance door (e.g. stained carpet, hardboard, floorboards, etc);
- when carpets/mats are lifted, the presence of moisture-loving insects such as silver fish and wood lice;
- in the worst cases any adjacent areas may be affected by dry or wet rot.

If a wood rot outbreak has occurred it is vital to inspect the timber joists below. If these are not easily accessible then further investigations may be required. Any defective timber will need to be replaced and appropriate water bars and weatherboards provided.

8.3 Other joinery items

There are a large number of other external timber components and some are more accessible than others. These include the following.

8.3.1 External features – canopies, porches and conservatories

Many of these features may have been constructed at the same time as the main house and so are probably of good quality. Others may have been added at a later date with variable standards. External features can add amenity and value to a property but if built to a poor standard they can have a significant impact:

- they may be unsightly and detract from the aesthetic qualities of the dwelling;
- the expense of either repairing them or demolition and clearing away can be very high;
- in a poor condition they can pose a danger to the occupants especially during strong winds when glass can be blown around.

These features should be assessed on a similar basis to doors and windows described above but also include some particular problems (HAPM 1991, section 5.3):

- The foundations and supporting structure are less effective than the main walls. Timber columns in particular should be well supported in appropriate shoes at ground level.
- They can have poor water-shedding properties which can lead to increased rates of deterioration of the structure below. Formal rainwater drainage systems should be provided for all features apart from those with very small roof areas.
- Timber gallows bracket supports to bays can suffer from rot where not isolated from damp masonry. Checks should be made for signs of movement in the window.

- In a similar manner to dormer windows, small pitched roofs over porches and bay windows, etc. can be difficult to form using the same tiles as on the main roof. Smaller plain tiles are often more effective. Standard ridge and hip tiles may also be inappropriate.

8.3.2 Fascias, eaves and barge boards

Most roofs will have a number of timber components at or above the guttering level. This is especially true around dormer roofs where it might be difficult to see fascia boards never mind inspect them. On some classically designed older houses, barge boards and fascias might be large and very ornate. The cost of repairing these features could be very high. Where the building is listed then the features might have to be replaced exactly.

Inspecting these features can often be difficult. Key inspection indicators could include:

- Where the features are less than 3 m off the ground they should be probed at regular intervals (say 1 metre) and at vulnerable positions (joints for example).
- At higher levels, the features should be assessed by careful visual inspection via binoculars. Key features would be peeling and missing paint, obvious signs of rot or a 'waviness' to painted surfaces and any open or sprung joints between components.

8.3.3 Timber cladding and boarding

A number of dwellings, especially the large number of timber-framed dwellings built between 1960 and the early 1980s, have external horizontal or vertical timber boarding. Inspection and assessment of these claddings are similar to any other timber components. Special attention should be given at vulnerable points such as:

- along the bottom edge of the cladding where it terminates against a sill or some other form of construction;
- at any external corner junction where one board meets another;
- around window and door openings where the timber is relied upon to seal the junction with the window frame and sill;
- in any areas where excessive rainwater run-off is evident e.g. beneath a leaking gutter or a defective overflow. The risk of rot in these locations is high and so should be inspected carefully.

8.4 External decorations

8.4.1 Assessment

If the external decorations are not regularly renewed then the durability of the windows and doors can be reduced. Consequently it is important that the state of the external decorations is assessed so an accurate estimate of when they need to be repainted can be given.

8.4.2 Mechanisms of deterioration

Paint failures are usually associated with inadequacies in the undercoat or primer (Richardson 1991, p.126) and most failures occur at the junction between components. There are many different ways a paint film can breakdown but the main ones include:

- **Blistering** – blisters in the paint film are usually caused by moisture that is trapped underneath or within the paint film.
- **Brittleness or flaking** – flaking and cracking of the paint film is usually caused by internal stresses set up when drying. Another cause could be dimensional changes in the substrate, i.e. moisture movement in timber.
- **Chalking** – this is where the paint loses its gloss leaving white or slightly tinted powder on the surface. This is caused by photochemical breakdown of the surface layer releasing the pigment.

A more complete description has been given by Cook and Hinks (1997, p.417).

If any of these features are noted then the affected area of paint should be removed back down to the substrate and the decorations renewed. Figure 8.5 attempts to describe a range of typical conditions of painted surfaces that links with client advice. This can help this assessment.

Overall	*Condition of the decorations*	*Advice to the client*
Good (satisfactory repair)	Looking at the property as a whole, the external painted surfaces are in a good condition. There are few areas of loose and flaking paint and glazing beads/putties are in good condition. The paintwork looks 'fresh' and relatively unblemished suggesting it has been recently done.	The decorations are in sound condition and should give a number of years good service. The surfaces should be decorated within the next 4–5 years.
Fair (minor maintenance required)	Although much of the external decorations are in a fair condition, a significant percentage (say 40%) have developing defects. These typically include small areas of rotten timber, cracked or missing glazing putties and beads, flaking, peeling or powdering paintwork.	The decorations are in a fair condition only. Painted surfaces will need redecorating within the next year at the very most. Some of this work may include extensive preparation such as burning off defective paint work, etc.
Poor (significant or urgent repair needed)	Most of the external decorations are in a poor condition with a number (say 55–60%) showing significant defects. These might include several rotten areas of timber; cracked and missing glazing putties and large areas of missing, peeling, flaking or powdering paint.	The decorations are in a poor condition and are affected by a number of serious defects. They will require a significant amount of preparation work to bring them back to an acceptable condition. This work should be carried out as soon as possible.

Figure 8.5 Table linking condition of the external decorations and client advice.

8.4.3 Other decorative problems

Painted wall surfaces

The external wall surfaces of many dwellings may have been painted in the past to either enhance the appearance of the building or offer added protection against the elements. In some cases these coatings might have been proprietary applications containing additives such as mineral substances or silicones. They are often thicker than normal paints and if the surface was not properly prepared, these coatings can soon blister and flake. In the worse cases they can actually hold moisture in the wall preventing it from drying. One of the drawbacks about painted walls is that the owner has to regularly redecorate large external areas of the dwelling to keep up appearances! This can be expensive and clients should be made aware of this additional liability.

References

BRE (1985a). 'Preventing decay in external joinery'. *Digest 304*. Building Research Establishment, Garston, Watford.

BRE (1985b). 'External walls: joints with windows and doors – applications of sealants'. *Defect Action Sheet 69*. Building Research Establishment, Garston, Watford.

BRE (1985c). 'Inward-opening external doors: resistance to rain penetration'. *Defect Action Sheet 67*. Building Research Establishment, Garston, Watford

BRE (1994). *BRE Housing Design Handbook: Energy and Internal Layout*. Building Research Establishment, Garston, Watford.

BRE (1995). 'PVC-U windows'. *Digest 404*. Building Research Establishment, Garston, Watford.

BRE (1997). 'Repairing timber windows: investigating defects and dealing with water leakage'. *Good Repair Guide 10, Part 1*. Building Research Establishment, Garston, Watford.

Cook, G.K., Hinks, A.J. (1997). *Appraising Building Defects*. Longman Scientific and Technical. Harlow. Essex.

HAPM (1991). *Defect Avoidance Manual – New Build*. Housing Association Property Mutual Limited, London.

Melville, I.A., Gordon, I.A. (1992). *Structural Surveys of Dwelling Houses*. Estates Gazette, London.

Richardson, B.A. (1991). *Defects and Deterioration in Buildings*. E&FN Spon, London.

Chapter 9

Internal matters

9.1 Introduction

Defects to the internal elements of a dwelling are often overshadowed by the problems associated with the major elements of a building such as the external walls and the roof. Yet internal defects are important for two reasons;

- they may give an indication or confirm possible defects that exist in the main superstructure of the dwelling; and
- they may be significant defects in their own right for which the remedial works would be very costly.

This section will concentrate on defects that are closely associated with internal features. Signs associated with defects such as subsidence and rising dampness are included in the appropriate chapters.

9.2 Walls and partitions

9.2.1 *Types of partitions*

The main function of internal walls within a dwelling is principally to divide space. Over the years a wide range of materials have been used:

- **Solid brick walls** – these can vary between 265 mm (one brick thick) down to 150 mm (half a brick thick). These measurements include plaster to both sides. The walls will give out a solid sound when tapped.
- **Timber walls with brick infill** – a traditional variation in some areas. This type of wall usually consists of vertical timber studs and noggins with brick-on-edge infill panels plastered over both sides. Usual thickness about 140 mm.
- **Timber stud partitions** – in older properties two types may be encountered:
  - **Trussed partitions** – these can consist of a series of struts, braces and ties. They may have been designed to take a considerable load and may be a very important part of the building's structure. If any part of this is cut through during alteration work the whole dwelling could be affected.
  - **Conventional stud partitions** – still used widely and consists of vertical timber studs with horizontal noggins clad with either plasterboard or lath and plaster finish.

Both of the timber partitions will sound hollow when struck.

- **Blockwork partitions** – a variety of blocks have been used since the 1920s including breeze blocks and hollow clay versions. Modern blockwork is made from concrete with a wide variety of aggregates that produce a range of products from a lightweight non-load-bearing block 75 mm thick to more substantial dense load-bearing versions 215 mm thick.

9.2.2 Load-bearing partitions

One of the most critical judgements to make is whether the partition is load-bearing. For example, an internal partition may take the loads from (see figure 9.1):

- struts and other components from a traditional 'cut' or purlin roof;
- water tanks in the roof space;
- the self-weight and imposed loadings from intermediate floors and staircases;
- the weight of the partitions directly above.

In some circumstances they may also provide bracing and lateral stability to the whole structure. As a general guide:

- older properties that have traditional roofs often have load-bearing partitions at all floors;

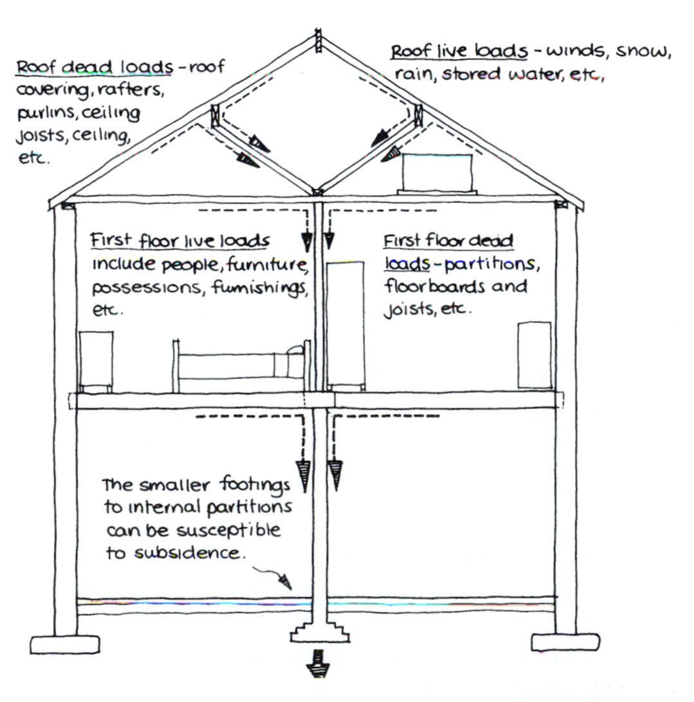

Figure 9.1 Load paths through an internal partition in a typical domestic dwelling.

- newer dwellings with trussed rafter roofs may have load-bearing partitions at ground floor only. Where properties are three storeys or more, load-bearing partitions may extend higher than the ground floor.

9.2.3 Structural failures in load-bearing partitions

There are a number of reasons for a partition to develop structural defects. They include:

- **Poor foundations** – in some cases the loads carried by internal walls can exceed those supported by some external elements. Yet they can have minimal foundations much shallower than the footings of external walls. This can result in differential settlement between the external envelope and the internal structure.
- **Overloading from new roof covering, etc.** – a change in the weight of a roof covering or a change of use in the building as a whole can cause the partition to subside through overloading.
- **Structural alteration of internal partition** – including through lounges, new door openings, etc. This can interrupt the existing load paths and cause a corresponding deflection at a higher level.

Key inspection indicators

- **Distorted door frames** – doors will often have to be continually planed down and adjusted so they can close properly. Typical signs will be sloping transoms and door heads (assessed with hand-held spirit level), top rails to doors that are out of shape because they have been frequently planed down by the owner. Evidence of scuff marks to lining or frame where the door has been binding also suggests problems.
- **Sloping floors** – floors that slope towards the subsiding partitions (tested by bricklayer's spirit level minimum 1 m in length).
- **Cracking** – that can radiate from the corners of door openings in the affected partitions.

In many older properties, these distortions might have occurred many years ago. Buildings often reach an equilibrium albeit a little out of true. The key factor here is how recent these movements have been. The following signs can indicate whether the movement is contemporary or not:

- if the door shows clear signs that it has been recently adjusted and planed. This could include freshly exposed timber, edges of the door recently touched up with paint, etc.;
- cracking that extends through decorations that are relatively new (i.e. less than 3 years old);
- evidence of building alterations that match the age of the movement.

If these triggers are present then the principles of 'follow the trail' may suggest that further investigations are appropriate.

9.2.4 Non-load-bearing partitions

This type of partition mainly divides spaces and rarely will carry any formal loads apart from their own self-weight. Key issues will include the following.

Structural stability

Although non-load-bearing partitions may support nothing but their own self-weight they must be properly supported by the structure beneath especially on upper floors. A common problem is associated with timber intermediate floors that support block partitions. Ideally where these partitions run parallel to the floor joists, they should be positioned directly above a larger single or double joist. Even then it may be necessary to calculate the adequacy of the support just to make sure. If not, then deflection of the floor can occur. Where partitions run at right angles to the joists, a line of solid noggins should be provided. In this case it is vital that the block partitions are not built directly off the flooring (HAPM 1991, section 3.2).

Plaster surfaces to block partitions can also develop defects especially if the blocks are to dry when plastered. In this case the water can be drawn from the plaster resulting in poor adhesion characterised by map cracking to the surface.

Rising dampness in internal walls

Rising dampness was discussed in some detail in chapter 5. Because the subsoil adjacent to external walls is generally wetter than that beneath the building itself, rising dampness in internal partitions is not a widespread problem. Even so, to make sure that dampness does not affect internal surfaces, partitions still have to have effective damp-proof courses that are properly linked to adjacent damp-proof membranes. If not then rising dampness can occur.

A particular problem that can occur with ground floor timber-framed partitions especially in older properties is wood rot in the sole plate. If this is not protected by a dpc or rubble has breached the existing one, conditions for the development of wet or dry rot may exist. The position where a suspended timber floor changes to a solid floor is particularly vulnerable (see figure 9.2). In these locations special care must be taken to test the lower walls and skirtings for dampness and visual signs of wood rot.

9.3 Fireplaces and chimney breasts

Most properties will have at least one chimney breast. In older dwellings there may be a considerable number with flues serving every room in the house. The primary function of the chimney breast is to provide a passage for the products of combustion but many often have an informal structural role. When they are properly bonded into the main walls chimney breasts can have two opposite effects:

- they can provide a measure of stability and restraint to the main structure;
- because of their mass and self-weight they can result in subsidence if the foundations are not adequate.

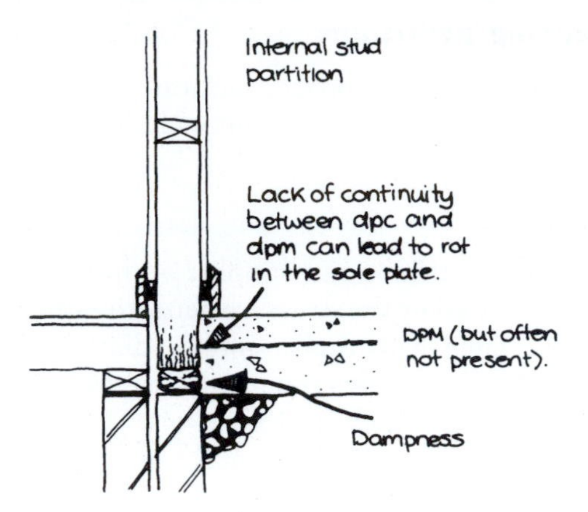

Figure 9.2 Sketch section showing junction of a solid and suspended floor with an internal partition.

Because of their structural influence any modifications to a chimney breast in a dwelling may have serious ramifications (see Smith v. Bush 1990) and so these must be inspected very closely during a survey. The main problems associated with chimney breasts are outlined below.

9.3.1 Adequacy of flues

The adequacy of flues for the purposes of fuel-burning appliances is discussed in more detail in section 10.5.1. For older properties it is likely that the chimneys have been used for decades and so can be in a very poor condition. The internal lining or parging may have deteriorated and if the chimney breast has an external face, sulfate attack may be active. The testing of flues is not part of any standard survey and so little comment can be offered but there are a number of visual signs that could indicate that significant problems exist:

- evidence of soot falls in the fireplace area;
- brown watery stains to the surface of the chimney breast within the room;
- deteriorated mortar joints and brickwork to the chimney stack in the loft space. White salt crystals may also suggest moisture-related chemical defects;
- old chimney pots and flaunching in a poor condition.

Even where these signs do not exist, it cannot be assumed that the flues are adequate for use. Clients should always be advised that flues should be tested and regularly cleaned if they are going to be used for an open fire.

9.3.2 Blocked fireplaces

The fireplaces may have long since been taken out and the chimney breast blocked. This is often done in brick or blockwork and plastered over but sometimes a less satisfactory method of sheet boarding on a timber framework is used. If it is not done correctly there could be a build-up of moisture in the redundant flue leading to damp staining on the internal wall surfaces. The BRE (1987b) have identified a number of different situations (see figure 9.3):

- **Chimney on an internal wall** – the chimney stack should be taken down below roof level and capped. The flue should not be ventilated.
- **Chimney on external walls (eaves elevation)** – the chimney should be taken down to below roof level and capped. Ventilation to external air provided at head and foot of the stack.
- **Chimney on external wall (gable elevation)** – the flue should be ventilated at the head and foot to external air. The top of the stack can be vented by providing a proprietary venting hood or capping the stack off and venting via air bricks.
- **Disused chimney stack that has been left in place above the roof line** – the following advice should be given:

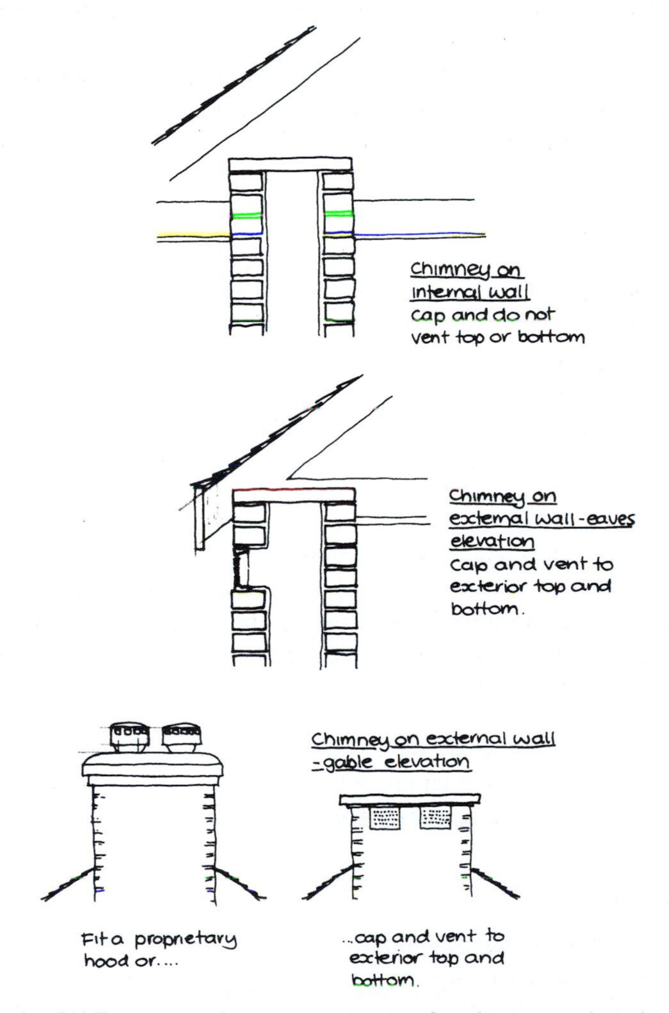

Figure 9.3 Sketch of different ventilation arrangements for chimneys when they are capped off (reproduced courtesy of BRE).

- the existing flue should be vented at the top with a proprietary venting hood that excludes moisture;
- a 225 × 225 mm vent with appropriate fly screen and cover should be provided to the room the fireplace used to serve;
- ideally the chimney should be swept and cleaned prior to this work being completed.

Other problems to be alert to:

- Make sure that a flue ventilated to an internal room does not conduct water vapour into the roof space. This will cause condensation in the roof void. Check the chimney flue in the roof space.
- A disused flue that is part of a larger stack that contains others still in service should not be capped. There is a risk that gases may leak into the disused flue and spill into the living accommodation.

9.3.3 Structural alterations

To create extra living space, many owners remove existing chimney breasts. The structural implications can be serious. As well as the Smith v. Bush (1990) case, some interesting discussion about the removal of chimney breasts occurred in Sneesby and another v Goldings (1995). In this case a surveyor carrying out a mortgage valuation failed to notice that where a chimney breast had been removed the structure above had been inadequately supported. The surveyor had realised the chimney breast was missing and even checked in the room above for signs of structural distress but found none. The inadequate structural support had been boxed in by new kitchen cupboards. If the cupboard doors had been opened, inadequate and shoddy work could have been seen. Soon after the plaintiffs had moved in, large sections of ceiling plaster began to fall down. The chimney breast had to be rebuilt to prevent further structural damage. The court held that the surveyor was liable despite the checks that he had made. This was because the removal of the chimney breast had created a 'trail of suspicion that should have been followed and resulted in a more complete inspection at the site of the chimney breast itself to see what was done there'. The doors of the kitchen cupboard could have been quickly opened so revealing signs of the defective work. There are a number of lessons from this case. When inspecting a property:

- All the chimney stacks should be located. This can be done externally but care must be taken where chimneys have been removed to below the level of the roof covering. Always check for a chimney from and old kitchen or 'scullery'. Smaller chimneys may have served an old 'copper' or water boiler that has long since been removed.
- Track each of the chimney stacks through the property to identify any missing or altered chimney breasts.
- Where chimney breasts have been removed:
    - Check in the room or space above for any signs of distress or structural movement.

- Open any cupboards, wardrobes, go into accessible roof spaces to see if there is any evidence of the support work that has been completed. Going beyond the usual remit of a standard survey may be necessary in these cases.
- Ask the vendor if they have any details of the work that was carried out or copies of guarantees, building regulation approvals, etc. Most building control authorities would classify such an alteration as 'notifiable work' and require building regulation approval.

Where a chimney breast has been removed and the supporting work can be inspected, a substantial construction should be expected. For example:

- A large heavy gauge set of 'gallows' brackets properly welded together, supporting the chimney above via a concrete padstone and all bolted into the wall. Some authorities do not allow this form of support as it can impose additional stress on neighbouring properties (see figure 9.4).
- A substantial concrete or steel lintel or beam properly pinned up to existing chimney breast spanning between adequate supports such as a load-bearing internal and an external wall.

Anything less than this could justify further investigation. If a chimney breast has been removed more than 5 years before, the supporting work is concealed and there is no evidence of adverse structural movement, then it could be safe to assume the work is adequate. Such opinions are a matter of judgement and as Hoffman LJ suggested in the Sneesby case, it could also be a matter of luck!

9.3.4 Rising dampness to chimney breasts

The chimney breast is a continuation of the main wall of the property and so should be properly isolated with a damp-proof course. Below the ground floor level, brickwork is usually built up and backfilled with rubble to form a base for the hearth. In these areas, poor constructional techniques often result in the dpc being breached. Therefore, it is important to check the cheeks and the face of the chimney breast at low level with a

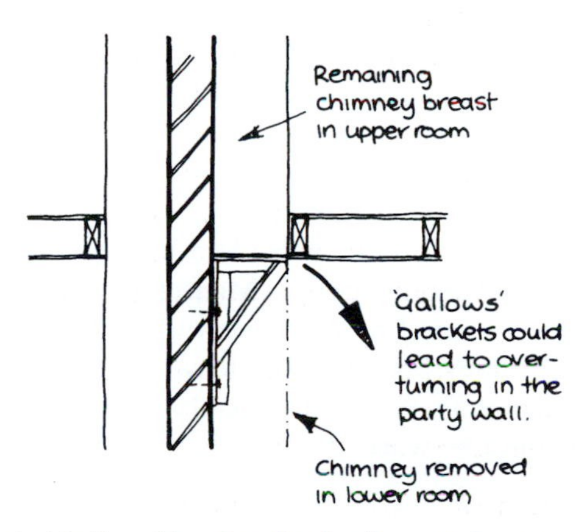

Figure 9.4 Sketch of a typical 'gallows' bracket that is often used to support the upper portion of a chimney breast.

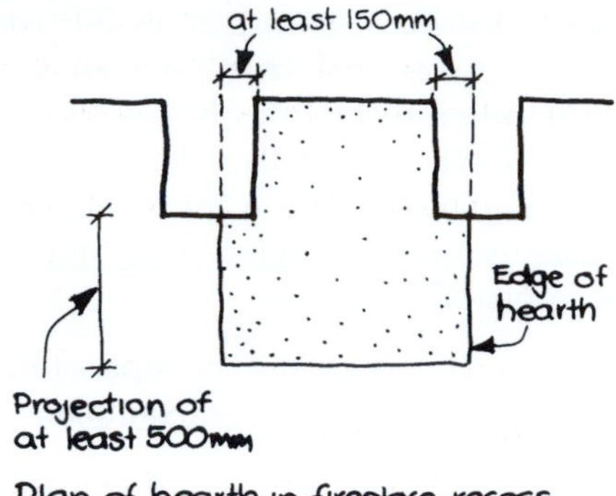

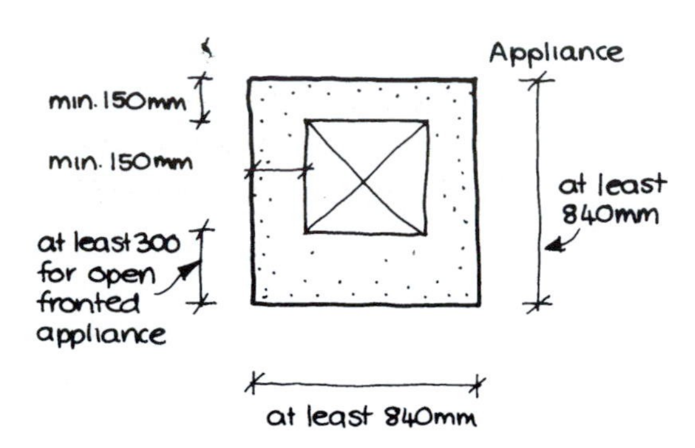

Figure 9.5 Minimum sizes of different hearth constructions (Crown copyright. Reproduced with the permission of the Controller of Her Majesty's Stationery Office).

moisture meter. Some of the timber joists and floorboards bearing on to the hearth may be in contact with damp materials. Where possible a moisture meter should be used in these locations.

9.3.5 Hearths

The primary function of a hearth is to provide a non-combustible area should any spark or burning fuel fall out from the fire. Many hearths are not big enough. The current Building Regulations, Part J require constructional hearths to extend 500 mm beyond the face of the chimney breast and at least 150 mm beyond the sides of the fireplace opening (see figure 9.5). Where it is less than this and the fireplace is used for an open fire the client should be warned about the of the dangers of such a feature.

9.4 Floors

Floors have been mentioned in two previous sections:

- section 3.4.3 where the lifting of floor coverings was discussed;
- section 6.2.5 which identified the key indicators that could help assess the condition of a timber ground floor even though the space beneath could not be inspected.

These aspects will not be repeated here. This section will focus on those defects not previously covered.

9.4.1 Lower floors

The first fact to establish is whether the floor is solid or not. If the floor is timber then it will require a very different surveying routine to a concrete one. The usual test is to 'jump' on the floor to see if it gives out a hollow sound. If it is timber the level of vibration may also give an insight into the floor's stability.

Timber lower floors

The information that can be gathered will depend on whether it was possible to visually inspect the floor space beneath. The main concerns will be with stability, rot and beetle infestation. For a more comprehensive guide to assessing timber floors see BRE (1997a) but the main points are summarised below.

TYPICAL DEFECTS

The following indicators may suggest that the floor has structural problems:

- excessive vibration and rattling of ornaments, etc. when jump tested;
- perceptible slope to the floor that can be detected with a spirit level.

There could be a number of causes of these problems:

- **Inadequate size of floor timbers** – undersized timbers may have been used in the floor resulting in excessive deflection. A traditional rule of thumb is illustrated in figure 9.6. This will give a rough assessment of the floor's adequacy. Another method is to compare the floor joists with table A1 of Appendix A of Part A of the Building Regulations (HMSO 1991a, p.44). But remember that the span is measured between effective supports (e.g. honeycombed walls) and is not the total length of the joist.
- **Poorly constructed floors** – much of the mass-produced Victorian housing suffered from variable standards of design and workmanship (Douglas 1997). Suspended floors could also have been altered over the years resulting in joists that are poorly seated on wall plates. Sleeper walls have often been removed for access purposes and not properly rebuilt.
- **Floor timbers weakened by rot and woodworm** – older timber floors are very vulnerable to outbreaks of wet and dry rot and attacks of wood-boring beetles. Joist ends were often built into supporting walls without adequate dpcs and debris may have built up over the years to partially bury timber components. The under-floor surface was rarely sealed with concrete but left in contact with bare earth. This can result in moisture levels building up encouraging both rot and beetle attack. This is often compounded by poor ventilation to the sub-floor area.

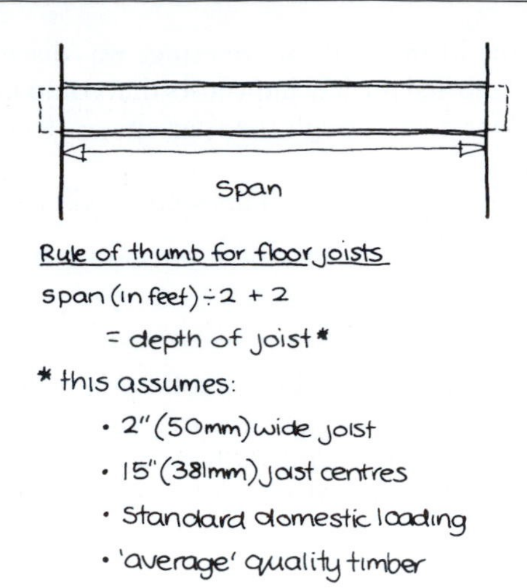

Figure 9.6 Rule of thumb for sizing floor joists.

Assessing the adequacy of ventilation to sub-floor areas

When deciding whether there is enough sub-floor ventilation, there are three separate issues to consider:

- Are there sufficient air bricks in total? – Part C4 of the 1992 Building Regulations states that current suspended floors should be vented as follows:
  - 1500 mm^2 for each metre run of wall;
  - the space between the ground and the underside of the joists should be at least 150 mm.

This would equate to approximately one 225 × 150 mm air brick at every 1.5 m centres around all accessible sides of the building.

- Are these airbricks properly distributed to prevent stagnant air pockets where moisture might build up? – Even if there are enough airbricks they may not be in the right position. Melville and Gordon (1992, p.195) identify the typical problem of a terraced house with a back addition or 'offshot' that has a solid floor (see figure 6.2). Although there are adequate airbricks front and back, there is a whole area of the sub-floor void that is stagnant. It is these areas that are at risk from rot and beetle attack.
- Are the sleeper walls adequately perforated to allow good cross-ventilation? If not, bricks may need to be removed (BRE 1986b).

A more detailed review of the moisture that collects beneath suspended timber floors has been produced by Harris (1995, p.11).

Solid lower floors

Concrete is often regarded as being a very durable material. Yet not all concrete floors are defect free and many older solid floors contain little concrete. For example, solid floors in older rear kitchens and sculleries may be no more than a thin screed laid over ash or clinker hardcore to provide a level bed for a quarry tile surface (BRE 1997a, p.1). The main defects include:

- **Settlement of the slab** – this is where the support from beneath is removed causing the concrete bed to settle. Because most concrete floors are unreinforced they crack haphazardly as they settle down to their new level. This might be caused by:
    - consolidation of the hardcore and subsoil below;
    - subsidence caused by external factors such as shrinkable clays, mining activities, etc.

KEY INSPECTION INDICATORS

- Gaps below skirtings (see figure 9.7; BRE 1990);
- cracks or unevenness in floor finishes that suggests cracks in the slab below;
- sloping floors (detectable with a large spirit level) and dished or isolated depressed areas of floor surface;
- ill-fitting doors in partitions that are built off the defective slab.

- **Swelling of the slab** – in some cases the concrete floor may swell upwards. This will cause the concrete slab to crack in a similar manner to that described above but for opposite reasons. This may be caused by:
    - sulfate attack to the underside of the concrete slab – hardcore or subsoils that are rich in sulfates can combine with moisture and bring about chemical change to the concrete. The main effect is to cause the concrete to expand and the floor to swell upwards. For more information on this aspect see BRE 1996;
    - swelling of the ground and hardcore – where a tree has been removed, clay subsoil may swell as moisture returns. A similar effect can occur where moisture affects some hardcores such as steel slags and colliery wastes although this is rare.

KEY INSPECTION INDICATORS

- A clear rise in the level of the floor especially (though not always) towards the centre of the slab;
- concrete slab or floor finish showing signs of random cracking;
- binding of doors that open into the room;
- distortion of internal partitions that are supported on the slab;
- external walls pushed outwards by the force of the expanding slab. The wall tends to slip at the dpc level (see figure 9.8);
- is it known locally that the subsoil or ground water is high in sulfates? Or did local building techniques often incorporate unstable hardcores below ground-floor slabs?

Category of damage	*Description of damage*	*Approximate: a) crack width b) gap[1] mm*
0	Hairline cracks between floor and skirting	a) NA b) up to 1
1	Settlement of the floor slab, either at a corner or along a short wall, or possibly uniform, such that a gap opens up below skirting boards which can be masked by resetting the skirting boards. No cracking in the walls. No cracks in the floor slab. Slab reasonably level.	a) NA b) up to 6
2	Larger gaps below skirting boards, slight slope of slab and some local rescreeding may be necessary. Fine cracks appear in the internal partitions and slight distortion of door frames and some 'binding' of the doors. Slab reasonably level.	a) up to 1 b) up to 13
3	Significant gaps below the skirting boards especially at the corners where there might be slight cracking to the floor slab. Sloping of floor is clearly visible in these areas (1 in 150). Some disruption to drains and other services. Damage to internal walls more widespread with some requiring crack filling or part replaster. Doors may have to be refitted.	a) up to 5 b) up to 19
4	Large, localised gaps below skirting boards; possibly some cracks in the floor slabs with sharp fall to edge of slab. Inspection reveals voids of over 50 mm below the slab. Local breaking out and part refilling and relaying of slab required. Damage to internal walls may require part replacement of bricks or blocks and re-lining of internal stud partitions.	a) up to 15 but may also depend on number of cracks b) up to 25
5	Very large overall settlement with large movement of walls and damage at junctions with first floor area. Voids exceeding 75mm below slab. Risk of instability. Most of the ground floor requires breaking out and relaying or grouting of fill. Internal partitions need replacement.	a) usually greater than 15 but depends on how many b) greater than 25

1 'Gap' refers to the space between the skirting and finished floor caused by the subsidence.

Figure 9.7 Table describing typical signs of solid floor affected by subsidence (based on BRE *Digest 251* and reproduced courtesy of BRE).

- **Failure of the damp-proof membrane** – the problems associated with dampness have been outlined in general terms in section 5.4. This section looks at those problems closely related to solid floors. The floors to many older houses simply do not have a damp-proof membrane and those built between 1950 and 1966 may rely on thermoplastic tiles stuck down with a bitumen adhesive (BRE 1997b). Therefore where the solid floors to older properties have not been replaced they must be viewed with caution. For newer construction, Part C4 of the Building Regulations (HMSO 1991b) states that a damp-proof membrane must be provided within the thickness of a solid floor. This may be above or below the concrete floorbed. Because well-laid good-quality concrete is virtually waterproof it can

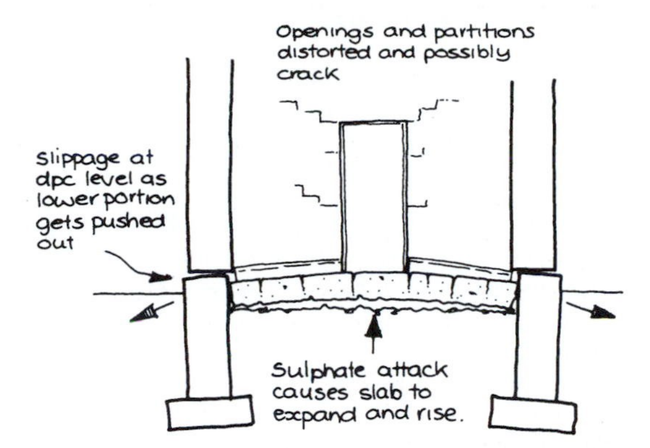

Figure 9.8 Sketch showing effects of sulfate attack to a solid floor.

cope with a few minor problems with the damp-proof membrane. Where the defect is significant, high levels of moisture can enter the floor slab and cause serious deterioration. Moisture can affect a floor by a number of routes:

- missing or partially missing dpm;
- badly punctured dpm;
- poor link between the dpm of the floor and the dpc of the wall.

In each case both the slab and any overlying screed can act as a huge reservoir for moisture. This may take many months to dry out even when the defect has been rectified.

KEY INSPECTION INDICATORS

- Deterioration or staining to the underside of the floor coverings and finishes;
- high moisture meter readings to the surface of the floor;
- in some cases where the plaster is in contact with the floor surface, high moisture meter readings to the base of the plaster with characteristic tide-mark staining;
- high moisture meter readings to any skirting boards.

The main problem with this type of defect is that it can easily be confused with others. For example:

- low levels of rising dampness in the wall;
- cold bridging in uninsulated slabs causing condensation dampness on the floor surface and heating; and
- hot and cold water pipes that are leaking into the screed (BRE 1988).

Recognising that there is a problem is straightforward, finding the true cause and organising the correct remedial work is far more complicated. The duty of the surveyor in the first instance is to recognise that a dampness problem exists. The potential effect can be more pronounced if the floor covering is timber block or boarding which could lead to the development of wet or dry rot.

Ground floor suspended concrete floors

Many newer properties now incorporate suspended concrete flooring systems. These usually consist of concrete beam and infill blocks with various forms of concrete screed laid over. Generally speaking in older dwellings the only suspended concrete slab floors were associated with system built houses (BRE 1997b, p.4). Distinguishing this type of floor from a suspended timber floor is relatively straightforward (see section 9.4.1 above) but it can easily be confused with a solid ground-bearing alternative. Because both can suffer from different types of defects, proper identification can be important. Here are a few tips:

- If the house is relatively new try and obtain the drawings and/or specification.
- When 'jump' tested (see section 9.4.1 above) a suspended concrete floor can still give a slightly hollow sound when compared with one fully supported from beneath;
- Some suspended concrete floors may have air vents to the floor space beneath. Part C4 of the 1991 Building Regulations (HMSO 1991b, p.15) only requires ventilation where the sub-floor void is not effectively drained or is likely to fill with explosive gases.

The main defects that can affect suspended concrete ground floors are related to movement. The main causes and associated signs include (BRE 1997b, p.4):

- inadequate bearing to the pre-stressed concrete beams that support the infill blocks. This can result in displacement at the perimeter;
- poor connections and lack of lateral restraint to the wall where it runs parallel to the span. Gaps and cracks at the floor and wall joint are typical signs although these can be difficult to see because of skirting boards;
- where the beams and blocks have not been 'locked' together properly by the grouting laid over the top. This can give rise to cracking to the finishing screed in line with the span direction.

Another defect can occur where the surface of the floor is uneven causing any pre-formed insulation or chipboard finish laid over to 'rock' when walked on.

9.4.2 Upper floors

In the vast majority of cases the upper floors of most residential properties will be timber. The exceptions are purpose-built flats and maisonettes which may have solid concrete party floors.

Timber upper floors

In many ways, intermediate timber floors are constructed in a similar manner to the ground floor equivalents. Consequently the approaches to assess the condition of the two are similar. This is summarised below.

- **Deflection check** – the standard 'jumping up and down on the floor' test is appropriate for upper floors. Any excessive deflection may also be evidenced by:

- cracking to the ceiling and ceiling/wall junction in the room below;
- distortion of partitions supported by the defective floor.

The causes of this could be:

- **Inadequate joist size for the span of the floor** – this can be checked by using the rule of thumb described in figure 9.6. The depth of the joists can be estimated by measuring the depth of the whole floor where it is revealed at the staircase opening. Deductions are then made for floorboarding (say 13 mm) and the ceiling (allow 20 mm maximum). If a carpet can be lifted and the floorboarding exposed, the spacings of the joists can be estimated by measuring the distance between the nails in the floorboarding. This is not a certain way of assessing the adequacy of the floor. Hidden notching or other constructional faults may still give problems. Despite this, it can help a surveyor add value to the assessment of condition.
- **Lack of strutting at mid-span** – most suspended timber floors will require some form of strutting at mid-span to stop the joists twisting and to stiffen the floor up. Without lifting several floorboards it is impossible to tell whether this is in place.
- **Inadequate end support** – joists in older properties were built into solid walls without any sort of restraint or positive fixing. If the wall moves out of plumb (see section 4.7) the joists can be left with very little bearing. In new properties, where joists have been hung on joist hangers, the end bearing can be badly made offering little effective support to the joist (BRE 1984).
- **Inadequate support around openings in the floor** – where staircases and chimneys pass through the floor, a series of trimming joists are often used. They normally consist of wider joists or two standard joists bolted together. These will then connect onto two other trimmers that collectively define the opening. Because the trimmer joists are carrying much higher loads their design, sizing and jointing is far more critical than for other joists. Landing areas should be closely inspected for signs of deflection to the floor.
- **Deflection due to excess notching** – many older properties may well be on their second or third 'replumb' and 'rewire'. Each time contractors may have notched or drilled the joists inappropriately (see figure 9.9). In some cases this might lead to deflection especially where a lot of pipes pass through the floor in a small area (BRE 1987a). Special checks should be made around hot and cold water tanks where a high number of pipes, cables and associated notches can be anticipated.

There are a number of other defects that could affect intermediate floors. These are related to problems with other parts of a building emphasising that any deficiency should be looked at holistically rather than just in isolation:

- **Wood rot to joist ends** – joists built into older solid walls may be as little as 100 mm away from the external face (see figure 9.10). If the bricks are porous or the pointing is in a poor condition then dampness can affect the timber and cause dry or wet rot. Leaks from WCs, baths and washbasins can also cause damage to floors. Indicators would include:

  - poor pointing and deteriorating masonry externally;
  - evidence of wood rot to skirtings along with high moisture meter readings;
  - water stains and/or mould growth to adjacent walls and ceiling surfaces.

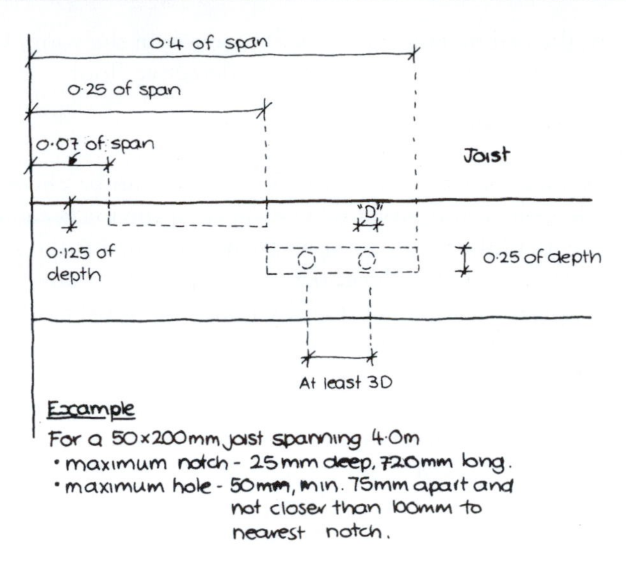

Figure 9.9 Recommended notching sizes and spacings.

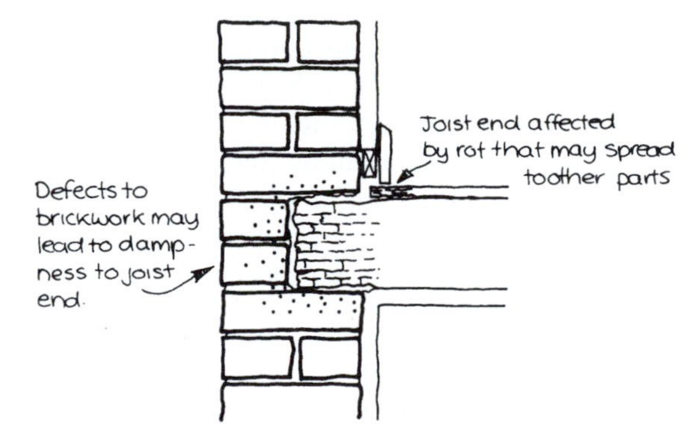

Figure 9.10 Section through external solid wall showing the bearing of internal floor joists. If the external face is exposed to driving rain and the pointing is in a poor condition then the joist ends are vulnerable to wet or dry rot.

- **Beetle attack** – the effects of wood-boring beetle are discussed in detail in section 6.3. Upper floors can be affected so the floorboarding should be inspected wherever floor coverings, furniture and possessions allow.
- **Lack of lateral restraint** – it is important that the floor and the external walls of a building are structurally connected to ensure the wall is laterally stable (see section 4.7). Many older properties will not have this and the wall may bulge outwards. The presence of strapping cannot be deduced without taking up the floorboards but a number of key indicators may suggest lateral instability can be a problem:
  - gaps between the underside of the skirting and the floorboarding. This can often be masked by timber beading;
  - cracks at the junction of the affected external wall and the ceiling of the upper floor;

  - the presence of tie bars externally can indicate remedial work has been carried out to the floor.

- **Defects with panel flooring** – chipboard flooring panels are commonly used in new housing for the floor deck of intermediate floors and can be very satisfactory if the correct type is used. But the investigations have shown that this is not always the case (BRE 1983). Using the wrong type in areas of high humidity, not allowing space around the edge for expansion and inadequate nailing, gluing and edge support (noggin) can all lead to problems. Typical indicators can include:
  - sagging and buckling panels;
  - loose and squeaking areas of flooring;
  - dramatic loss of strength in wet areas.
- **Incorrect thickness of timber boarding for the spacing of the joists** – although this is not such a serious defect it can give rise to excessive vibrations. The thickness will be difficult to determine without lifting a floorboard or panel, but the BRE (1983) suggest::
  - 16 mm boarding for 450 mm joist centres;
  - 19 mm boarding for up to 600 mm joist centres.

Concrete upper floors

Most properties will have timber intermediate floors. However a number of purpose built residential blocks may have concrete 'party' floors constructed using a wide variety of materials and techniques. These may include (Marshall et al. 1998, p.116):

- reinforced concrete that may incorporate secondary steel beams and/or a variety of hollow clay or concrete pots;
- pre-cast concrete beams with concrete blocks or panels laid between;
- pre-stressed concrete structures where the reinforcing wires are put under tension and sealed within the concrete.

Because the concrete is usually not exposed to external influence, the likelihood of serious defects is much less than for external concrete (see section 4.7). Despite this, a number of problems can occur:

- Stability problems at the end bearing – this is where the floor bears onto the wall. Examples could include:
  - lateral instability in the wall that reduces the end bearing of the edge of the floor;
  - differences between the expansion rates of the concrete slab and the wall construction can lead to cracking and spalling of adjacent brickwork.
- Deterioration in the concrete where it extends to the external face of the building (see section 4.7.9 for more information).

- Problems where the floor extends to form a balcony. Poor design, detailing or material breakdown can result in the failure of the cantilevered balcony. The signs can include cracking to the top side of the balcony adjacent to the external wall.
- Although dampness defects are not common, moisture can affect intermediate floors in a number of ways:
  - through faulty damp proof course and cavity tray detailing at the junction of the floor, balcony and wall (see figure 7.21). Water staining to internal surfaces may be an indicator;
  - leaking water and heating pipes, faulty sanitary fittings, etc. that may result in large amounts of water entering the screed above the main slab. This can give rise to staining to the ceilings below. In some cases 'tide marks' to internal partition walls can occur and resemble typical rising dampness signs. The authors know of defects like this that have led to the injection of a chemical damp-proof course in to the walls of a 3rd floor flat!
- General structural problems with the floor as a whole that may result in distortion. Although not common, some pre-1975 buildings may have been constructed with High Alumina Cement Concrete (HACC). This was used to develop a high early strength in the concrete but often resulted in a dramatic loss of strength shortly afterwards. Stability problems can occur in warm, humid environments where the strength of the concrete can be further reduced because of chemical change. Dramatic building collapses in the 1970s highlighted the danger posed by these types of structures. Based on a visual survey, it is virtually impossible to identify whether HACC has been used. However the following points could be useful:
  - It was often used in concrete buildings constructed before 1975 so buildings later than this time will not usually be affected.
  - Wide-span pre-cast concrete sections are the most at risk.
  - Warm, humid environments can accelerate the deterioration of the concrete. Therefore, high risk uses include swimming pools, sports facilities or even where a concrete element regularly gets wet through other defects (e.g. leaking flat roof, faulty water pipes).
  - Deflection in the slab can suggest problems. Signs could include:
    - sloping floors;
    - gaps between the floor and skirtings; and
    - distorted internal door openings and binding doors.

Where these factors are noted, the surveyor should consider advising further investigations.

9.4.3 Ceilings

Ceilings have been included at this point because their condition is linked with that of the floor that they are attached to. Older ceilings can pose many serious problems as they can collapse suddenly without warning causing injury to the occupants and damage to the possessions. The two main types of ceiling are:

- **Lath and plaster ceilings** – where three coats of lime plaster were applied to a base key of thin strips of softwood or laths nailed to the joists above. The plaster was squeezed between the laths to form a key (see figure 9.11).
- **Plasterboard ceilings** – the modern version used on most dwellings built since the Second World War.

Many years of continual vibration and changes in moisture content, temperature, etc. can lead to the failure of the plaster key of the ceiling. Whole sections of the ceiling can bulge downwards, sometimes only kept in place by the decorative lining paper. When a ceiling has reached its limit of stability, one more slammed door or heavy-footed child running up the stairs can lead to quite dramatic collapse. These failures can be helped on by:

- water leaks from roofs and sanitary fittings above;
- pugging or 'deafening' (for our Scottish readers!) material placed on the back of the ceiling for sound insulation purposes;
- heavy ornamented centre roses and cornices that add to the weight of the ceiling.

Key inspection indicators

- Large cracks and bulges to the ceiling sometimes showing through any lining paper;
- water staining to the ceiling;
- evidence of patch repairs to the ceiling suggesting problems in the past.

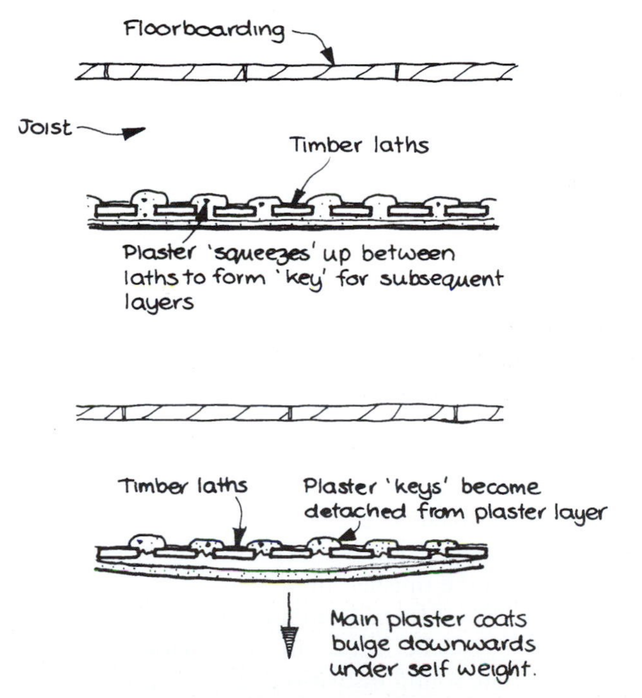

Figure 9.11 Sketch section through a lath and plaster ceiling. The lower drawing shows what can happen when there is a failure of plaster 'key' to the ceiling.

Where these indicators are present the client should be warned of both the danger the ceiling may present and the high cost of replacing it. It is rarely possible to successfully patch repair a lath and plaster ceiling. If the dwelling is a listed building, then the local planning authority might require the owner to apply for listed building consent to replace it.

Plasterboard ceilings are generally more reliable. Some of the common defects include:

- uneven ceilings, cracking at the plasterboard joints and at junctions with the walls. One of the main causes is where the appropriate noggins (timber bearers to the free edges of the board) have not been included or the plasterboard is too thin for the spacings of its supports (BRE 1986a). If this occurs in ceilings of the uppermost floor then it might be possible to inspect the support arrangement from above in the loft space.
- nail heads have popped out where they have not been properly driven home.

These faults may be minor defects but occasionally major building movement may also result in similar signs. The surveyor should always look broadly at problems like this.

9.4.4 Sound insulation

Sound insulation is not usually a problem in a single-family dwelling. In conventional two-storey single-family properties there are no special requirements under the current building regulations. The issues become more critical in flats where the lack of sound insulation can be very unpleasant for all concerned and the source of many neighbour disputes.

Where one flat is provided above another either in a purpose-built block or in a conversion scheme, the level of sound insulation is outlined in Part E of the Building Regulations (HMSO 1991c). Although these standards are not retrospective they can give a good measure of the level of provision that should be aimed for. There are a number of different ways of achieving this but one common method, shown in figure 9.12 is used in many conversion schemes. Figure 9.13 shows typical sound proofing to a staircase where it separates two dwellings. Establishing whether any floor has an adequate level of sound insulation is beyond the scope of the standard survey but a few key indicators can help give an insight.

Key inspection indicators

- If the flat has been recently converted (i.e. since 1985) was it granted the appropriate building regulations approval?
- At the entrance to the upper flat, is there a timber threshold piece that marks a change in level between the finished floor level of the flat and the common area? This could suggest that some form of sound insulation has been laid over the floor.
- If a corner of a carpet can be lifted can floorboards be seen? If yes, then it is unlikely that appropriate sound insulation has been provided.
- During the survey, can sounds be clearly heard from the property above or below? This will depend on whether the dwellings are occupied at the time and also on the surveyor's judgement.

Figure 9.12 Section through an existing timber floor showing retro-fitted sound insulation.

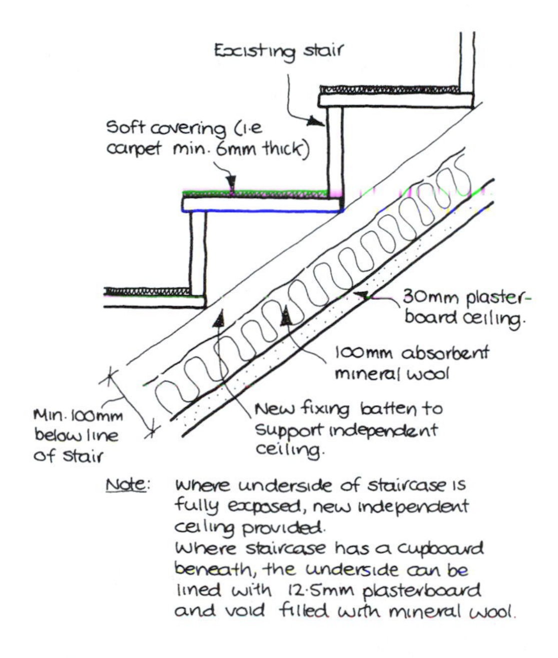

Figure 9.13 Typical sound proofing to existing staircase between dwellings.

If doubts are raised by any of these questions the client should be advised that noise nuisance is a possibility especially with conversions carried out before 1985.

9.4.5 Fire resistance

The ability of the internal walls and floors to resist the passage of fire and smoke may mean the difference between life or death for the occupants. Most standard surveys do not include an assessment of the adequacy of the fire resistance of the property. In fact there are few special provisions for fire precautions in single-family two-storey houses. The regulations mainly apply to flats and maisonettes.

When looking at this type of accommodation there are a number of key features that can help the surveyor identify any clear deficiencies. These indicators are a summary of what can be seen and do not necessarily represent a breach of the regulations. A more detailed account of the requirements is contained in Part B of the Building Regulations:

- For single-family dwelling:
  - smoke alarms should be fitted in the circulation areas close to the kitchen and living room;
  - houses that have a storey more than 4.5 m above ground level (typically a three- or four-storey house) should have a protected staircase (see below for definition) down to the final exit;
  - loft conversions should be as described in section 7.6.6.
- For flats and maisonettes, although the regulations are very complex, the main features should include:
  - smoke detectors in the circulation areas of the individual flats;
  - fire doors with self-closing devices to all internal flat doors and the flat entrance door itself. There should always be two fire doors between an individual room and the common staircase or landing. The doors to any large cupboards or storage areas should also be half-hour fire resisting;
  - a protected staircase from the flat door to the main entrance door of the whole building. A 'protected staircase' is defined as a 'fire sterile' area which leads to a place of safety away from the building. The enclosing walls, floors and staircases should be able to resist the passage of fire and smoke for at least 30 minutes. Any cupboards or storage areas off the main staircase should also have a 30 minute fire rating;
  - a block of flats can be served by a single staircase providing that the top floor of the building is no more than 11 m above ground level and there are no more than three storeys above the ground storey. Above this level the regulations become very detailed and specific but usually involve an alternative means of escape.

It must be stressed that these indicators are very broad way of attempting to give an initial assessment of the level of fire safety in a building. If the inspection reveals significant deficiencies then the Building Regulations should be consulted in detail or the matter referred to an appropriate professional.

9.5 Internal joinery

9.5.1 Staircases

The staircase in a dwelling should provide a safe and serviceable means of access from one floor to another. Safety is the most important aspect and it has been estimated that over 20,000 accidents that occur in the home are associated with staircases (Melville and Gordon 1992, p.209). Defects are not the only cause of the danger as the original design can create a number of other hazards. This is a problem for the surveyor. Many staircases in older dwellings will not conform to the current building regulations yet they might be safe to use. On the other hand, winding and narrow stairs that lead up to attic bedrooms will be very difficult to use safely. This is another area where the surveyor must use sensible and well balanced judgement.

Design of the staircase

A safe staircase should have the following features:

- For a private stair in a dwelling the maximum pitch should be 42 degrees with a maximum rise of 220 mm and a minimum going of 220 mm.
- For a staircase in a common hallway of a block of flats the maximum rise is 190 mm and the minimum going is 250 mm.
- There should be at least 2 m headroom above the flight of the staircase.
- There is no minimum width for a staircase in a single dwelling but a common stair in a block of flats should be at least 1 m wide.
- There should be no more than 16 risers in any one flight.
- open treads on the staircases are potentially dangerous. The current building regulations state that the gap should be blocked.
- There should be a landing at the top and bottom of each flight with at least 400 mm of clear space (see figure 9.14).
- There should be at least one handrail to a flight.
- All staircases, landings, balconies, etc. should be protected by appropriate guarding. This should be between 900–1100 mm depending on its location. To protect young children, a 100 mm ball should not be able to pass between the balustrades. This guarding should be robust and be able to resist lateral pushing pressure.
- Any tapered treads should conform to figure 9.15.

Condition

Many of the defects to a staircase will originate in faults to other parts of the adjacent construction including the walls and floors. Failure to comment on the condition of staircases has been frowned upon in the courts. Although it was not the main cause of action in Sneesby and another v. Golding (1995), the surveyor failed to properly report

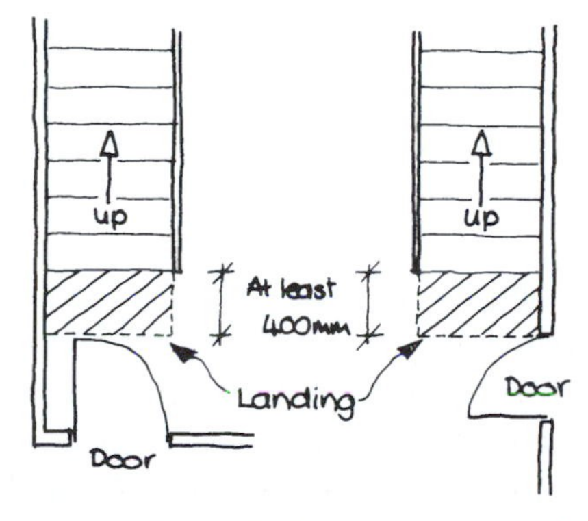

Figure 9.14 Safe arrangements for landings and doors on staircases (Crown copyright. Reproduced with the permission of the Controller of Her Majesty's Stationery Office).

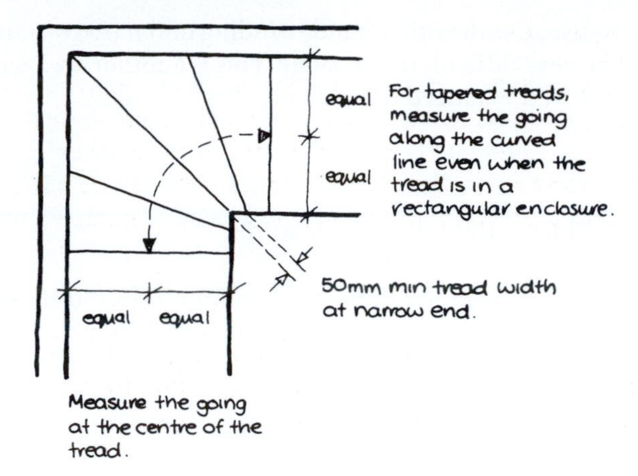

Figure 9.15 Guidelines for assessing suitability of tapered treads on a staircase. It is important to note that these measurements apply to new work. Many existing staircases will not conform to these strict rules. It will be a matter for the surveyor's judgement as to whether the staircase is suitable (Crown copyright. Reproduced with the permission of the Controller of Her Majesty's Stationery Office).

on the implications of removing a supporting partition to a staircase. When the wall was removed, the newel post was left standing independently. Because it was not designed to do this the whole balustrade became unstable.

KEY INSPECTION INDICATORS

- Check all treads and risers for any loose, split, damaged or unsupported timber. The condition of the nosings is particularly important. Where the timber is exposed, look for signs of wood-boring beetle.
- Carefully test (by shaking using a gentle pressure) all handrails and balusters and note any balusters that are missing or more than 100 mm apart. Special attention should be paid to the newel post. If this is loose then the whole of the balustrade may be unstable.
- Check whether the treads are horizontal and do not dip from one side to the other. If they do it could indicate some form of settlement in the adjacent construction.
- Where the side strings are attached to the internal faces of external walls, test the timber for signs of dampness.
- If the underside of the staircase is open to view, inspect the following:
    - check that all the wedges and blocks are well glued and fixed into position so the treads and risers are secured;
    - look for evidence of wood-boring beetle that is common in this location;
    - check the rear face of the bottom riser and the base of the adjacent strings for wood rot and moisture content. Rising dampness often occurs in these locations.
- If the underside of the stair has been plastered or boarded check the soffit for soundness. Where the staircase is in a common hallway serving two or more flats, the underside of the staircase should be fire protected. Where there is an understairs' cupboard, it should be fitted with a fire-resisting door (see section 9.4.5 above).

9.5.2 Other joinery items

This will include a wide range of components including internal doors and linings, skirtings, architraves, window surrounds, dado panels, picture rails, fitted cupboards, etc. Many of these have been mentioned in previous sections when specific defects were discussed. Because all are made of timber special care should be taken to look for signs of wood rot and beetle especially on the internal faces of external walls. The level of detail reported will depend on the type of survey but the following tips might be useful:

- Open and close all internal doors to check for warps, twists, binding and general operation. Check ironmongery to see if it is working properly.
- Check all door, window and other openings for squareness. Distortion might suggest structural movement.
- Are there any joinery items that have special aesthetic or historic features? For example panelled doors or original dado features might not only be attractive but also protected through the listing process.

9.6 Internal decorations

9.6.1 Introduction

The condition of the internal decorations rarely affects the value of a property. In most cases it will have little impact on the dwelling's condition. There are a few general aspects to account for when giving advice on internal decorations:

- Most new owners will want to redecorate to their own tastes even if the vendor spruced up the house just before it went on the market. This decoration is usually done on a DIY basis and many clients will see this as almost a 'fixed' cost that they will have to meet when they move into a property.
- The style, form and condition of the internal decorations are features that many lay people feel able to partly assess themselves when they first look at the house.
- 'Taste' is a very subjective concept and should be left to the client to judge. What may be disturbing to one person might be another's idea of style!

The surveyor's role should therefore be focused on the following:

- What general condition are the decorations in?
- Does the type of decorations affect the safety of the occupants?
- Will the costs of redecorating be high due to the particular nature of the existing decorations?

These are discussed in more detail below.

9.6.2 General condition of decorations

The quality of the decorative finish mostly depends on:

- the quality of the original materials used; and
- the standard of workmanship.

Although many clients will be able to make this assessment, an objective view will help them. They will expect to do some redecoration, the key question is how much. The answer may affect their decision to buy. A room-by-room schedule would not be appropriate for most standard surveys. Instead an overall judgement should be given. Figure 9.16 may give some useful descriptors that can help in this process.

Particular comments should be made where the decorations have been affected by other building defects such as damp staining, mould growth, cracking, etc. These signs are possible 'trails of suspicion' in themselves and the surveyor should make sure they have been properly followed. Cross-reference to other parts of the report might be appropriate.

9.6.3 Safety implications of decorative finish

Over the years, a number of different finishes may have been applied that can threaten the health and safety of the occupants. This could include:

- surface finishes that allow excessive fire spread and release toxic gasses when ignited. The most common example would be polystyrene tiles (especially when painted with a gloss finish) but could also include fabric wall hangings, etc.;
- textured finishes that may contain harmful substances such as asbestos, lead, etc. Older 'artex' finishes may be particularly problematic.

Where these are suspected, removal should be recommended with appropriate safety precautions.

9.6.4 Relative costs of redecoration

Where decorations need to renewed, the relative cost will depend largely upon the nature of the existing finish. For example, imagine a magnolia and white painted room that is a little grubby. If the new owner wishes to repaint it in similar colours then a quick wash down and one coat of paint might be all that is needed. On the other hand a room that has been poorly decorated with a heavy lining paper painted in a deep colour will require much more work. It is only reasonable to try and give the client an impression of what work might be involved. The following factors may help:

- Dark colours will be far more difficult to cover up with lighter shades and so may take many coats of paint.
- Where wallpaper is to be removed, the condition of the plaster beneath becomes very important. If the surface of the existing wall shows signs of unevenness, cracks and loose and bulging sections of plaster then these areas are likely to be damaged even more when the wallpaper is removed. A large amount of filling or even plaster repairs might be required.
- Heavily textured surfaces will be very difficult to remove. Artex and sculptured plaster may have been fashionable several years ago but now most people would want to remove them. This can be very costly and time consuming.

Category	*Description*	*Action*
Good (satisfactory repair)	The standard of internal decorations is generally good. Good quality materials have been used and many have been recently applied. Surfaces are generally clean and well maintained.	Little work is required apart from small scale repair/redecoration or decoration to suit individual taste.
Fair (minor maintenance)	Although the standard of decoration is reasonable, a number of surfaces are showing signs of aging and deterioration (say 30–40%). In some areas, wall paper is beginning to peel and some paint surfaces are chipped and scratched.	A number of the surfaces require redecoration to restore an acceptable aesthetic appearance and/or to suit individual taste.
Poor (significant or urgent repair needed)	Most surfaces are in a poor condition showing signs of age, wear and tear (say 55–60%). Many areas of wallpaper are loose and paint finishes are extensively chipped, scratched and peeling.	Most surfaces in the dwelling will require redecoration in the short term to restore their aesthetic appearance. This may involve a considerable amount of preparation. This could be both time consuming and expensive.

Figure 9.16 Table linking the condition of internal decorations to client advice.

These are just a few examples that are commonly encountered. People do weird and wonderful things in the privacy of their own homes. Although it might suit their unique tastes it could be very off-putting to those with more conventional opinions. One of the most dramatic examples encountered by the authors was where the teenage daughter of the previous owner had created a work of art in the corner of her bedroom. Although the sculpture was very artistic not everyone appreciates over 300 cola tins stacked and glued from floor to ceiling in a black-painted bedroom!

9.7 Basements, cellars and vaults

9.7.1 Introduction

These are terms that are applied to the lowest space in a dwelling. The nature of this space will vary from something that is little more than a sub-floor void to a full-height room with natural daylight that has potential for conversion to a habitable room. Because they are either totally or partially below ground the walls can face particular problems from lateral forces imposed by retained ground and high levels of dampness. Key indicators of potential problems can include the following.

9.7.2 Strength and stability

- Where the basement walls are retaining is there evidence of any horizontal displacement such as bulging or cracking to the masonry?
- Is there evidence of relatively recent extensions or other buildings close to the basement walls? The loads from new foundations may put pressure on old basement walls.
- Do internal load-bearing partitions appear stable and well founded? Do they line up with the internal partitions above? Have they been altered in a way that might

affect their stability? Internal partitions at basement level may be taking a considerable load from above and play an important role in helping the external walls resist the thrust of retained ground. It is essential that any alterations do not affect their stability.

- Is there evidence that the lowest floors have been renewed and possibly lowered? Look at the lower part of the existing walls – are there still signs of a former floor level? Renewing and lowering a floor may increase the head height and cut down on dampness levels but it may also affect the stability of the walls by reducing the effective depth of the foundations.

9.7.3 Dampness, beetle infestation and wood rot

Basements can be poorly ventilated, damp places where the conditions are often perfect for wood rot and insect attack to begin. The following checklist might be helpful:

- Is the basement adequately ventilated? Use the rule of thumb described in section 9.4.1 as an initial guide. Pay special attention to stagnant areas as many basements are sub-divided into small spaces.
- Are there any timber components built into damp walls? Major timber supporting beams, trimmers and joists may bear into damp brickwork. Wet rot can cause a slow deterioration and compression of the timber leading to settlement of any supported structure above. It is important to test these features with the moisture meter and prodding with a sharp implement for signs of rot. Other joinery item can also be vulnerable such as door frames and linings, cupboards, shelf brackets and fixing plates. Stored timber can also start a dry rot outbreak.
- Is the basement ceiling underdrawn? If yes, what is its condition? Any finish to the underside of the floor above will restrict the inspection that must be reported to the client. It will also restrict ventilation to the floor structure and increase the risk of wood rot occurring. It is common practice to recommend that underdrawing be removed. This has two main implications:

  - It may affect the fire resistance of the building as a whole. This is important for flatted accommodation and houses that have three storeys or more. If a fire starts in a basement it can quickly spread to the ground floor and prevent escape for occupants at higher levels. Therefore, in these cases, removal of any underdrawing needs to be carefully considered.
  - The amount of uncomfortable drafts and heat loss may increase with the removal of the underdrawing. It is sensible to advise draughtproofing the floor with hardboard and insulating from beneath.

- What is the level of dampness? *All* basements are damp. Very few are 'dry' and fewer still have been adequately waterproofed. It is a matter of assessing the scale of the dampness. In the worse cases the walls and floors may be very damp with many salts showing on the surface. If the basement is below the level of the water table there may be standing water or staining to the floor that suggests water occasionally collects there. Adjacent drains might also leak allowing running water to enter the basement. In some cases this can be foul water that will produce very unhealthy conditions.

- Has the basement been previously treated for dampness? Physical tanking may have been provided to the internal faces of the basement. In some cases this may be a proprietary waterproof cement system. Is this effective? Are there signs of dampness? Do any guarantees or details of the work exist? If yes, is it backed by an insurance scheme?

9.7.4 Access to services

In most basements, piped and cabled services will be visible. It is also the most common location for the electric and gas meters. This provides an excellent opportunity to gain an insight into the condition of these elements. See chapter 10 for more details.

References

BRE (1983). 'Suspended timber ground floors: chipboard flooring – specification'. *Defect Action Sheet 31*. Building Research Establishment, Garston, Watford.

BRE (1984). 'Suspended timber ground floors: joist hangers in masonry walls – specification'. *Defect Action Sheet 57*. Building Research Establishment, Garston, Watford.

BRE (1986a). 'Plasterboard ceilings for direct decoration: nogging and fixing – specification'. *Defect Action Sheet 73*. Building Research Establishment, Garston, Watford.

BRE (1986b). 'Suspended timber ground floors: remedying dampness due to inadequate ventliation'. *Defect Action Sheet 73*. Building Research Establishment, Garston, Watford.

BRE (1987a). 'Suspended timber floors: notching and drilling of joists'. *Defect Action Sheet 99*. Building Research Establishment, Garston, Watford.

BRE (1987b). 'Chimney stacks: taking out of service'. *Defect Action Sheet 93*. Building Research Establishment, Garston, Watford.

BRE (1988). 'Solid floors: water and heating pipes in screeds'. *Defect Action Sheet 120*. Building Research Establishment, Garston, Watford.

BRE (1990). 'Assessment of damage to low rise buildings'. *Digest 251*. Building Research Establishment, Garston, Watford.

BRE (1996). 'Acid resistance of concrete in the ground'. *Digest 363*. Building Research Establishment, Garston, Watford.

BRE (1997a). 'Domestic floors – assessing them for replacement or repair'. *Good Building Guide 28, Part 3*. Building Research Establishment, Garston, Watford.

BRE (1997b). 'Domestic floors – assessing them for replacement or repair: concrete floors, screeds and finishes'. *Good Building Guide 28, Part 2*. Building Research Establishment, Garston, Watford.

Douglas, J. (1997). 'The development of ground floor constructions: Part II'. *Structural Survey*, Vol. 15, No. 4, pp 151–56.

HAPM (1991). *Defect Avoidance Manual – New Build*. Housing Association Property Mutual, London.

Harris, D.J. (1995). 'Moisture beneath suspended timber floors'. *Structural Survey*, Vol. 13, No. 3.

HMSO (1991a). Building Regulations 1991. Approved Document A, *Structure*, p.44. HMSO, London

HMSO (1991b). Building Regulations 1991. Approved Document C, *Site Preparation and resistance to moisture*. HMSO, London.

HMSO (1991c). Building Regulations 1991. Approved Document D, *Resistance to the passage of sound*. HMSO, London.

Marshall, D., Worthing, D., Heath, R. (1998). *Understanding Housing Defects*. Estates Gazette, London.
Melville, I.A., Gordon, I.A. (1992). *Structural Surveys of Dwelling Houses*. Estates Gazette, London.
Smith v. *Bush* (1990). 1 AC 831.
Sneesby and another v. *Goldings* (1995). 36 EGCS 137.

Chapter 10

Building services

10.1 Introduction

The building services within a domestic dwelling receive little attention from most surveyors. Lack of knowledge and confidence results in referrals to so-called 'specialists'. Yet in modern houses up to 25% of the value of the property may be attributable to services alone and defects in these systems also have an immediate impact on the building user. Lights that do not work or boilers that fail to warm the house will get new owners critically thumbing through a survey report quicker than most other defects.

The condition of the service systems can also affect the health, safety and well-being of the occupants. Gas explosions, fires, carbon monoxide poisonings and electrocutions can kill and so a surveyor has a moral duty (as well as a legal one) to give appropriate warnings where the signs can be detected.

Despite this importance few surveyors give more than superficial advice to their clients. This section aims to challenge that approach. Within the professional context of building inspections and surveys the following chapter gives targeted guidance that will help surveyors give practical advice to their clients.

IMPORTANT ADVICE: This chapter contains a number of summaries of complex guidance and advice relating to the health and safety of service systems in residential buildings. The objective is to help surveyors identify key and obvious defects. Care must always be exercised when giving advice to clients using the information in this chapter because of its general nature.

10.1.1 Extent of inspection

Because of the specialist nature of services within buildings, the standard guidance given by professional institutions restricts the extent to which surveyors should comment.

Mortgage valuations

Under this inspection the surveyor should identify whether or not there are gas, electricity, central heating, plumbing and drainage services to the dwelling. It is clearly stated that the testing of services is not undertaken. Because only factors that materially affect value should be reported, unless comprehensive replacement of a service system is required then many repairs will not be mentioned. Even so, where the cost of repair is minimal there may be considerable risk to the health and safety of the user. A faulty light switch is just one example. Is this an issue for the lender or the buyer? This will be a matter of professional judgement for the surveyor.

Homebuyer survey and valuation

Taking an overview of the most recent version of this type of survey, the extent of inspection includes the following:

- Services are inspected but no tests to assess the efficiency of the systems are carried out.
- Only significant defects readily apparent from a visual inspection should be reported.
- Leisure facilities such as swimming pools and saunas are not inspected.
- In flats drainage, lifts and security are not looked at.
- For drainage systems:
    - the chambers should be inspected for damage and blockage;
    - a note should be made of trees close to the drain runs;
    - no comments are made on the condition of cesspools and septic tanks but customers should be warned to ask the local authority about the cost of emptying. If the condition is a cause for concern then the tank will need emptying and testing;
    - the surveyor should determine whether surface water drains to a soakaway or main drains.

The guidance is more helpful on issues of health and safety. Under what constitutes an urgent repair the guidance refers to '... matters that are defects judged to be an actual or developing threat ... to personal safety'. Mention is also made of urgent issues that present a danger to safety but are not materially significant. A broken power point is specifically mentioned here (RICS 1997, p.15).

These descriptions can begin to identify the limits of services inspections under each type of commission. The rest of this chapter gives more practical examples.

10.1.2 Problems in reporting on services

The introduction to this chapter suggested that many reports offer little or no comment on the services. Many surveyors seem to prefer to leave it all to the specialists. With mortgage valuations this might be acceptable but for a Homebuyers survey and building survey a more proactive service is required. Many customers complain that reports already contain enough referrals without adding a long list of plumbers, heating and electrical engineers that will result in additional expense and considerable time delays. When a customer has paid out a lot of money for a *survey* rather than an inspection (which they usually get *free* anyway), they may find this situation unacceptable. See section 3.8.2 for a broader discussion on the role of specialist referrals.

If surveyors do take a more proactive approach to reporting on building services, defining their own skill boundaries becomes critical. Deciding what can and cannot be said and properly identifying the point where further investigations are needed is an important judgement.

10.1.3 The specialists

Once a decision has been taken to call for further investigations the next step is to decide who to advise the client to go to. Although this is not part of a surveyor's duty it is an important role. This is because almost anyone can call themselves an engineer, surveyor, electrician, plumber, etc. This can expose the customers to exploitation at the hands of unqualified and sometimes unscrupulous operators. A few examples of reputable organisations are outlined below:

- **Electricians** – National Inspection Council for Electrical Installation Contracting (NICEIC); Electrical Contractors Association (ECA).
- **Heating Engineers** – Confederation for the Registration of Gas Installers (CORGI); Heating and Ventilating Contractors Association (HVCA).
- **Plumbers** – Institute of Plumbing: The National Association of Plumbing, Heating and Mechanical Services Contractors.

An inspection by a Building Services Engineer could also be recommended. They have a professional status similar to Chartered Surveyors but only dwellings with complex or high value services would justify the fee involved. All of these named organisations exercise a certain amount of control over their members and have limited compensation schemes.

10.1.4 What the specialists do

When a specialist investigation is recommended in a report most surveyors are unaware of the outcome both in terms of the scale of the deficiencies or the repair costs. Awareness of the type of testing procedures that the specialists carry out helps to give surveyors a broader appreciation of the issue. For example take gas and electrical services.

Electrical services

The Institute of Electrical Engineers (IEE 1992) recommend a visual inspection and testing of an electrical system. As some of these tests might cause damage to vendor's property (especially where socket covers are removed) it is important to gain prior approval from the vendor. The IEE goes on to state that periodic testing should be considered in the following situations to verify continued compliance with regulations:

- where there has been a change of ownership or tenancy;
- after a change of use;
- following alterations or additions to the original installations;
- a change in electrical loading, i.e. more electrical equipment added;
- reason to believe that damage has been caused to the system, e.g. by a fire or flood.

The guidance note also suggests that domestic properties should be tested every 5 years even where none of the above changes have occurred.

The inspection that the electrician carries out takes into account:

- safety
- wear and tear
- corrosion
- damage
- excessive loading
- age
- external influences, e.g. building alterations.

The IEE also issue a number of recommended forms and schedules that are used to record the outcome of these investigations. These should have been retained by the vendor as proof that the work has been done. This criteria suggests the testing of the electrical systems should be done more regularly than most surveyors would normally recommend. The biggest change would be the need for tests on the change of ownership. If this advice was to be followed an electrical test would have to be recommended in every survey report. Taking a balanced view of this and applying the philosophy developed in this book, an electrical test should only be recommended if:

- any 'trails of suspicion' have been spotted during the survey; and
- the system has been installed over 5 years ago and has not been tested since.

In all cases the client should be advised to have a test done if they want to be 100% sure that the electrical system is safe.

Gas services

Gas pipes are usually tested by subjecting the system to twice operating pressure and monitoring with a pressure gauge. If any drop in pressure is revealed then the leak must be traced and repaired. This could be difficult if gas pipework is in concealed spaces requiring floorboards to be lifted or boxing to be removed and so has implications for the vendor. A test for gas soundness must be carried out:

- whenever a smell of gas is noticed;
- whenever work is carried out on a fitting that might affect its gas soundness;
- if a gas escape is suspected;
- before restoring a gas supply once it has been cut off;
- on installation pipework before fitting of a gas meter;
- on the original installation prior to fitting any extension;
- on completion of a major extension to the system.

The rest of the chapter will look at the main service systems common to domestic dwellings and will highlight some of the critical features that will help with the inspection process. A basic knowledge of the building service systems is assumed.

10.2 Cold water supply systems

The regulations governing the installation of piped water services in dwellings are the Model Water Bylaws 1986 Edition that apply throughout the UK. To ensure a uniformity of approach a guide to the bylaws has been produced by the Bylaws Guidance Panel (WRC 1989).

10.2.1 Mains supply systems

Most dwellings are supplied with drinking water by a supply pipe that is connected to the mains of one of the water companies in this country. The normal method of connection is via a 13 mm communication pipe up to the boundary of the site where a stopvalve or stopcock is located. This is usually in a box or chamber with a small cast iron hinged lid at ground level. This is where a water meter will be located. From this point onwards, the supply pipe usually becomes the responsibility of the owner. It continues across the property and rises up just inside the dwelling where a second stopvalve should be fitted (see figure 10.1).

Key inspection indicators

- The main stop valve should be located so that it can be 'readily examined, maintained and operated' (WRC 1989, p.161). The hinged lid should work and the chamber should be clear of all debris so the handle of the stopvalve can be seen. If the stopcock cannot be located and turned off quickly then a small leak in the house could easily become a disaster.
- The stopvalve should be at a depth of 750 mm or more to avoid the effects of frost action. If it is less than this then the supply pipe should be insulated.
- The supply pipe should not pass through an inspection chamber, refuse chute, sewer, drain or cesspool. This could lead to possible contamination of the water supply.

10.2.2 Water from springs and wells

Dwellings in remote areas may take water from springs, wells or private boreholes. Where this is taken for the sole use of the household and no water comes from water company mains then the national Water Bylaws do not apply. Where dwellings use

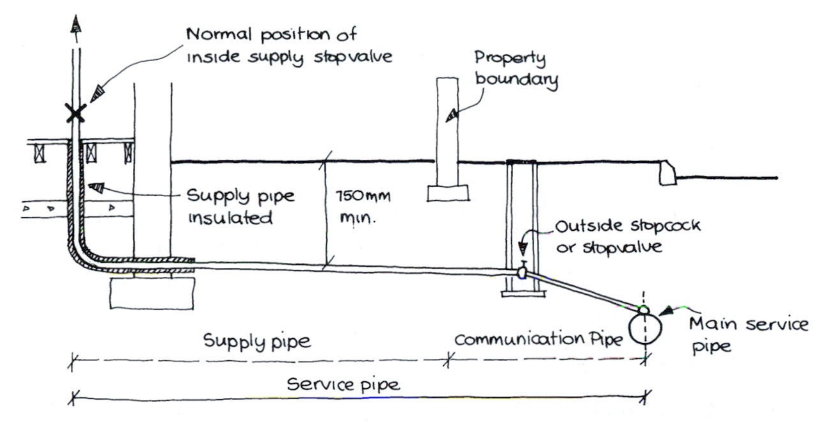

Figure 10.1 Diagram showing rising main from the boundary of a property to the main stopcock.

spring or well water and mains water the Water Bylaws would apply to both systems. Assessing the adequacy of the water supply from these alternative sources is a specialist skill requiring additional investigations by appropriate organisations.

Key inspection indicators

- Contamination of the supply is the main problem and in this respect water collected on the surface is more problematic than from a well or a borehole.
- Continuity of supply is another concern especially during periods of prolonged drought. Is there any evidence or information that suggests the supplies run short?
- Even if supplies are adequate, some form of local water treatment may still be required. This could involve small scale settlement tanks or ultra violet treatment equipment. These are specialist installations that require an electrical supply.

Further advice on the suitability of private supplies can be obtained from:

- the Environmental Health department of the local authority in relation to testing of the water supply for potability;
- specialist contractors who sink wells, boreholes and provide purification systems;
- the local water company if a back-up mains supply is provided.

The National Trust has published a useful guide entitled *Private Water Supplies – a Guide to the Design, Management and Maintenance* (National Trust 1996) that surveyors may find useful. The statutory instrument no. 2790 'Private Water Supplies Regulation' (HMSO 1991) outlines the legal requirements.

Private water supplies can have significant financial and health implications for an owner. Because the assessment of suitability of the supply is difficult to make without specialist knowledge, referrals for further investigations might be necessary in most cases.

10.2.3 Rising main – internally

The supply pipe usually rises up in a corner of the dwelling and the owner's stopvalve should be positioned just above the ground level. Ideally the main should be insulated down to a depth of 750 mm although in older properties this is rare (see figure 10.1).

Key inspection indicators

- Is the stopvalve in a convenient position and easy to operate in an emergency?
- The supply pipe should be insulated where it passes through an unheated space such as a sub-floor void or a cellar where there is risk of frost damage (BRE 1987a).
- Is there any evidence of leaks to the valve or associated pipework? Small leaks often occur along the spindle that connects the handle to the main body of the valve. This can lead to an outbreak of dry or wet rot in adjacent timbers.
- Is there any evidence of excess condensation run-off? Supply pipes are very cold for most of the year and condensation can occur on the surface of the pipe. Rot can occur in this situation as well.

- Is there a change of materials in the piped water systems? Dissimilar metals connected in the same water system may lead to corrosion through galvanic action. The following sequence is acceptable as long as it follows the direction of flow:
    - galvanised steel
    - uncoated iron
    - lead
    - copper.

  If three or more of these are present, further investigations may be warranted.

10.2.4 Lead pipes in buildings

In many older properties, lead is found in a number of locations (see figure 10.2):

- lead mains
- lead-lined pipes
- lead-lined water cisterns
- lead/tin solder in joints of capillary pipe fittings.

Lead pipes can be identified by the following features:

- a dull grey when unpainted;
- bulky pipes large in overall diameter when compared with copper pipework of the same bore;
- bulbous joints where lengths of pipes are connected;
- because lead pipes are heavy they will often sag between supports. This can put a strain on connections and joints that can result in leaks;
- because it is a soft metal, lead pipes can be easily scratched with a screwdriver.

Figure 10.2 Lead water pipes are clearly thicker than other types. The bulbous joints shown in this photograph are typical of this material.

Apart from problems of durability, lead can be harmful to health. Water can dissolve the lead allowing it to be ingested into the human body. The Water Bylaws (WRC 1989, p.155) do not require the removal of lead pipes but does control the repair and alteration of them:

- Lead should not be used in either the repair or replacement of lead pipe systems. This includes lead solder as well as pipework.
- Copper pipe should not be used in the repair of lead systems because this will result in galvanic action. Where a lead pipe system has to be repaired, the National Water Bylaws recommend the use of plastic pipes with appropriate connectors.
- If water storage tanks are lead lined then they should be replaced as soon as possible as these do pose a high health risk.

If a customer is concerned about lead levels in their water supply, the local water company has a duty to respond. In most cases they will test the lead levels and report back on the outcome. If the customer is still concerned then the water company should agree to replace the service pipe from the road to the boundary as long as the owner replaces the remaining lead supply pipes up to the main stop valve inside the property.

In summary, although lead does not legally have to be replaced many customers may be concerned about the effects on their health. These worries combined with the durability problems of lead systems makes it sensible to advise:

- where lead piping forms part of the internal cold water system it should be replaced as soon as is practically possible;
- where it is restricted to the main communication pipe the priority would not be so high.

In either case if the customer is concerned by the presence of lead they should be advised to contact the local water company.

10.2.5 Cold water storage systems

Once in the building, the rising main should then serve the cold water supply system. There are two types:

- **The direct system** – where all the cold water draw-off points are fed directly off the main. The only cold water storage tank is usually one that supplies the hot water storage tank (see figure 10.3)
- **The indirect system** – where only the kitchen sink has a mains connection and the rest of the cold water draw-off points are supplied from a much larger cold water storage tank (see figure 10.4).

Both are still allowed and the indirect system is the most common modern system. The relative merits of each system are compared below.

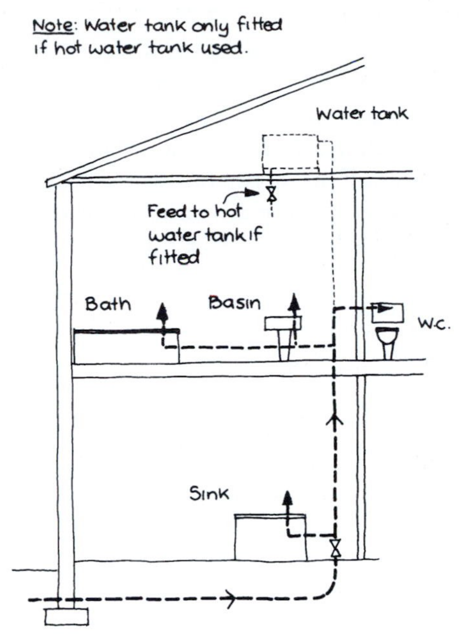

Figure 10.3 Diagram showing direct cold water system from the stopcock to cold water storage tank (if fitted).

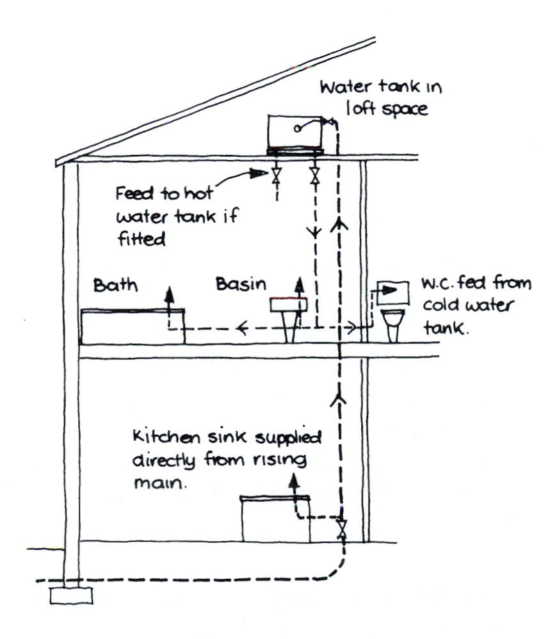

Figure 10.4 Diagram showing indirect cold water system from the stopcock to cold water storage tank.

Direct system

- Drinking water at every cold water draw-off point.
- Smaller diameter cold water pipes, shorter runs and so less pipework.
- No water storage if the water supply is interrupted.
- Water pressure will drop when different draw-off points are used at the same time. This is especially important if instantaneous water heaters are used. In these cases the pressure of the hot water can drop dramatically when more than one tap is used.
- The higher pressure will result in excessive noise and wear on taps and valves.

Indirect system

- Usually only one drinking-water point.
- Water storage during interruption of supply.
- Less noisy systems because of reduced pressure.
- A need to accommodate and support a large storage tank.
- Larger diameter pipes and longer pipe runs.

The different component parts of a cold water system are described below.

Cold water storage tank

Both the direct and indirect systems will usually have a cold water storage tank. The difference is capacity. Normally the indirect system will have a 227-litre capacity with the direct system matching the size of the hot water storage tank (usually 112 litres).

The tank is often located in the loft space although it could be contained in a cupboard at high level in one of the upper rooms. Assessing the adequacy and condition of this tank is important. A flood of 227 litres through a house can cause extensive damage!

KEY INSPECTION INDICATORS

- Tank material:
  - **Galvanised steel** – these tanks may be badly corroded and at the end of their useful life. Look for rust around pipe connections and within the tank itself.
  - **Asbestos cement** – used in older installations. In water asbestos poses no danger. The outside face of the tank should be checked to ensure it is not breaking up or 'dusting' (see later section 11.5 for more guidance).
  - **Plastic** – modern types of tank that can be either round or rectangular. Although these are more durable they still have to be properly installed. One of the most common defects is a lack of rigidity to the tank wall where the rising main is connected to the ball valve. A metal reinforcing plate is usually required at this point.
- The tank should be properly supported on a plywood base that is at least 19 mm thick. Additionally:
  - trussed rafter roofs – there is a specific arrangement of timber bearers that spreads the load over at least four trussed rafters (see figure 7.13);

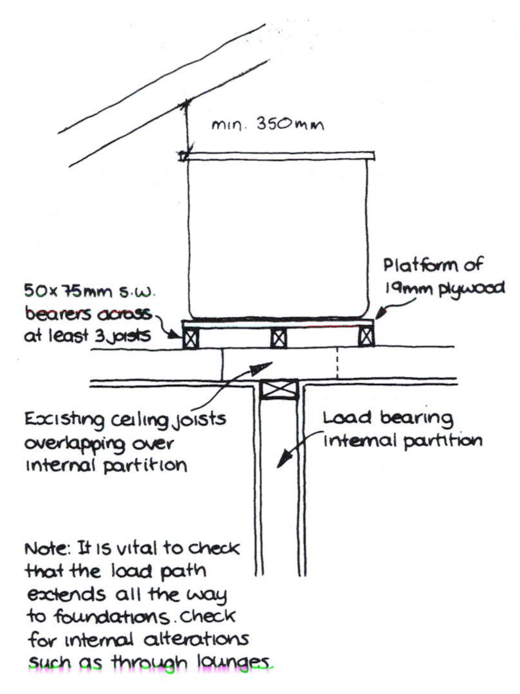

Figure 10.5 Support for a cold water storage tank in a traditionally built roof.

- traditional roof structures – 50 × 75 mm bearers spread over at least three or four ceiling joists and positioned as close as possible to a supporting wall below. Make sure that this wall has a load path down to the foundations. Partitions removed at ground level can often result in problems higher up! (See figure 10.5.)

- The tank should be insulated against frost and heat:
  - all sides and the top of the tank should be fully insulated to prevent freezing. Where the tank is over a heated space the insulation should be omitted from beneath;
  - all pipes to and from the tank should be insulated;
  - as a general guide the insulation should be at least 40 mm thick.

- The tank should have a tight-fitting lid to stop any debris getting into the water. There should be at least 350 mm clear space above the tank to allow for maintenance and cleansing (see figure 10.5).
- The tank should have a warning pipe made of rigid material (usually plastic) that discharges in a conspicuous position externally (BRE 1990a).
- If easily accessible the ballcock should be checked to see if it is working properly. This can be done by pressing down the float to below the water level and releasing. This will show how quickly the flow is stopped when the float is let go.
- The pipes to and from the tank should be properly supported. If not, they can vibrate and possibly damage connections and joints.

10.2.6 Cold water pipework

The cold water distribution pipework can be in any of the following materials:

- copper – the most common type of pipework. Joints can be mechanically made (compression joints) or soldered. New joint fittings should be stamped with 'SS' and the British Standard Kite mark to show that silver solder has been used (i.e. no lead);
- galvanised steel – generally no longer used on domestic installations;
- unplasticised PVC – becoming more common in modern housing. Joints are solvent welded;
- polyethylene – usually used for the mains supply from the outside stopcock into the house. This type is blue in colour. This can often be seen in basements.

Key inspection indicators

- The pipework should be properly clipped and supported to prevent water hammer. This occurs when water pressure rises or falls sharply (when a tap is turned on and off quickly) and sets up a reverberation in the pipes. In the worst cases it will cause them to repeatedly 'bang' against neighbouring pipes or floor joists, etc. The noise can be very loud and carry on for several minutes. Water hammer can often be cured by fitting special mechanical water hammer arrestors. These are small cylinders the size of a cricket ball that can absorb any pressure surges.
- The pipework should be properly insulated where it passes through unheated areas;
- Servicing valves should be fitted to the pipework serving float operating valves. These are 'in-line' valves that are operated with a screwdriver rather than a handle and allow the cistern or tank to be repaired. It is good practice to have these servicing valves on all sanitary fittings and appliances so taps and valves can be easily maintained.
- Always check carefully for leaks especially at the connections with cold water tanks, hot water cylinders, WC cisterns etc.

10.3 Hot water supply systems

The hot water in a dwelling can be supplied by a number of different methods.

10.3.1 From the main heating boiler

This can be through either the direct or indirect hot water systems.

Direct systems

These are generally associated with older installations and were often used to heat one or two radiators as well as the hot water. The main limitations include:

- Because the hot water circulates directly between the heating boiler, the storage cylinder and in some cases the radiators, the pipes can easily become furred up reducing the effectiveness of the system.

- When hot water is being used the amount of heat going to the radiators is correspondingly low. This makes the space heating erratic and often inadequate.
- Direct systems are outdated and have a very limited useful life. Clients should be advised to budget for replacement in the short to medium term.

Indirect systems

These are more common with newer installations and overcome the disadvantages of the direct system through the separation of the space heating and water heating circuits (see figure 10.6):

- The incidence of furring is greatly reduced because the hot water between the boiler, radiators and the hot water tank (called the primary system) is continuously circulating and sealed.
- The space heating is unaffected by hot water draw off and so is more effective.
- To keep the primary system topped up with water, a smaller feed and expansion tank is provided. This is usually found close to the main cold water storage tank and should have similar features, i.e. rigid cover, ball valve, insulation, warning pipe, adequate support, etc. To allow for expansion of water and steam from the primary circuit an expansion pipe discharges over the feed and expansion tank.
- The energy efficiency of the system can be improved by insulating the pipework between the boiler and the hot water tank although this is rarely done in practice.

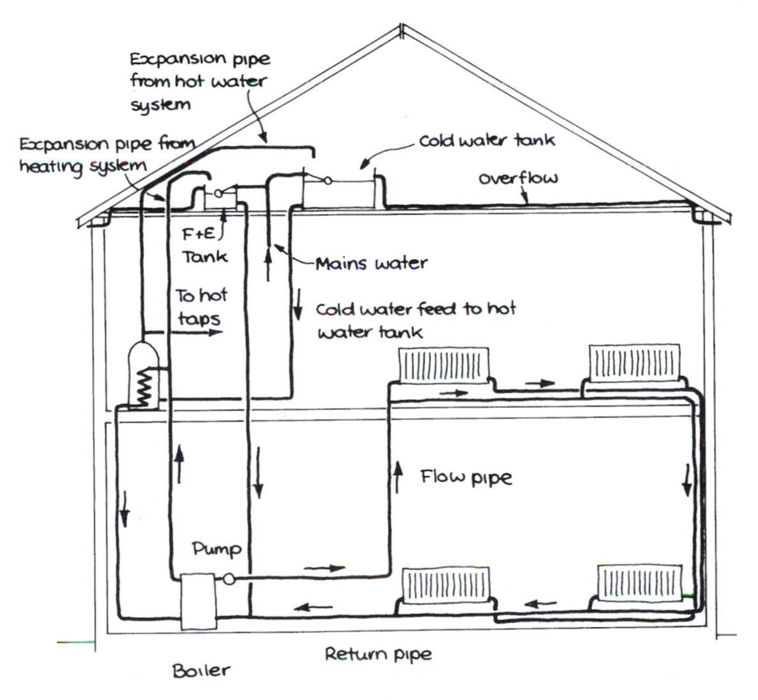

Figure 10.6 Indirect hot water and central heating system.

HOT WATER CYLINDER

One of the main components in the indirect hot water system. The key factors to investigate include:

- It should be close to the hot water heating source and be properly insulated. Sprayed foam insulation that is factory applied is the most effective. If not then a well-fitted insulating jacket a minimum of 40 mm thickness should be provided;
- An open expansion pipe should be taken off from the top of the cylinder and discharged over the cold water storage tank through a hole in the insulated lid. It is vital that this expansion pipe is well insulated as water has been known to freeze within it.
- It is important to check that the right expansion pipe discharges over the correct tank. For example the expansion pipe from the primary circuit of an indirect system should only discharge over the smaller feed and expansion tank. This is because the water will contain a rust inhibitor that should not be allowed to contaminate water in the main storage tank.
- Most hot water cylinders have an electric immersion heater to provide hot water when the boiler is switched off. It is better to have two: a smaller one towards the top for short draw-offs with a larger one towards the bottom for baths, etc.
- Where possible cylinders should be checked for leaks especially where the pipework connects to the tank and around the immersion heater seals.
- Cylinders should be fitted with a thermostat that regulates the temperature of the hot water by controlling the main heating boiler. This is normally positioned at the front of the cylinder about half way down and will have an electric cable connection.

10.3.2 Independent gas water heaters

These are separate gas heaters that provide hot water and can be classified as follows:

- **Small gas water heaters** – These are usually installed near a kitchen sink or a wash hand basin and are flueless. The combustion gases discharge into the room itself. Although these are still permissible they are rarely used. Where local water heaters are required instantaneous electric types are most commonly used.
- **Large gas water heaters** – normally intended to supply a bath and a number of hot water draw off points. Sometimes called a 'multi-point', these have to be fitted with a flue that discharges externally and are usually room-sealed. These should be assessed in a similar manner to heating boilers.
- **Storage gas water heaters** – these are usually described as gas circulators and provide a hot water supply either directly or indirectly to a thermal store. To make sure that the system is protected a thermostatic control must be provided. They are rarely used on new properties.

10.3.3 Combination boilers

This is the term usually applied to main heating boilers that also supply instantaneous hot water on a similar basis to a gas multi-point. This system is described in more detail in section 10.8.

10.3.4 Electrical storage systems

Two main types are most common:

- **On-peak electric water heating** – this is usually via an immersion heater(s) in a hot water cylinder. Better installations will have two immersion heaters. This type of hot water heating is very expensive and it essential that the tank is well insulated (BRE 1987b).
- **Off-peak electric water heating** – time controlled and linked to overnight cheap-rate electricity. The tanks are usually much bigger and must be well insulated to ensure that hot water is available throughout the day.

10.3.5 Hot water pipework

In an effort to conserve energy, hot water pipes in new installations should be insulated. This is laid down in Part L of the Building Regulations (HMSO 1991a). Although these provisions do not apply to existing systems they can serve as a benchmark. The main requirements are summarised below:

- All pipe insulation should be at least 15 mm thick.
- Hot water pipes connected to hot water storage tanks should be insulated for at least 1 m back from the tank.
- Hot water pipes in unheated areas should be insulated against freezing.

One of the most useful tests of any hot water system is to run the hot water tap at the fitting that is furthest from the heating boiler or storage cylinder. Care must be taken when making judgements but factors could include:

- Is the flow of water adequate? Does it come out at a reasonable pressure?
- Does it reach an adequate temperature within a reasonable period of time (say 15 seconds)?
- Is the water discoloured? If yes, this could indicate corrosion problems with the heating boiler especially if it is a direct system.

10.4 Sanitary fittings

Surveyors can be forgiven if they sometimes think clients are only really interested in the colour of the bathroom suite of their future home. Yet the condition and suitability of the sanitary fittings can be very important. It is an area that many past owners may have 'done it themselves' and consequently left disastrous legacies! Although sanitary fittings are of secondary importance, replacing defective suites can be very expensive. For the purposes of this section 'sanitary fittings' include WCs, wash basins, baths, showers, bidets and even the kitchen sink.

Key inspection indicators

- Are they good-quality matching fittings that have a long life or are they the cheaper versions that will quickly become tarnished? For example cast iron or heavy gauge steel baths are often more robust than many light steel or plastic alternatives.
- Are the fittings in good condition? Are they chipped, cracked and/or crazed? What is the condition of the vitreous enamelling? Cracked and chipped WC pans will harbour germs and will be unhealthy. The enamelling to baths and basins may have been badly stained by leaking taps, etc. The enamelling to baths and wash basins may have been renewed in situ by the application of a resin coating by a specialist contractor. Although these often come with guarantees the long-term durability will always be in doubt.
- Are the fittings well fixed? Are they properly supported? Basins, baths and WCs have to support human beings of all shapes and sizes which can lead to intense loadings! If the fittings deflect or move they can impose a strain on the water and waste connections resulting in leaks and damage.
- Check the adequacy of the water seal between the fitting and the splash-back tiles. This is especially important around showers where large amounts of water can seep behind the tray and cause wood rot. Carefully inspect the ceiling and wall finishes **below** the bathroom for signs of leaks. Be suspicious if ceilings have recently been redecorated.
- Carefully check all the plumbing and waste connections for leaks. If time is available fill the fittings with water and pull the plug out to see if there are any leaks on the trap and wastes. Flush the WCs and carefully watch for any leaks at the junction of the pan and the outgo. Often a carefully placed 'drip catching' plastic tray can be seen! (See figure 10.7.)
- Check sanitary fittings that have been fitted in bedrooms. Showers and vanity basins are particularly problematic in this respect. These can be remote from the main drainage points resulting in long runs of waste pipes with an inadequate fall which can easily lead to blockages.

10.5 Heating systems – general guidance

10.5.1 Introduction

Being warm in a dwelling is a high priority for most people. This is an area fraught with difficulties for a surveyor as assessing the adequacy of any heating system is often beyond skill boundaries. Section 5.8 looked at this issue from an energy efficiency point of view; this section will give advice on the condition, functionality and safety of the heating system.

10.5.2 Need for a specialist inspection

Clients should always be advised to have the gas services professionally tested when they move into a property. This is because:

Figure 10.7 Leaks behind the WC are surprisingly common. A variety of 'drip bowls' can give vital clues to developing problems. The risk of wet or dry rot in this type of location is very high.

- the service may have been turned off for a period resulting in malfunctioning of mechanical parts;
- changes may have been carried out by the previous owner in a DIY fashion;
- to ensure the health and safety of the occupants are protected.

Accepting this as general advice there are a number of other features that a surveyor may be expected to report on. Before considering these in detail it is important to review two key issues that determine whether a gas system is safe or not – flues and ventilation. If the products of combustion are not removed effectively and replaced by fresh air the appliance will not work properly and will also pose a danger to the occupants. The client will be generally be concerned about health first and then cost implications of putting the system in working order. The lender's level of interest on the other hand will be governed by other criteria. If there is no impact on value then the lenders will have no interest. The surveyor has to balance the moral duty of helping to protect the health and safety of their clients with the duty to the lender.

10.6 Flues

10.6.1 Inspection of flues

Although surveyors are not expected to check the '... internal condition of any chimney, boiler or other flue' (RICS 1997, p.33) a background knowledge of the different types of flue will help in the assessment of the various systems.

10.6.2 Types of flue

There are a number of different kinds:

1) Individual open flues – this is the simplest type that gas appliances can discharge into:

   - A traditional chimney system – this must be swept and free of all dampers or restrictors, etc. Before a traditional chimney can be used it has to be checked visually and subjected to smoke tests to ensure that combustion products will not leak from the chimney into other spaces or spill back into the room. This can only be carried out by a CORGI registered contractor. The opening in the fireplace around the appliance needs to be properly sealed.
   - Flue liners – where the chimney is not suitable, a preformed metal flue liner is installed in the chimney. The decision whether one is required depends on:

     - type of appliance
     - condition of the chimney and its height
     - manufacturer's requirements.

A flue liner should be made of flexible stainless steel. It must be in one complete length, top supported and be fitted with an appropriate terminal fitting. It should not be used external to the chimney.

NOTE – If a gas appliance has to be replaced then it is likely that the existings stainless steel liner will have to be replaced as well. This is because the life of a liner is limited to 10–15 years and most new gas appliances will last much longer than this.

It is not always possible to assess whether a gas appliance is fitted with a flue liner. Even if it has been then the adequacy of the installation cannot be assumed. If the gas appliance has not been fitted with a liner then the key question is, does it need one? In either case the key features to look for include:

- Is there a metal terminal or cowling on the chimney stack? This could provide some evidence that work has been undertaken.
- Are there any brown watery stains on the chimney breast within the individual rooms? This could indicate that condensation of the combustion gases is occurring in the chimney flue.
- Inspect the chimney in the loft space. If there are large amounts of white crystal deposits on the face of the brickwork that could indicate condensation in the flue. Check for any cracks, gaps or missing bricks in the chimney that might allow gases to escape.

2) Flue pipes – some appliances are fitted with flue pipes. These have spigot and socket joints that are sealed and caulked using fibre rope and fire cement. These flues are often made from a cement based materials and in older properties the likelihood of an asbestos content is high. Key inspection indicators will include:

   - Check that the flue is properly supported at approximately 1.8 m centres (maximum) throughout its height.
   - Is there evidence of cracks or gaps at the joints. Is there any evidence of flue gases escaping, i.e. staining or scorch marks?

- There should be at least 25 mm clearance between the flue and any combustible material, e.g. floorboards, joists, etc. Where the flue passes through storage cupboards (e.g. airing cupboards) there should be a protective guard that gives 25 mm clearance around the pipe. Double-walled flue pipes are better protected so can have smaller clearances.
- Is the flue pipe likely to contain asbestos? Some older flues may have been insulated with an asbestos based quilt insulation often held in position with chicken wire. The risk from this type of asbestos will be high. See section 11.5.2 on dealing with asbestos (see figure 10.8).
- Flue pipes should terminate in positions where the effects of the wind will not affect the performance of the flue.

3) Pre-cast flues – these are constructed out of pre-cast concrete blocks built into the main structure and are slimmer in profile than a brick chimney. Not all appliances can be used with these flues and so manufacturer's instructions need to be checked. Their useful life tends to be similar to conventional chimneys.
4) Balanced flues – these flues are usually fixed directly to the appliance and includes a terminal in the design. It can be natural or fanned draught. The most important feature of balanced flues is the external position so that combustion products can disperse safely at all times. Figure 10.9 illustrates some of the main requirements. It is important to note that this is a summary of guidance given by a number of

Figure 10.8 A flue passing through the loft space of a non-traditional dwelling. It has been insulated with some form of quilt held in position with a wire mesh. The possibility of this being asbestos is high.

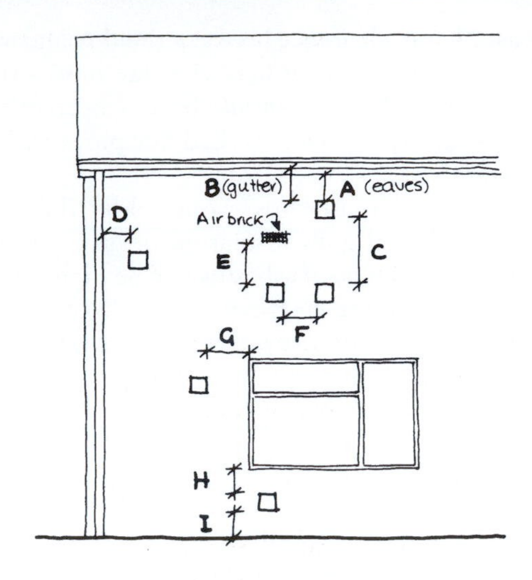

Location	Natural draught	Fanned draught	
A*	300	200	Below the eaves.
B*	300	75	Below a gutter.
C	1500	1500	Above or below another flue.
D	75	75	From a downpipe.
E	300	300	Above or below an airbrick.
F	300	300	From another flue.
G	See manufacturers Instructions		
H	300	300	Below a window.
I	300	300	Above the ground.

Dimensions shown are in mm and are minimums.

*Note: If less than 1000mm away from gutters and 500mm from painted eaves then heatshield to be provided.

Figure 10.9 Positioning of balanced flues on the external wall of a typical dwelling.

different organisations and should be used to judge the general adequacy of the flue position. For more precise guidance see the current CORGI regulations. In addition:

- flues can only discharge into a car port that is open on at least two sides
- flues cannot discharge into covered passages between dwellings;
- any flue that is less than 2 m above a pavement, ground level, balcony, etc. is to be fitted with a wire guard;
- heatshields should be provided where the flue is within 0.5 m of painted surfaces or 1 m of a plastic gutter.

5) Condensing appliances – the development of high efficiency appliances results in lower flue gas temperatures. The combustion products of condensing boilers (including condensed water) are low temperature and mildly acidic and so plastic flue pipes are most common. They are usually 22 mm in diameter and fixed to a slight slope away from the appliance. The pipes should drain to a hopper, soakaway,

gulley or stack pipe (SVP or RWP). If they do they should have a proper trap on the pipe.

10.7 Combustion ventilation

10.7.1 General provisions

This is a very important part of the Gas Safety Regulations (CORGI 1997) and Building Regulations (HMSO 1991b, Part J). They apply to new, existing and replacement installations. Adequate ventilation to spaces where gas appliances are located is necessary for the safety of the users and to ensure the appliances works properly. When assessing the adequacy of ventilation there are a number of general principles:

- Air vents should not be located in areas that are likely to be blocked by leaves, snow, etc.
- The free ventilation area should be measured and not just the overall size of the ventilator itself. For example a 220 × 220 mm terra-cotta airbrick (total area 48.4 cm^2) might only have 23 cm^2 of free ventilation.
- Ventilation to an appliance can be provided through a loft space or underfloor area as long as the space does not communicate with another premises.
- Air vents must be at least 300 mm away from the nearest flue terminal.

10.7.2 Ventilation for open-flued appliances

It is not the job of the surveyor to check whether the amount of fixed ventilation is correct but developing a feel for what is appropriate can help in the assessment process.

Types of ventilation

Open-flued appliances get combustion air from the room. If that room has not got enough free ventilation then life threatening situations can easily develop. It is with these types of appliances that a surveyors judgements can be most critical. There are two types of ventilation:

- **Adventitious ventilation** – a room or space can never be made completely airtight. There will always be leakage through cracks and gaps in the construction. It is safe to assume that this adventitious ventilation is equivalent to a 35 cm^2 of free ventilation in all habitable rooms (CORGI 1997, p.84). This is enough ventilation for a small domestic gas fire. Usually there are no special ventilation requirements for standard appliances that have less than a 7 kW input rating.
- **Purpose provided ventilation** – where gas appliances are rated higher than 7 kW then additional ventilation to a room has to be provided. The general rule is that 4.5 cm^2 of free ventilation is required for every kW of total rated input of the gas appliance above 7 kW. The problem facing the surveyor is that assessing the adequacy of purpose provided ventilation is a skill that is beyond the scope of the standard survey. One of the key difficulties lies in being able to accurately determine the

rating of the gas appliance. Therefore, key 'trails of suspicion' should be looked for. If spotted, then further investigations should be called for.

Open-flued appliances in a room

These types of appliances are the most dangerous to occupants. Therefore particular care should be taken when assessing them:

- A standard gas fire or heater does not require any special ventilation to outside BUT make sure that the appliance does not conceal a more powerful back boiler behind.
- Where there is a water or space heating back boiler in the room then there should be an air brick or another form of permanent ventilation. If there isn't then further investigations should be recommended.

Open-flued appliances in compartments

Gas appliances are often installed in compartments or enclosed spaces such as airing cupboards or understairs spaces. Some owners might have built a small timber-clad structure to hide an ugly boiler! There are special ventilation requirements for these locations that are designed to:

- make sure that the supply of combustion air remains adequate;
- prevent residual heat build-up within the enclosure and keep the surface temperature of the boiler within safe limits;
- stop electronic circuitry from malfunctioning because of high temperatures;
- allow servicing engineers to dismantle the boiler during maintenance works.

Where an open-flued appliance is located in a compartment the requirements are for ventilation to outside air at both high and low levels. Where the compartment is ventilated to the room then a similar arrangement is required but with a much greater free-ventilation area. The air vents in this case must not communicate with bathrooms, bedrooms, bedsitting rooms, showers or private garages.

Many DIY constructed compartments will not meet with these standards. The compartment may also need to be adequately fire-proofed in communal areas. In airing cupboards special separation will be required to ensure stored clothing does not get too close to hot surfaces.

10.7.3 Ventilation for room-sealed appliances

Where the gas appliance is room sealed (i.e. does not get its combustion air from the room but from the outside via a balanced flue) then the following rules apply:

- Where the appliance is positioned in a habitable room then no additional permanent ventilation to the outside is required.

- Where the appliance is positioned in a compartment then permanent ventilation to the external air needs to be provided at both high and low levels.
- As above but where the compartment is vented from a room or an internal space then a greater amount of permanent ventilation is required at high and low levels.

In this context 'compartment' has a similar meaning to that described for open-flued appliances.

10.7.4 Ventilation of flueless appliances

Flueless appliances include gas ovens, hobs, instant water heaters, gas fridges, gas-powered tumble driers, etc. All these are ventilated to the room or space that they are in. Generally speaking the room only needs permanent ventilation if it is below a certain size which is usually taken as 20 cubic metres.

10.7.5 Extract fans and open flues

Studies (Shepard 1993) have shown that extract fans can affect the operation of open-flued appliances when they both serve the same rooms or spaces. Part F of the Building Regulations (HMSO 1991c) does not allow a fan that has an extract rate of 60 litres per second to be in the same space as an open-flued appliance. The BRE suggests if the two have to be in the same space then the extraction rate of the fan should be a maximum of 20 litres per second. As most kitchen fans are rated at over 60 litres per second this may be difficult to achieve. Where this occurs one solution is to provide an additional 50 cm^2 of permanent ventilation over and above that normally required. This sort of detail may be difficult to establish in practice. The main trigger is where an open flued boiler (i.e. not a balanced flue) is positioned in the same room as an extractor fan. In these cases the surveyor will have to investigate further or refer the issue to a heating engineer.

10.8 Location of gas boilers

10.8.1 Typical locations

Boilers can be found in a wide variety of locations in a dwelling. Apart from those normally expected, gas appliances can be fitted in the following locations as long as the described conditions are met:

- **Roof spaces** – room-sealed and open-flued boilers are allowed as long as:
  - There is sufficient vertical head to the feed and expansion tank as specified by the manufacturer.
  - There is adequate and safe flooring to allow for maintenance and servicing.
  - The fixing of the boiler is sufficient to support its weight.
  - Floor mounted boilers are set on at least 12 mm thickness of non-combustible material (but not asbestos!).

- A fixed retractable but permanent loft ladder is fitted with adequate guarding to protect from falling.
- There is fixed lighting in the roof space.
- Stored items are prevented from falling against the boiler by appropriate guarding.

- **external locations** – boilers that are outside the main dwelling must be in a weather-proof compartment that has the following features:

  - The enclosure must meet the construction requirements of the building regulations.
  - The boiler must be able to be isolated by a double-pole switch.
  - The ventilation must match that for a normal compartment or enclosure but with the additional provisions:

    - low-level ventilation must be at least 300 mm above ground level to prevent splash back and flooding;
    - air vent holes must be between 5–10 mm across to prevent blockage or entry by vermin and birds.

  - The door to the enclosure must be lockable to prevent unauthorised access.
  - The gas appliance and associated pipework should be protected against freezing and corrosion.
  - The appliance should be fitted with a frost thermostat that operates the boiler when the temperature falls below 4°C.

It is not suggested that these regulations should be checked off on every occasion but taken as a guide to the standard required for external compartments.

10.8.2 Banned locations for gas appliances

There are a number of locations where the installation of gas appliances is not allowed:

- **Sleeping areas**

  - Appliances over 14 kW must always be room sealed (balanced flue) but because of noise nuisance associated with central heating and water boilers it is always better to site these elsewhere if possible.
  - Below 14 kW appliances (normally gas fires) can be open flued but must be fitted with an automatic shut-down valve that will switch the appliance off if there is a build-up of combustion products. Such devices are rarely found in practice.

Therefore, if normal radiant gas fires are observed in bedrooms then it is likely that they contravene regulations and so should be referred to a specialist.

- **Bathrooms and showers** – only room-sealed appliances are allowed in these areas and even then any electrical switch or control must not be accessible to someone using the bath or the shower.
- **garages** – only room-sealed appliances are recommended for use in private garages.

10.9 Gas heating systems

There is a variety of different types of heating systems fitted in domestic dwellings.

10.9.1 Different types of gas heating systems

These normally include:

- Traditional fully pumped open vented – conventional system where an expansion pipe is taken over the feed and expansion tank located at the highest point of the system (see figure 10.6).
- Sealed systems – this is similar to the above but has a sealed system doing away with the need for the cold feed and expansion components. The expansion vessel is usually separate from the main boiler.
- Thermal storage systems – this is another type of sealed system where the boiler is used as a heat source for a storage vessel that holds a quantity of hot water ready for immediate dispersal into the radiator system. The heat-up times for the radiators can be very quick.
- Combination boilers – sealed systems with integral expansion vessels that provide both space heating and instantaneous hot water.
- Condensing boilers – an increasingly popular system that uses two heat exchanges to extract extra heat from the flue gases. This increases the efficiency of the system and because the combustion gases are cooled down, condensation of the flue gases can occur. This drains out from the boiler through the plastic flue via a waste pipe into the drainage system.
- Warm air heating – these were common during the 1960s. If one is encountered it will either be at the end of its useful life or a replacement unit. The air is warmed via a heat exchanger and fanned around a duct system within the dwelling. The stub stack system is more restricted and will heat only those rooms directly adjacent to the boiler itself.
- Individual gas room heaters – individual gas fires can be either open flued or room sealed (balanced flued). Condensing gas fires are also becoming more common.

10.9.2 Suitability of gas heating systems

The detailed assessment of a heating system can only be carried out by an appropriately qualified heating engineer. Despite this there are still a number of key features that a surveyor can comment on.

Key inspection indicators

The most important factors are:

- **position of the flues** – are these located in the correct positions, etc.;
- **provision of combustion ventilation** – does the appliance have the right sort of ventilation;
- **adequacy of the heating system**.

One of the key questions is whether the heating system is capable of raising the internal temperatures to appropriate levels. Detailed information about the construction and exposure of the house, climate, etc. will all influence how big the boiler has to be. The aim is to produce the following average internal temperatures:

- lounge 21°C
- bedrooms 18°C
- bathroom 18°C

Anything below these levels will be considered too cool for most people. A standard survey will not reveal enough detail to judge whether the system will be adequate or not. Radiator size for example will influence the temperature achieved in a space even if the overall capacity of the boiler is adequate. A robust rule of thumb involves matching heating capacity of the boiler to common dwelling types (see figure 10.10). It must be emphasised that this is an approximate guide only to give the surveyor a 'feel' for what is adequate. If the boiler is clearly below these levels then there is clear justification for referring the matter to a heating engineer.

10.9.3 Control of gas heating systems

The more efficient heating systems are fitted with controls that ensure the whole installation heats up quickly when heat is required and shuts down when it is not. That is one reason why electric storage radiators and solid fuel systems cannot match the efficiency of gas. There has been a number of new developments in this area as heating systems are becoming more sophisticated:

- **Programmers** – these vary from simple time clocks with four on and off settings through to digital versions that can be programmed to suit a wide variety of lifestyles. The key issue here is usability. Some programmers can be far too complicated for the users.
- **Thermostats** – a room thermostat, usually located in the hall, makes sure the heating system shuts down when a pre-set temperature is reached. Because this is based on temperatures in just one part of the house, the system often continues to produce heat in the more remote parts of the house even if it is not needed.
- **Thermostatic radiator valves** (TRVs) are a better way to ensure that the temperature in individual rooms does not rise too high. Sensors in the valve shut the radiator down at preset temperatures avoiding waste of heat.
- **Boiler managers** – these systems are in effect mini-computers that monitor external and internal temperatures. They are programmed with the vital statistics of the

House type	*Exposure*	*Space heating only*	*Heating and HW*
3 bedroom semi	Normal	10.5 kW	14.5 kW
3 bedroom semi	Exposed	13.5 kW	17.5 kw
4 bedroom detached	Normal	14.5 kW	17.5 kW

Figure 10.10 'Rule of thumb' for relating boiler capacity to typical dwelling type.

dwelling so they are able to calculate how long the heating needs to be switched on to allow the internal temperatures to reach the desired level by a certain time.

As technology progresses, more of these features will become standard provision and expected by home owners.

10.9.4 Condition of gas heating systems

Although heating systems vary, a number of key features can give an insight into the condition of the system as a whole:

- **Operation of system** – If the services to the dwelling are still connected does the system work? Does the boiler operate when switched to 'constant' on the programmer? Does the water run hot at a sink or a basin? Do the radiators feel warm? Do the TRVs operate easily or are they stuck?
- **Leaks** – Are there any leaks on the system? These might be obvious but vulnerable locations include:
  - the various joints and water seals within the boiler itself. Look for corrosion, water stains, etc. around the boiler casing;
  - at the pipework connections to the radiators, hot water tanks, etc.;
  - if the pipework is buried in the screed of a solid floor and the floor covering can be easily lifted, test the surface of the screed with a damp meter. Leaks from any buried pipework joints are common.
- **Safety issues** – Are there any signs to suggest that the system may be unsafe? For example:
  - evidence of staining caused by the leakage of combustion gases from the boiler especially where the flue joins the boiler casing;
  - defective or damaged flue terminals, etc.;
  - open or exposed electrical wiring and switch gear;
  - with unvented domestic hot water systems, faults in the safety devices could, in the worst cases, lead to explosion. The BRE (1989) suggest an inspection checklist which is summarised below:
    - Was the operative who installed the system British Board of Agrement (BBA) approved?
    - Was the firm BBA approved?
    - Is the system itself BBA approved?

    It may be impossible to answer these questions on a standard survey even with a helpful vendor. Therefore, when sealed systems are encountered a stronger recommendation for specialist testing may be justified even where the system appears to be relatively new without obvious faults.
- **maintenance history** – all gas installations need to be serviced and maintained at least once a year. If not then potential faults may develop that could render the system dangerous and reduce its effective life. Ask the owner for the maintenance record for the appliance.

Category	*Description of condition of heating system*	*Advice to client*
Good (satisfactory repair)	The system appears to be relatively new (5–7 years old) and in good condition. There are no signs of defects (corrosion, flue gases escaping etc.) and no obvious features that give rise to safety concerns. The system is effectively controlled and there is clear evidence that the current owners have had the system serviced regularly.	The visual inspection has revealed that there are no obvious defects or deficiencies that could suggest that there might be problems with the central heating system. Because of the specialised nature of the installation, if the customer wants to be sure then a test should be arranged before a legal commitment to purchase is entered into.
Fair (minor maintenance)	The system may be of intermediate age (5–10 years old) and in fair condition. There are a few defects (small amount of leakage, corrosion, etc.) and the possibility that a few features may not conform to the current gas regulations. The system has few controls and appears to have been serviced sporadically.	Although the visual inspection has shown the system to be of a serviceable nature, defects and deficiencies have been identified which suggest that the system requires further inspection and testing by an appropriately qualified specialist. This should be arranged before any legal commitment is entered into as it may result in additional expenditure.
Poor (significant or urgent repair needed or threat to health and safety)	The system appears to be old (over 10 years old) and in poor condition. There are a number of defects (leakage, scorch marks around the flue etc.) and there are serious concerns about the safety of the system (not enough ventilation, spillage of flue gasses, etc.). There are few controls on the system and there is no evidence to suggest that the system has been serviced.	The visual inspection has revealed considerable defects and deficiencies indicating that the system will require extensive repair or replacement. Further inspection and testing by an appropriately qualified specialists should be arranged before any legal commitments are entered into. Although the cost of the work will only be known once you receive a report, it is likely that the system will require significant investment to render it safe and efficient to use.

Figure 10.11 Table linking the condition of the heating system and client advice.

Like other judgements about service systems, only a qualified engineer can properly assess the condition of a heating system but figure 10.11 attempts to provide descriptors that may help surveyors give more targeted advice to their clients. This approach must be applied with care. In some cases where new heating systems have been installed wrongly they may present a danger to the occupants requiring immediate attention. Such a situation will cut across these common descriptions.

10.9.5 Replacement of gas heating systems

If a surveyor judges a heating system to be at the end of its useful life, replacing like with like is rarely appropriate. Heating engineers will be able to give more targeted advice but some general guidance might be appropriate (CORGI 1997, p.141):

- Always convert to a pumped system if the existing system is gravity fed.
- Fit TRVs to give more control.
- Choose a timer that is suited to the user.
- Install condensing boiler to gain maximum efficiency.
- Consider replacing the entire system including the pipework and the radiators. Not only will this prolong the life of the boiler but also it will be more efficient.
- Always use a room-sealed appliance (i.e. balanced flue), it is safer.

10.10 Oil-fired central heating

10.10.1 Introduction

As with liquefied petroleum gas (see section 10.14.4), oil-fired central heating is common in areas that are not served by mains gas. The system should be very easy to identify; the large fuel-oil storage tank in the garden is the clearest clue and the supply pipe between the tank and the dwelling is usually thinner than a typical gas pipe (only 10 mm diameter). There are two types of fuel oils:

- 35 secs – this is quite a viscous oil
- 28 secs – this is a much thinner oil, sometimes called kerosene.

Common boiler models include Glowworm Thermglow and Stelrad Ideal Buccaneer. Oil condensing boilers are also available and are very energy efficient.

10.10.2 Storage of oil

The storage of the fuel oil should conform to the following (see figure 10.12):

- for most single family dwellings a maximum tank size of 3,400 litres will probably be sufficient and can be contained in either a galvanised steel or plastic storage tanks;
- the tank should be supported by brick or block piers built off an adequate concrete base. There should be a dpc between the piers and the base of the tank;
- oil tanks are sometimes provided with a 'catchpit' or bund wall that will retain the oil if the tank should leak. BS 5410 (HMSO 1977) states that a catchpit only needs to be provided in hazardous situations such as tanks near watercourses or a sewer;
- the tank should be at least 760 mm from any boundary and 1.8 m away from the building. If the tank is closer than this than radiation barriers have to be provided;
- the oil tank can be stored in a building or underground but must be enclosed in a 1 hour fire resisting enclosure, have a catchpit and the tank should be resistant to corrosion.

10.11 Coal-fired central heating and room heaters

Coal has been used for centuries to heat dwellings and is still very common in former mining communities. There are a number of solid fuel heating systems and some of these are described below. The relative energy efficiency of the appliances have also been included for reference:

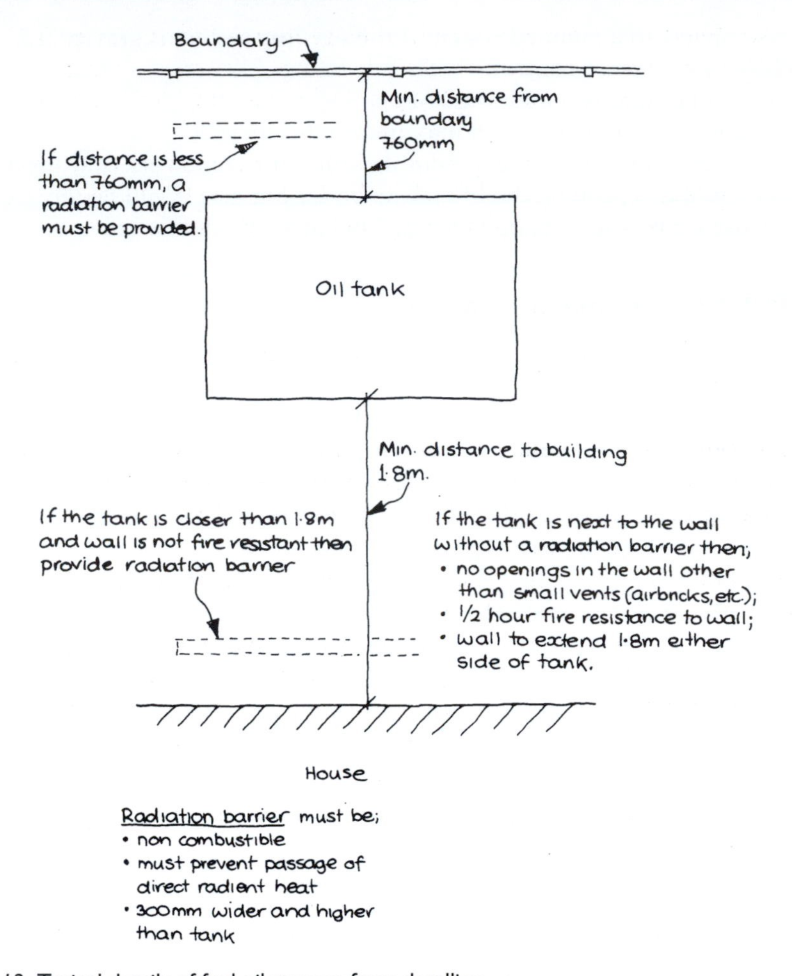

Figure 10.12 Typical details of fuel oil storage for a dwelling.

- **Traditional open fire** – contained within a fireplace with chimney breast and grate. Very inefficient way of heating a space (32% efficiency). Some versions may have a handle near the top that operates a throat restrictor which reduces heat loss up the chimney (42% efficient).
- **Open fire and back boiler** – feeds a hot water cylinder and possibly a few radiators. Look for water pipes coming out the chimney breast (55% efficient).
- **Closed room heater** – glass door to the front but ensure that it is not a gas coal effect fire! Common models include Rayburn and Parkray (60% efficient).
- **Closed room heater and back boiler** – a combination of the above systems (65% efficient).
- **Solid fuel central heating boilers** – these include:
  - manual feed – this has a single tub and is loaded by the users;
  - auto-feed – these are often called coal flow boilers. They generally have twin tubs side by side. One is the combustion chamber and the other is the feed and storage chamber. A room thermostat controls the flow of coal.

The central heating boilers can supply enough heat for the heating and hot water for an average 3-bedroomed semi-detached dwelling. The systems can be used with all the usual heating controls with the auto-feed freestanding boilers being the most powerful.

A coal storage bunker should be provided that is a maximum of 45 metres away from a delivery point. For an average-sized appliance it should be able to store 1.5 tonnes of coal which can usually be accommodated in a 2.25 m^2 bunker.

10.12 Electric heating

Although the 1973 oil crisis made electricity an expensive heating fuel, many systems are still used. These include:

- **Drycore boilers** – this is usually a radiator-based wet system. It draws its heat from a large heat store made from storage radiator bricks. The heat store is warmed up using cheap rate electricity. Usually in the kitchen and could be the size of a large fridge.
- **Storage radiators** – uses off-peak or Economy 7 electricity at night to heat up storage radiator bricks that release the heat during the day. Older types are very bulky, often 225–300 mm deep. Newer versions are much slimmer. Common models include Creda, Dimplex, Storad, Unidare. Some more sophisticated and responsive systems are fan assisted to allow greater and more flexible release of heat. Ideally storage radiators should be positioned in the following locations (HAPM 1991):
    - on internal walls to minimise heat loss;
    - because of their weight they should not be placed mid span position of floor joists
- **Underfloor heating** – electrical elements warm a large proportion of the dwelling. Common in purpose-built flats of the 1960s and 1970s. Partial systems only heat the lounge and hallway.

10.13 Electrical systems

In section 10.1.3 the specialist assessment of the electrical installation in a dwelling was described. This section identifies the signs that can enable a surveyor to judge whether further investigations are justified or not. As with other services, a little knowledge is a dangerous thing so this guidance should not be seen as detailed interpretation of the regulations. Instead it should be taken as a broad framework against which an electrical installation can be assessed. The objective is to advise whether the system is likely to be safe and give an indication of its probable performance and life.

10.13.1 Main intake and fuse board

The main intake and fuseboard must not be sited in:

- a bathroom or toilet
- an inaccessible position that would make it difficult to read or repair.

Fuse boards that are old or obsolete metal types may indicate that the whole system needs upgrading.

10.13.2 Presence of earth bonding

This is a skilled judgement and just because green and yellow wires are spotted it does not mean that the system is adequately earthed. A properly installed system will show the following characteristics:

- mains water pipe – earth connection on the customer's side within 600 mm of the stop tap. This could have been altered if supply pipe has been replaced by plastic main;
- mains gas – on the customer's side within 600 mm of meter;
- pipes from oil tanks, LPG cylinders, etc. – as close to point of entry as possible;
- any structural steelwork in contact with earth;
- fittings such as sink tops, baths, pipework to sanitary fittings, radiators, etc.

If earth connections are seen in the majority of these locations then confidence can be expressed about the electrical system. BUT the client should always be encouraged to have the system properly tested before commitment to purchase if they want to be sure.

10.13.3 Wiring

There are many different types of wiring that were used prior to PVC. These include:

- vulcanised rubber and cotton braid;
- lead sheathed cable;
- tough rubber insulated varieties.

Because of the age of these installations most will be at the end of their safe and useful life and will need rewiring. Key locations to spot older sections of wiring include in floor voids, loft spaces, fuse boards, etc. Other characteristics of an adequate installation include:

- wiring neatly clipped to the sides of ceiling joists rather than the top to avoid damage;
- wiring should be clipped at 225 mm centres;
- wiring should not be covered by thermal insulation. Electrical wiring will give off heat when used normally and if power cables (i.e. those serving power points, water heaters and showers) are covered by thermal insulation then the resulting overheating can damage the cable. Lighting circuits are not so badly affected so the key is to look for the thicker cabling.

10.13.4 Equipment and fittings

There are a number of other clues that can indicate that a system may be below standard:

- The age, type and condition of the switches and light fittings. Round pin plugs and old circular light switches for example (see figure 10.13).
- New switches and sockets may suggest recent rewiring but they may simply connect to old wiring behind.
- Switches and sockets may be too close to sinks, basins, showers etc. and present a danger of electrocution.
- Surface-run wiring in most locations is not recommended.
- Surface-fixed sockets on skirtings.
- Large numbers of appliances plugged into the same socket via double and triple plug adaptors suggests too few power points.
- DIY alterations should be viewed with particular caution. Examples could include spurred sockets into extensions, conservatories and informal loft conversions.
- Look at the mains fuseboard for evidence of older wires.

Pulling together these indicators can give a feel for the adequacy of the system. Figure 10.14 attempts to present descriptors that may characterise electrical systems in a range of different conditions. Caution must be exercised when applying this approach. Assessment of electrical systems is a complex and specialist process that can only be properly done by a qualified electrician. Therefore, any advice given by a surveyor must be carefully given.

Figure 10.13 Old or unusual switch gear may give important clues about the suitability of the system.

Category	*Description of condition of the electrical system*	*Advice to client*
Good (satisfactory repair)	The system appears to be relatively new and in good condition. The observable wiring is in PVCu and there are no obvious features that give rise to safety concerns. The fittings are modern, the fuseboard is fitted with MCBs and earth wires were seen in most of the expected locations. There are no obvious DIY or substandard additions to the system.	The visual inspection has revealed that there are no obvious defects or deficiencies that could suggest that there might be problems with the electrical system. Because of the specialised nature of the installation, if the customer wants to be sure then a test should be arranged before a legal commitment to purchase is entered into.
Fair (minor maintenance)	The system may be of intermediate age and in fair condition. Although the wiring is largely in PVCu the switches and sockets are of older designs. The fuseboard may not be fitted with MCBs and some earth wiring does not appear to be present in the expected locations. The system appears to have been altered in the past, although the standard appears reasonable.	Although the visual inspection has shown the system to be of a serviceable nature, defects and deficiencies have been identified which suggest that the system requires further inspection and testing by an appropriately qualified specialist. This should be arranged before any legal commitment is entered into as it may result in additional expenditure.
Poor (significant or urgent repair needed or threat to health and safety)	The system appears to be old and in poor condition. Although some of the wiring is in PVCu there is some braided or sheathed varieties. There are not enough power points and some are of the round pin variety. There appears to be little earthing and the fuseboard is an older type. The system has been altered a number of times resulting in spurs and extensions of a doubtful DIY quality.	The visual inspection has revealed considerable defects and deficiencies indicating that the system will require extensive repair or replacement. Further inspection and testing by an appropriately qualified specialist should be arranged before any legal commitments are entered into. Although the cost of the work will only be known once you receive a report, it is likely that the system will require significant investment to render it safe and efficient to use.

Figure 10.14 Table linking the condition of the electrical system and client advice.

10.13.5 'Plug' testing of electrical systems

Although the testing of electrical systems can only be done by a qualified electrician many surveyors may use a proprietary 'plug' tester. These can be plugged into a ring main and depending on the combination of indicator lights they claim to identify certain deficiencies. Although some commentators claim that plug testers can enhance the ability of the surveyor to comment on the effectiveness of electrical systems, care must be exercised in using them. It will not test the whole of the system, just that bit it happens to be plugged into. On balance they are probably more misleading than helpful.

10.14 Gas service

The supply of gas to domestic dwellings is currently in the process of change. It is no longer a monopoly; a number of different companies are offering to supply gas to domestic consumers. What this will mean even in the short term is not clear but it is assumed that the regulations governing installation and safety issues will ensure some level of consistency across the country.

10.14.1 Gas meters

These should be as close as possible to the point of entry of the gas main. A meter control valve should be fitted as near as possible to meter inlet. The meter should be must be accessible for:

- installation and maintenance
- meter reading
- operating pre-payment mechanisms
- emptying any coin boxes.

The meter must not be sited:

- where food is stored
- under a sink
- in damp areas (although they often are!)
- not on an escape route unless it is in a fire proof compartment.

The meter can be outside as long as it is appropriately protected from the elements, usually in a standard plastic box.

10.14.2 Gas pipework

- Gas pipes in domestic dwellings can be:
  - copper – with capillary soldered joints or compression fittings;
  - steel – usually screwed joints.
- Polyethylene pipe – only used below ground externally apart from where the pipe bends up to connect into the meter.
- Pipe routes – gas pipes should be separated from electrical wiring by at least 25 mm and from fuseboards by at least 150 mm.
- Pipe fixings – gas pipes should be properly supported along their length. Typical spacings are as follows:
  - copper – 2 m vertically and 1.5 m horizontally
  - steel – 2.5 m vertically and 2 m horizontally
- Where the pipes are fixed to the inside faces of external walls, stand-off clips should be used so that the pipe does not touch the wall.

- Lead pipes – it is common to find lead gas pipes still incorporated in older systems. Although there are no regulations preventing this the lead should be properly supported to prevent 'sagging' of the pipe under self-weight. Any such movement could put a strain on the joints with other pipework.

10.14.3 Gas in flats

The gas service in flats up to 15 m high (usually five storeys) are covered by the general gas regulations. For buildings higher than 15 m, installations are covered by additional regulations that require specialist advice. Even for flats up to five floors, certain precautions have to be taken. For example, gas risers in communal areas will have to be run in a fire-protected and ventilated duct. Gas meters should only be installed on escape routes in properly protected compartments with any doors to meter compartments properly secured and fitted with appropriate signs. Therefore, it is beyond the skill level of any surveyor to comment adequately. If the client wants to be more fully advised about the gas service they should refer the matter to an appropriate heating engineer.

10.14.4 Liquefied petroleum gas (LPG)

This is a liquefied type of gas (usually butane or propane) that is stored near to the dwelling in a large metal tank. It is common in areas that are not connected to mains gas. Although many believe that natural gas and LPG are the same, they differ in some important aspects which can give rise to dangerous situations. Some key points to consider include:

- **Tank storage** – LPG is stored in a liquefied form under pressure in a tank usually provided by the gas supplier. This should conform to the following:
    - The tank should be a minimum of 3 m away from the dwelling and all boundaries.
    - The pipe between the tank and the dwelling should be buried.
    - A control valve should be provided at the point of entry into the building.
    - The tank should not be encased or painted by the occupier. Grass and shrubbery should be cut back.
    - The tank must be visible by the installer from the tanker position just in case of spillage, etc.
- **Cylinder storage** – some dwellings may be supplied by smaller, separate cylinders. Some general points to consider include:
    - The cylinders must not be stored in a basement or sunken area. This is because the gas is heavier than air and can accumulate in lower areas.
    - The tanks are usually connected via flexible hoses but these should not pass through walls, partitions, ceiling or floors.
    - The cylinders should be stored upright.

10.14.5 Timber-frame dwellings and gas services

Because of the special nature of timber-framed dwellings, some aspects of a gas service are critical. It is beyond the scope of a survey to take into account this level of detail but some general issues are described below:

- Check that any gas supply service is properly secured if run within the cavity between frame and outer cladding.
- The gas appliances should be properly supported on the walls of timber structures.
- There should be an appropriate non-combustible separation between hot surfaces and timber components.
- The vapour barrier in the wall construction must be continuous where penetrated by flues, service pipework, meter boxes, etc.

10.15 Below-ground drainage

10.15.1 Introduction

Such a high proportion of the underground drainage system is concealed that it is difficult for a surveyor to give anything but a qualified report to the client. Like other parts of this book, this section attempts to highlight those visible signs that will enable the surveyor to make assumptions about the condition of the drains. Where this is not possible, the call for further investigations can be made confidently for good reason in the best interests of the client. Some of these indicators may be more relevant for a full building survey but are included here to give the reader a broader view.

10.15.2 Definitions

Before the faults of drainage systems are described in detail it will be useful to clarify some basic definitions:

- **Drain** – the outfall pipe that serves one dwelling.
- **Common drain or private sewer** – several houses may be served by a common drain that crosses a number of properties before entering the main public sewer. Every aspect of maintaining, renewing and cleansing is the responsibility of the owners concerned. The proportion of the liability is usually fixed at the time of construction. It will be important to bring this information to the client's attention so their legal representative can check the deeds and any repairing liabilities.
- **Public sewer** – often under the main roadway and is adopted by the local authority. In many cases, the property owner(s) is responsible for the drain until it enters the public sewer even if this extends beneath the public highway for a considerable distance.

A few other definitions might also be useful:

- **Foul water** – water that is discharged from WCs, baths, showers, handbasins, sinks, sluices, washing machines, dishwashers, etc.

- **Grey water** – the term some people apply to waste from sinks, baths, handbasins, dishwashers, etc.
- **Surface water** – water that is discharged from rainwater pipe systems, yard gullies, etc.

10.15.3 Types of below-ground drainage collection systems

There are three main types of collection systems:

- the combined system where foul and surface water are combined in the same drains;
- the separate system where foul and surface water are kept apart. There is usually double the amount of inspection chambers and drain runs on the property;
- the partially separate system where surface and foul water are kept separate on estates, infill housing, etc. but combined just before they enter the main sewer.

During the urban developments of the 1960s and 1970s it was the intention to always keep the two separate. The surface water was to be collected separately and run directly to the nearest water course. The foul water would be taken for treatment but because there was much less of it (i.e. no rainwater) it was more economical. In many areas this was achieved but in some established urban areas it was impossible to separate the two. Here it was accepted that the sewerage will always be combined. Consequently there will be a variety of practice across the country and even within the same water company area. This is because the building control authorities act as the agents for the water companies. In some council areas, the building control department will insist on separate systems right up to the point of entering the main sewer while others will allow combined drained runs within the boundaries of even new properties.

In areas where there are separate systems there is a problem of contaminated water entering the surface water system. There are two main sources of this:

- 'normal' pollutants that are washed into the system, e.g. petrol, oil, dog faeces, etc.;
- foul and grey water that are mistakenly or purposely connected into surface water systems on domestic properties.

A typical example is a DIY installed wash hand basin that drains into the RWP pipe of a separate system. Such an arrangement would lead to higher than acceptable contamination of the water course. In the future these contravention's could be traced more actively and action taken against the offending owners. In the meantime it is most important to identify that this problem exists and the situation is contrary to regulations.

10.15.4 Typical faults in underground drainage systems

The commonest types of defects to underground drains include the following:

Cracked and leaking drains

Most of the problems are associated with ceramic pipes with rigid joints. The failures are caused by:

- thermal and moisture movements in the pipes themselves;
- overloading of the pipes especially below vehicular access routes;
- ground movements associated with subsidence.

Once cracked or fractured, the implications can be serious. Leaking drains can have a number of effects:

- **Subsidence** – this can be caused in two ways:
    - water can leach out into the ground and soften the subsoil beneath nearby foundations until the weight of the building overcomes the bearing capacity of the soil;
    - fine-grained material (i.e. sands and gravels) can be washed into the drains creating a void. This can result in the subsidence of nearby structures.
- **Root damage** – the roots of nearby trees can seek out the moisture leaking from cracked pipes. In some cases, large mats of finer roots can enter the drains and cause regular blockages.
- **Localised flooding** – if the leak is bad enough, the result water may permeate through the subsoil and leak into nearby basements and sub-floor voids.

Insufficient fall

Drains have to be laid with adequate falls to be self-cleansing. Sluggish flow will lead to the build-up of solids and eventually a blockage will occur. The amount of fall on a drain depends largely on its diameter and the number of appliances it serves. For example, a drain run serving several houses might be allowed a shallower gradient because of the higher volume of water flowing down it. According to Melville and Gordon (1979, p.837), a robust rule of thumb for determining an appropriate slope on a drain is to take the diameter of the pipe in inches and multiply this by ten. This will give the following:

- 4-inch pipe (100 mm) will require a gradient of 1 : 40
- 6-inch pipe (150 mm) will require a gradient of 1 : 60.

When compared with the values stipulated in the Building Regulations, Part H (HMSO 1991d) these values are a little steeper than they need to be with good reason. Many of the older drainage systems were laid at a time when accuracy levels did not match those of the modern construction industry. Also movements in the subsoil may have also created a variety of backfalls and misalignments. Therefore a cautious approach would be appropriate.

Inadequate access

Even well-designed and installed drains will become blocked from time to time. When this does happen there must be sufficient access for the blockage to be cleared. Inspection chambers and rodding eyes should be provided at every change of direction of the drain, changes of gradient and at regular intervals along longer drain runs. If not then the client may face a number of consequences:

- when a blockage does occur, a contractor could have to trace the drain and excavate to break into the drain to clear the problem. There will also be the cost of making good afterwards;
- to prevent it happening again in the future the client would have to have a rodding access point or a chamber constructed over the existing drain. This could be very expensive.

Even where there is not a blockage problem, should the owner want to add additional sanitary appliances in the future (i.e. a new bathroom, WC, etc.) then the local building control authority may insist that the existing drainage system is brought up to standard. In these cases the client should be clearly warned of the possible implications and higher costs it could involve.

Blockages

Drain blockages are often the result of other problems associated with the drainage system, especially if they occur regularly. The most common causes of blockages include:

- inappropriate solid objects accidentally passing down drains and lodging on bends, etc.
- tree roots partially blocking the drain;
- poor construction including:
  - badly aligned pipes causing a projection at the joint;
  - settlement where a pipe passes through a wall;
  - poor-quality bedding/support of pipes allowing them to become mis-aligned.
- design failures such as inadequate falls, insufficient size, etc.

Where there is clear evidence that the drains have been blocked on a number of occasions then it can be taken as evidence of something more serious being wrong.

10.15.5 Inspecting underground drainage systems

Within the limits of a standard survey there are key indicators that can give an insight into the likely condition of the drains below ground, as follows:

Inspection chambers

Looking inside one can give valuable clues about the condition of the drainage as a whole. As most types of standard surveys include the lifting of the cover to an inspection chamber it is worth reviewing this process.

- **Locating the chamber** – finding the inspection chamber may be the first challenge! It is possible that there never was one but most post-1900 dwellings did have at least one that is normally in the back garden somewhere. If it cannot be found quickly it might have been covered by:

- the surfacing of a new path laid directly over the cover;
- raised garden beds gradually enveloping the cover over the years;
- unauthorised rear extensions, patios and remodelled gardens directly over the chamber.

If the chambers cannot be found or are found to be covered then this should be reported to the client who should be advised to commission further work to expose and make the drains more accessible.

- **Lifting the lid** – even when the covers are located they are not always easy to lift. The cover and the frame may be rusted together or accumulations or dirt, grit, etc. may have made them impossible to prise apart. The holes for the inspection chamber keys may have corroded or the lifting hand holds may have broken. When faced with difficulties like this, a surveyor must decide what level of effort to get the cover open is reasonable. This will always depend on the circumstances but if the cover can not be moved with conventional lifting tools without damaging the frame, cover or surface of the path within a 3–5 minute period then the attempt should be abandoned and the matter reported to the client. Surveyors should be prepared to put some effort into it because most covers will require a little force to shift them. Therefore, it is likely that the courts would expected to see that a 'reasonable' surveyor made more than a token effort. Holden (RICS 1998) points out that little has changed even with the new form of Homebuyers survey 'There is no problem with using a tool to open an inspection chamber, provided that you don't cause damage'.
- **Internal condition of the chamber** – once the lid has been lifted, with the aid of a strong torch and an inverted 'head and shoulders' inspection, the internal condition should be assessed. Typical signs might include:
  - If the walls are rendered is there evidence of spalling, cracking or crazing? Sulfate attack could be the cause.
  - The concrete benching should in a smooth, clean condition without any cracks. Is the benching fouled suggesting that there might have been a regular blockage in the past?
  - If the chamber is deeper than 1.5 m then there should be galvanised access step irons at regular intervals in good condition. Do not go down into the chamber. There could be danger from poisonous gasses and only trained operatives should undertake such a task.
  - Are the half-round channels and channel bends at the base of the chamber in good condition or are they cracked and spalled?
  - Is there any gravel, sand or soil debris in the channels suggesting cracked drains further up the drain run? This would allow the subsoil to be washed into the drain.
  - Are all the drainage connections into the chamber properly made? Often new connections are poorly connected allowing haphazard water flow. This could result in blockages.
  - Are there any roots in the chamber from nearby trees or shrubs?
  - Has the chamber got the right sort of cover?

  - If it is internal it should have a screw-down lid with a double seal.
  - If beneath a vehicular access drive is the cover the appropriate strength? Has the existing one distorted under load?
  - Are the cover and frame properly fixed? Is the cover corroded so there is a danger that pedestrians could trip or fall down the chamber?
- If the chamber is the final one before the drain enters the sewer, has it got an interceptor trap with a rodding eye fitted with a plug. If the cover is missing then vermin will be able to enter the drainage system.
- Is there visual evidence that the drain runs flowing into or out of the chamber have been lined with a proprietary plastic lining system? If this is the case then the owner should be asked to produce details of this installation including any guarantees which should be insurance backed. Such a discovery should also generate a 'trail of suspicion'. Why was the drain lined in the first place? Has this measure resolved the deficiency?

Underground drainage runs

Although the condition of the drain runs cannot be assessed from an above-ground inspection, there are a number of things that can be deduced from visual signs. Firstly, locate all the inspection chambers and roughly establish the route of the main drains. Are there any of the following features apparent along this line:

- Are there any trees or large shrubs close by? The roots could be penetrating the drains.
- Is there a vehicular access above the line of the drain that could impose excessive loads? This is particularly important where the invert level of the drains is shallow. The Building Regulations, Part H (HMSO 1991) standard is that pipes should be at least 700 mm below the surface of 'light traffic roads'.
- Is there any new extension or other building over the line of the drain? If the new walls and foundations have not been properly supported over the drain differential settlement could damage the pipes. If there is a new building, the client's legal advisor should check whether the appropriate approvals have been granted.
- Is there any depression in the ground, paths or driveway above the line of the drain? Are any adjacent buildings showing signs of subsidence damage? Any of these signs could indicate leakage of the drain.

Combined with the information gained from the inspection of the chamber, an accurate picture of the condition of the drainage system can be formulated without sophisticated tests.

Gullies

Gullies to some extent represent the transition from above-ground to below-ground components of the drainage system. There are a number of different types:

- **Open gullies** – these are used externally for surface water run-off from hard-standings, yards, etc. They should be fitted with a grating.

- **Back inlet gullies** – these provide a connection for either a rainwater or waste pipe connection into the drainage system. They are usually trapped and contain a water seal to stop smells from the main drains backing up.
- **Sealed gullies** – these are usually used within buildings and have a screw-down metal cover.

It is most important to see if the gullies are blocked or not. In some cases debris on the covering grate will give an indication of any recent blockage and flooding. Some gullies can become cracked leading to a loss of seal allowing smells to affect the property.

10.15.6 Giving advice about drainage problems – to test or not to test

It should be the intention of the surveyor to obtain sufficient visual clues and evidence to be able to confidently reach a clear conclusion about the condition of the drainage system. Where there is a cause for concern, a call for additional investigations would be appropriate. Unless the surveyor has been retained to organise this on the client's behalf most surveyors do not have any further involvement.

Many clients are often unsure what type of 'further investigations' are appropriate and what sort of contractor to appoint. Some will contact the surveyor for further discussion and advice. Reviewing the current thinking on the testing of drainage systems it is not surprising that clients are confused because most of the leading commentators seem to disagree. These issues are presented below so surveyors may be better placed to advise clients.

- **Water tests** – this is a traditional method of testing drainage systems. A length of drain is plugged and filled with water to the top of the nearest inspection chamber. This creates a pressure equivalent to about 0.007 N (see figure 10.15). After initial absorption, the drains are left for up to 30 minutes. Any drop in level is noted. In some drains the water can often drop quite quickly. This can be due to serious cracks in the pipework or poorly made joints that have never really been watertight. Also because of the pressure exerted by the water some commentators worry that water tests can actually **cause** damage to old and weak joints.
- **Air tests** – this follows a similar procedure but uses air to pressurise the system rather than water. Some commentators state that air tests can be more effective than water tests but subject the drains to a higher pressure and so increase the possibility of damage.

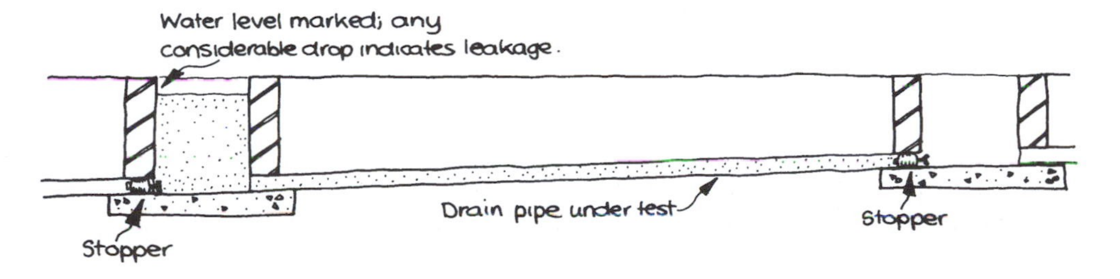

Figure 10.15 Typical arrangement for a water test of an underground drain.

- **Close circuit television inspections (CCTV)** – this is used to view and/or film the condition of the drain in question. The video camera is mounted on a small carriage and pushed down along the drain with specially adapted drain rods. The pictures are relayed to a monitor in the cab of the vehicle. These films are recorded on video and a copy is usually given to the client. The films can vary in quality and can clearly identify blockages, obstructions, cracks and other forms of damage to the pipes. These deficiencies can be precisely located as distances are measured and shown on the screen. The one drawback is that this method shows the damage but does not quantify the lack of watertightness.
- **Smoke tests** – this is of limited use on underground drains because it does not pressurise the pipe and so locate the precise source of the leak. Smoke tests are more appropriate for testing for leaks on above-ground drainage systems. The smoke is produced by a smoke rocket or pellet. The spread of smoke can be assisted by applying a low level of air pressure through a pair of bellows. Some contractors may use a special smoke machine. It is important to gain the owner's permission because smoke can often enter the dwelling especially if an old branch connection has not been properly sealed.
- **Drain run tracing** – there are a number of techniques used to find out the direction/position of drainage routes and connections. Two main methods are used:

  - **Drain dyes** – a powdered dye is added to test water and is poured down the drain that is being investigated. The appearance of the dye further along the system can reveal such things as how the drains are connected and which drains are causing leaks into basement areas, etc.
  - **Sonar sensors** – these small sensors are attached to the end of rodding equipment and pushed along a particular drain run. The high-frequency emissions are detected at the surface by special monitoring equipment and enables the drain runs to be precisely plotted.

Over the last few years the use of CCTV has become the predominant method of inspecting and assessing the condition of the drainage systems. Unless specific circumstances require other types it is safer not to recommend any form of pressurised test because of the possibility of damage.

10.15.7 Typical repairs to drainage systems

When a defective drain has been diagnosed then the scale of repair required will depend on the extent and cause of the problem. This decision is usually taken by the person who carried out the investigatory works and so is beyond the scope of the original appraisal. A knowledge of typical repairs will give an insight into the likely consequences of drainage defects and enable surveyors to better advise clients of the consequences. There are three types of drain repair:

- **Repairs to isolated drain pipes** – where only one or two individual lengths of pipe are defective and only a short run of drain will need to be replaced. The sound sections of pipes either side of the defective length may need to be disturbed for access purposes. The cost of this repair is largely influenced by the depth of excavation and the disturbance at the surface especially where it is beneath the dwelling itself.

- **Partial or complete replacement of drainage runs** – where whole drain runs have proved to be beyond repair then total replacement may be the only option. The cost of this can be considerable, running into several thousands of pounds. The work can also be very inconvenient for the occupier as the drains may be out of action for several days at a time. The work will also be complicated if the drain serves a number of other properties and the costs have to be apportioned.
- **Lining of an existing drain** – in some cases despite the existing drain being misaligned and damaged, a slightly cheaper and less disturbing option will be to line the existing drain with a fabric liner (usually polyester) that is coated with a special resin. The liner is pulled through the drain run and inflated to fit tightly against the drain walls. The resin sets hard and so seals any fractures or open joints that exist. This is a specialist operation and can be expensive but it is quicker and less disturbing. Many installers offer 15–20-year guarantees on their work.

10.16 Other methods of sewerage and surface water disposal

So far it has been assumed that the dwelling will be connected to the local authority sewer. In many rural areas sewerage has to be stored for regular collection or treated in a small domestic treatment facility and surface water drained away to the ground via soakaways. The assessment criteria are different to those for mains drainage.

10.16.1 Cesspools

A cesspool is a tank that is used to store untreated sewerage until a specialist company empties it on a regular basis. Although cesspools are cheaper to install than a small domestic sewerage treatment plant, the cost of emptying can be considerable. Cesspools could need emptying between 10–16 times per year at a total annual cost in the region of £3,000 – £5,000. As well as normal emptying, a cesspool will benefit from a periodic 'desludging' to remove the residue left at the base of the tank. This is important information that should be passed on to the client. The constructional standards of cesspools is governed by Part H of the Building Regulations (HMSO 1991d). Based on a visual inspection some of the key features include:

- The cesspool should have sufficient capacity to store foul water from the building it serves until it is next emptied. The aim is to give 45 days storage and the building regulations state that the minimum capacity should be 18,000 litres or 18 m^3 (although this will be impossible to check without emptying it!).
- The size of a cesspool will have been based on the level of usage estimated at the time of construction. Using an old rule of thumb of 30 gallons per day per person, the maximum number of people that a cesspool can economically and adequately serve will be 7 or 8. If the dwelling has been enlarged or extra dwellings added to the site then this figure might well be exceeded. If this is suspected further investigations should be recommended;
- It should be within 9 m of vehicle access so it can be emptied and desludged easily. This route should not go through the dwelling for obvious reasons!

- It should be as far as possible from a water course – a minimum of 10 metres. It should be at least 15 metres from the dwelling and downwind if possible.
- The cesspool should be properly and safely covered. Old timber boarding will not be satisfactory as it could easily rot and collapse. There should be an access cover that allows emptying, cleaning and maintenance.
- The cesspool should be ventilated via a 100 mm short vertical ventilating stack;
- There should be an intercepting inspection chamber on the drainage connection from the dwelling just before it discharges into the cesspool.
- only foul water should be drained to the cesspool. It is important to check that all surface water drains go to soakaways otherwise the cesspool will overflow during rain storms.
- The inspection cover of the cesspool should be lifted and the inside inspected visually. If there is a firm and 'crusty' layer on top of the pool's contents then this could be evidence that the cesspool has transformed into an informal septic tank (see below). This is where the walls of older installations have become permeable over time allowing liquid content to soak through. This then naturally disperses into the adjacent subsoil. In fact many older cesspools had an overflow pipe for this very purpose. According to Melville and Gordon (1992, p. 251) '... an overflow pipe from a cesspool taken on to a neighbouring fields ... was considered highly beneficial to the soil'. It goes without saying that this is not allowed under current regulations. This can be further assessed by:

  - asking the vendor for details of how regularly the cesspool is emptied including appropriate invoices;
  - inspecting the ground in the immediate vicinity for evidence of waterlogged ground, etc.

If there is any cause for concern then the cesspool should be tested. This usually involves emptying and cleaning the pool out; inspecting the condition of the walls and cover; checking for any overflows and subjecting it to a standing water test. This will reveal whether the cesspool is impervious or not.

If the cesspool is in a poor condition it might be more appropriate for the owner to consider installing a small domestic sewerage treatment plant in its place.

10.16.2 Small domestic sewerage treatment works or septic tanks

Rather than storing the sewerage, septic tanks are small-scale treatment installations. They usually consist of two chambers connected by 100 mm drainage pipe. The first chamber is called the settlement tank where the primary breakdown of the sewerage occurs. Further biological action occurs in the second tank before the effluent is disposed of in the following ways:

- discharged into a suitable water course;
- soakaway into the surrounding ground as long as it is not heavy clay or steeply sloping.

Whatever method is used, consent from the National Rivers Authority is required so the owners should be asked whether this has been obtained.

Key inspection indicators

- The positioning of the installation is very similar to that of cesspools described above.
- Both tanks should be covered, preferably with concrete planks set at a slight gap to provide ventilation.
- Although septic tanks treat most of the sewerage, both chambers have to be desludged once every 6 months or so to remove accumulated sediment. This will have a cost implication.

Modern septic tanks are usually factory made out of glass-reinforced plastic and installed in one unit to strict manufacturers specifications. In this case, the owner should be asked to supply details of the installation. A good source of detailed information on the provision of these types of facilities is BS 6297: 1983 Design and installation of small sewage treatment works and cesspools (BSI 1983).

10.16.3 Soakaways

Soakaways are used extensively where main drainage is not available or it would be uneconomic to provide an appropriate connection. Usually only rainwater pipes and surface water gullies are connected to soakaways. The size of the soakaway will depend on the amount of water that it has to deal with and the rate of percolation into the subsoil. It is difficult to assess the suitability of a soakaway as it is often impossible to determine where exactly it is. The BRE (1991) suggest that soakaways should have some sort of access cover or rodding point but on older properties these are rarely found. The following indicators might help in the assessment:

- The soakaway should not be less than 5 metres away from the dwelling. Where the subsoil is chalk or another type that might be vulnerable to instability caused by flowing water, further investigations might be required.
- Are the gullies connected to the soakaway in good condition? Are they clear, blocked or overgrown?
- Is there a visual water seal in the gulley? Is it full to overflowing or is there evidence of ponding around the gulley itself? This could indicate sluggish drainage. If time allows, a bucket of water could be emptied down the gulley to see how quickly it drains away.
- In the immediate area are there any areas of flooded or waterlogged ground that may suggest the soakaway cannot cope with that amount of run-off?

If there is any cause for concern then a recommendation for further investigations would be appropriate. This could include the tracing of the appropriate drain runs, location and excavation of the soakaway to assess its size and adequacy and absorption tests to assess the permeability of the soil. Repair works could include the enlarging or reforming of the soakaway in a new position.

10.17 Above-ground drainage systems

The above ground drainage system is defined as the pipework that conveys the soil and waste products from the sanitary appliances to the connection with the underground drainage system. Traditionally there are two different approaches:

- a two-pipe system where soil (i.e. WCs) and waste (i.e. washhand basins, baths, etc.) discharge into separate stack pipes. These are very common on older properties;
- a one-pipe system where soil and waste discharges into the same pipe but an additional and separate ventilating pipe was provided to prevent siphonage.

These have now been superseded by the single-stack system described below.

10.17.1 Single-stack system

Research revealed that as long as above-ground drainage systems were properly designed and installed, soil and waste can be discharged into the same stack without additional ventilating pipes with no adverse effects on the water seals to the sanitary appliances.

Although the single stack is well understood it is necessary to be aware of a few key characteristics. The position of the sanitary fittings in relation to the stack is regulated within very narrow limits. This is because effective performance can be affected by even the smallest of deviations from the building regulation standard. These are reviewed below:

- **Traps** – each sanitary fitting should be fitted with a waste trap that should be accessible for cleaning purposes. The sizes of traps should be as follows:
  - basins and bidets – 32 mm
  - sinks, baths and shower trays – 40 mm
- **Lengths of branch pipe connections** – the building regulations stipulate the following maximum lengths between the sanitary appliance and the connection with the stack pipe (see figure 10.16):
  - sink – 3 m max. for a 40 mm diameter waste and 4 m max. for a 50 mm pipe.
  - washbasin – 1.7 m max. for a 32 mm diameter waste pipe and 3 m max. for a 40 mm pipe;
  - bath – 3 m max. for 40 mm pipe and 4 m max. for 50 mm pipe;
  - WC – 6 m max. for single WC.

  If waste pipe runs are in excess of these distances then the integrity of the water traps could be affected by self-siphonage.
- **Positioning of branch pipe/stack pipe connections** – to prevent the loss of air seal of the sanitary appliances. Figure 10.17 shows the connection arrangements to the main stack pipe.
- **Ventilation of the stack pipe** – ventilation to the single-stack pipe system is usually provided by leaving the top of the main stack pipe open. This should be protected with a wire balloon or other special fitting to stop birds and vermin entering the system.

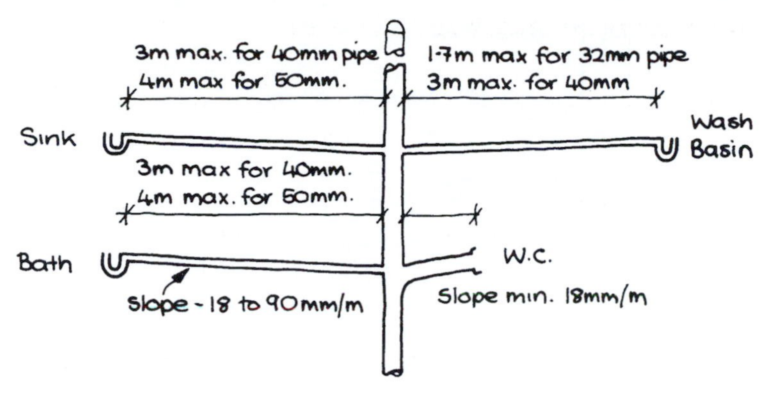

Figure 10.16 Length of branch pipe connections to a soil and vent pipe (Crown copyright. Reproduced with the permission of the Controller of Her Majesty's Stationery Office).

- Where there is an adjacent window opening, the stack pipe should be continued to at least 900 mm above the top of it.
- **Air admittance valves** – these are fittings that allow air to be drawn into a drainage stack but will not allow air to flow out. Therefore they can be located in a building without causing a problem of unpleasant smells. Potentially air admittance valves (AAVs) can avoid the need to take the soil and vent pipe up past the highest window and penetrate the roof covering. To work properly, AAVs have to be fitted properly (BRE 1990b):
  - For 1 to 4 dwellings of maximum 3 storeys an open vent on the system may not be needed.
  - For 5 to 10 dwellings all properties can be served by AAVs apart from the one at the head of the drainage run which will need to have an open vent.
  - For 11 to 20 dwellings an open vent will be needed at the midpoint and head of the run.
  - AAVs must not be used for venting cesspools or septic tanks.
  - If the AAV is housed within a duct:
    - the duct must be adequately ventilated;
    - access must be provided for rodding purposes;
    - sound insulation may have to be provided where the location is noise sensitive.
  - In a roof space the AAV must be insulated to prevent the flexible seal freezing and not working properly. The insulation must be placed so the seal is not stopped from working properly.
  - The AAV must be placed above the flood level of the highest sanitary appliance.
- **Fixing of the pipework** – all above-ground drainage pipework should be properly fixed by the appropriate brackets at the correct spacings. If not then wind loading can put a strain on the joints and possibly cause leaks. Another problem is that PVCu pipes can distort because of hot waste water running along the pipe. For 40 mm diameter pipes fixings should be at 500 mm horizontal centres and for 100 mm pipes this can be increased to 900 mm centres (BRE 1983).

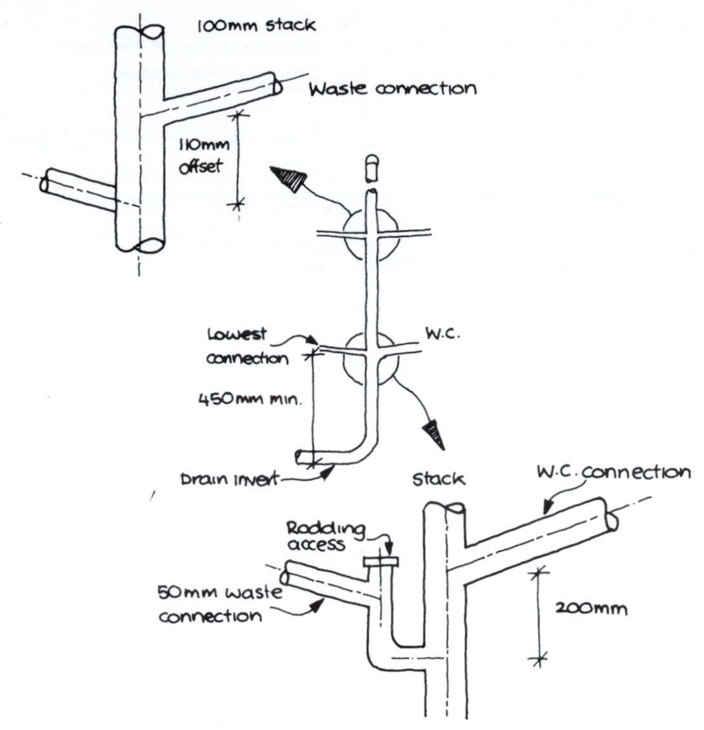

Figure 10.17 Offset connections to a soil and vent pipe to prevent siphonage and cross flow (Crown copyright. Reproduced with the permission of the Controller of Her Majesty's Stationery Office).

10.17.2 Common faults with above-ground drainage systems

Leaks to an above-ground drainage system will generally be obvious. Water staining and 'tide marks' of salt deposits, and the presence of moss and algae growth are all indicative of water flow. Leaks from WC branch connections will consist of foul water so there could be a threat to health.

Common problems include:

- leaks from the sanitary pipework especially at junctions and joints. These may be caused by:
  - inadequate provision for thermal expansion allowing 'push-fit' fittings to become disconnected;
  - poor workmanship where the junctions have not been properly made. This is especially common where 'strap-on' type connections have been used.
- difficult junctions between dissimilar materials. This is most common where PVC branch pipes have been connected to a cast iron pipe from a WC. Where water has been affecting a wall for some time, the dampness is a trigger for further inspection to make sure that it has not caused any defects internally.
- lack of access for clearing purposes. All pipework should be able to be cleared if a blockage occurs. Rodding access points should be provided at the end of runs, changes in directions, etc.

10.17.3 Rainwater pipes and guttering

In this country, a staggering amount of water falls on a typical domestic roof. The slope or fall of the roof ensures that all this water flows towards the adjacent guttering which in turn is channelled down two or three rainwater downpipes. Any faults or defects in this rainwater drainage system can result in high levels of moisture entering the building fabric. In wetter parts of the country this moisture will occur on a regular basis giving ideal conditions for dry and wet rot to develop. According to the BRE (1997), 1 in 12 houses have defects with their valley gutters and flashings and 1 in 5 have defects in the rainwater disposal system. Surveyors clearly need to be alert to this problem.

Inspecting gutters and downpipes

Unless it is pouring down with rain at the precise time the surveyor is carrying out the external part of the survey it will be very difficult to physically see the leaks themselves. On dry days some surveyors will often refuse to make any comment on the effectiveness of the systems at all. Depending on the circumstances there are a still number of visual clues that even on dry days could help give an insight into how the rainwater goods might be performing. These include:

- **Evidence of previous leakage** – this could include discoloured areas beneath gutter joints, rainwater pipe connections, etc. This staining could take the form of green algae and sometimes plant growth or even a washing clean of a previously dirty surface.
- **Older rainwater systems** – the gutters might be in such poor physical condition that if leaks are not occurring now they are likely to in the very near future.
- **Physical damage** – is there any visible damage or defects with the gutters or downpipes. Some key locations:
  - at low level, vandal and accidental damage to plastic pipes and at high level cracks or splits where ladders have been placed against the gutter;
  - at the junction between dissimilar guttering materials especially at the party wall positions;
  - to the rear face of cast iron downpipes where they have corroded because they have never been painted;
  - disconnected joints to PVC downpipes that have worked loose because of poor fixings;
  - sagging lengths of PVC guttering because of inadequate support (BRE 1984a, 1984b).
- **Downpipes blocked** – in some cases where there a blockage just below ground, the downpipe will fill up back to the next available joint and leak down the wall. Tapping the downpipe will produce a resonant note due to it being full of water.
- **Poor or inadequate design** – to properly drain, all the gutters and downpipes should be properly fixed and supported, set at the appropriate spacings and slopes. Obvious inefficiencies can sometimes be spotted.

Design of rainwater systems

- **Number of gutters and downpipes** – The design of rainwater pipes and gutters is based on the intensity of flow. For an average-sized semi-detached dwelling, 100 mm diameter guttering and two equally spaced 63 mm diameter downpipes should be able to cope with the overall rainwater flow. The key determinants are the positioning of the downpipes and how the gutters flow to them. If the downpipes are not equally spaced around the roof then water could overflow especially at changes in direction. The maximum length of gutter without a downpipe will be the region of 8 metres for most situations. Therefore, if two downpipes serve 8 metres of guttering to either side of each, they can properly serve about 32m of guttering. This can give a benchmark against which a system can be judged. For more detailed information see BRE *Defect Action Sheet 55* which gives advice on calculating the size of the collection system (BRE 1984a).
- **Fall on gutters** – Gutters should be set to a fall. The steeper the fall the more water the gutters can carry. Slopes should not be so steep that an excessive gap exists between the edge of the roof covering and the gutter itself. This would allow the rainwater run-off to miss the guttering altogether especially on windy days.

Guttering materials

Gutters can be half round or moulded in section. They can be in a variety of materials but the most common ones include:

- **Cast iron** – traditional material little used in recent times. Vulnerable to corrosion and leaking joints. Should be supported at every 600 mm. These are very heavy requiring good fixings.
- **Asbestos cement** – no longer used. Likely to be at the end of useful life. Joints formed with special sealing compound and sometimes coated internally with bitumen paint to extend life. They pose health and safety problems if they are replaced, repaired or repainted.
- **Metal guttering** – can be galvanised steel or extruded aluminium. Although commonly used on industrial buildings they are becoming more common as a modern replacements for 'ogee' cast iron guttering. Jointed with red lead or butyl rubber or other special compound.
- **Precast concrete eaves gutters** – a fashion in the 1950s. Very heavy section often supported on corbels or other parts of the building structure. Some are lined with non-ferrous metals or bituminous paint application.
- **Timber gutters** – in some areas of the country and especially in northern England, guttering is formed out of solid timber sections. Usually ogee in shape and formed out of a solid timber section. The joints are overlapping and sealed. The internal surfaces of the gutter have to be treated with bitumen paint every year or so. Vulnerable to leaks at joints and wood rot.
- **Plastic guttering** – most common current method. Jointed with special clips with integral gasket. This can be vulnerable to physical damage from ladders being put up against them, snow loads, etc.

References

Bowyer, J. (1977). *Guide to Domestic Building Surveys.* Architectural Press, London.

BRE (1983). 'Plastics sanitary pipework: jointing and support – installation'. *Defect Action Sheet 42.* Building Research Establishment, Garston, Watford

BRE (1984a). 'Roof: eaves gutters and downpipes – specification'. *Defect Action Sheet 55.* Building Research Establishment, Garston, Watford

BRE (1984b). 'Roof: eaves gutters and downpipes – installation'. *Defect Action Sheet 56.* Building Research Establishment, Garston, Watford.

BRE (1987a). 'Hot and cold water systems – protection against frost'. *Defect Action Sheet 109.* Building Research Establishment, Garston, Watford.

BRE (1987b). 'Domestic hot water storage systems: electric heating – remedying deficiencies'. *Defect Action Sheet 108.* Building Research Establishment, Garston, Watford.

BRE (1989). 'Unvented domestic hot water storage systems – installation and inspection'. *Defect Action Sheet 140.* Building Research Establishment, Garston, Watford.

BRE (1990a). 'Water storage systems: warning pipes'. *Defect Action Sheet 141.* Building Research Establishment, Garston, Watford.

BRE (1990b). 'Drainage stacks – avoiding roof penetration'. *Defect Action Sheet 143.* Building Research Establishment, Garston, Watford.

BRE (1991). 'Soakaway design'. *Digest 365.* Building Research Establishment, Garston, Watford.

BRE (1997). 'Repairing and replacing rainwater goods'. *Good Building Guide 9.* Building Research Establishment, Garston, Watford.

BSI (1983). *British Standard Code of Practice for Design and Installation of Small Sewage Treatment Works and Cesspools BS 6297:1983.* British Standards Institution, London.

BSI (1997). *Code of Practice for oil firing installations up to 45kW for space heating.* British Standards Insitution, London

CORGI (1997). *Gas installers Manual.* Council for Registered Gas Installers, Basingstoke.

HAPM (1991). *Defect Avoidance Manual – New Build.* Housing Association Property Mutual Limited, London.

Holden, G. (1998). 'New Homebuyer: more questions answered'. *Chartered Surveyors Monthly,* 8(2), October. RICS, London.

Hollis, M. (1986). *Surveying Buildings.* Surveyors Publications, London.

HMSO (1991) Statutory Instrument Number 2790 Private Water Supplies. HMSO, London.

HMSO (1991a). Part L. *Conservation of fuel and power.* Building Regulations 1991. HMSO, London.

HMSO (1991b). Part J. *Heat producing appliances.* Building Regulations 1991, HMSO. London.

HMSO (1991c). Part F. *Ventilation.* Building Regulations 1991, HMSO. London.

HMSO (1991d). Part H. *Underground Drainage.* Building Regulations 1991. HMSO, London.

IEE (1992). Guidance Note on Inspection and Testing Number 3. Institution of Electrical Engineers, London

Melville, I.A., Gordon, I.A. (1979). *The Repair and Maintenance of Dwelling Houses.* Estates Gazette, London.

Melville I.A., Gordon I.A. (1992). *Structural Surveys of Dwelling Houses.* 3rd edn. Estates Gazette, London.

National Trust (1996). *Private Water Supplies. Design, Management and Maintenance.* National Trust, Cirencester.

RICS (1997). *The RICS/ISVA Homebuyer Survey and Valuation. HSV Practice Note.* RICS Business Services, London.

RICS (1998). 'More homebuyers questions answered', *Chartered Surveying Monthly,* May, p.24. RICS, London

Shepard, A. (1993). 'Extra fans and open flues'. *Building Services Engineering and Technology*, 14(4), pp.143–49. Chartered Institution of Building Services, London.
WRC (1989). *Water Supply Bylaws Guide*, 2nd edn. Water Research Centre, Slough and Ellis Horwood, Chichester.

Chapter 11

External and environmental issues

11.1 Introduction

This chapter looks at those features on or around the site of the building being surveyed. These may not directly affect value but most will be important to the client. It is common for many of these elements to be looked at last when surveyors can often be tired, lacking concentration and thinking about their next appointment rather than carefully evaluating external features.

11.2 The site

11.2.1 Garages

The most important outbuilding on most properties is usually the garage. This has a direct impact on the value of the dwelling so its condition is important. The type of construction will vary and could include:

- a substantially built structure that uses the same constructional methods as the main house. This may include 225 mm brick walls, pitched and tiled roof with well-designed and manufactured doors and windows;
- an old sectional building made up of prefabricated metal and asbestos panels. Timber components are also common. Neglect over the years may have resulted in a poorly maintained structure that is close to collapse. If asbestos has been used, many of the components may pose a real danger to the occupants and so will need specialist removal. This can be very expensive.

Garages may have also developed other uses over time. Places to house animals, bird sanctuaries, DIY workshops, storage areas and even small offices from which the owner might work. They can present a variety of problems including unhealthy conditions, places where potentially explosive or hazardous materials are kept, through to serious breaches of planning regulations. The importance of correct reporting of garages was illustrated in Allen and another v. Ellis and Co (1990). In this case a surveyor failed to warn his client that the garage of a property had an asbestos roof. After moving in, the owner climbed on to the roof to carry out maintenance work, fell through and was injured. The judge felt that the surveyor should have told the client that walking on asbestos could be dangerous.

Key inspection indicators

- Ease of operation of garage door. Up-and-over types are prone to buckling.
- Condition of any inspection pits – are they properly and safely covered? Are they full of water?
- Are there of any boilers/gas appliances in the garage? If yes, assess as described in section 10.5.
- The standard of any electrical wiring and electrical equipment in the garage. The key is how the system is linked to the main house electrical system. It should have at least an earth leakage circuit breaker and any externally run cable should be properly protected.
- Has the use of the garage changed which might require planning permission?

11.2.2 Outbuildings

Although garages are variable in their construction they all have a similar function. Other outbuildings and structures that are found in the grounds of residential dwellings are much more variable and they are seldom maintained with the same degree of care given to the main building (Noy 1995, p.299). Consider the following range:

- outside toilets
- greenhouses
- summerhouses and 'follies'
- DIY built toolsheds and workshops
- old air-raid shelters
- animal cages and runs
- large timber-framed barns.

These may be in a very poor condition. Outbuildings are prone to rapid deterioration as few regulations control their design and construction. Most owners spend little time and money on preventative maintenance. Although the outbuildings are unlikely to affect either the value or whether the client buys the property or not, their condition will often impose significant liabilities on an owner. If the structures are unstable or contain hazardous materials then they could give rise to claims from neighbouring owners and third parties. Therefore, the outbuildings should be clearly described and their condition assessed. Clear warnings should be given if the structures could pose a danger to users. Clients should be made aware that demolishing and removing dilapidated and sometimes hazardous outbuildings can be very expensive.

11.2.3 Retaining walls

Many properties have retaining walls somewhere within their boundaries. Whether this is a small informal rubble wall holding back a flower bed or a tall, properly engineered retaining structure, it needs to be considered very carefully. This is particularly important where the retaining wall also marks the boundary line. Neighbouring properties may have acquired a right of support and should the wall collapse the owner may be liable for damages to their neighbour as well as for rebuilding a very expensive feature. Unlike other walls, a retaining structure has to resist very high lateral loading from the soil and water behind it. A rule of thumb for brick retaining walls is that the thickness at the base should be one third of its height (see figure 11.1). As retaining walls often reduce in thickness, the width of the wall at the top need not be the same as the width at its base. A well-designed wall should also be drained through the wall itself or via a land drain at its base. This is to prevent excessive loading during very wet periods.

The following indicators may give an insight into the effectiveness of the retaining-wall system:

- Are there any cracks to the wall? Does the wall appear to be vertical? Is there any apparent bulge? If so, these can be assessed using the middle third rule outlined in section 4.7 assuming that the thickness of the wall is as revealed at the top.
- What is the condition of the wall itself, i.e. what is the pointing and bedding mortar like? Are the bricks or stone breaking down through frost or chemical action? Defects like this can progressively weaken the wall.
- Does the wall appear well built using the same materials or is there evidence that the wall has been constructed using different methods at various times? This could suggest poor building practices and that the land has been further added to since the original retaining wall was constructed.
- Is there any provision for drainage through the wall? Reasonably sized drainage pipes at 1.5–2 m spacings would be appropriate. If they are present is there evidence that they are functioning properly? For example is there any staining or damp patch below them that suggest a flow of water passes through. If these drainage points do not relieve the build up of ground water behind the wall it could lead to a collapse. Some retaining walls may be drained by a land drain which cannot be seen so the absence of drainage holes should not be seen as a definitive indicator.
- Are there any buildings on the retained side close to the wall? As long as these are at least the height of the wall away from the edge then there should not be any problem (see figure 11.2). If they are closer they may be imposing an excessive

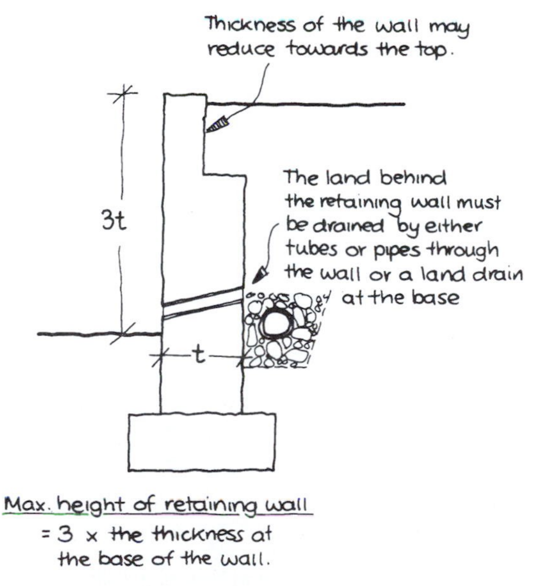

Figure 11.1 Sketch section through a typical retaining wall

load on the retaining wall and further advice may be needed. The same goes for stored materials. High loads can be imposed if there are piles of bricks, soil or other heavy materials on the retained side. If the neighbouring land can be inspected look out for any large ponds, streams or other water features. They may be the cause of potential problems – remember Rylands v. Fletcher!

If any of these signs raises concerns, further investigation by a structural engineer or other professional with geotechnical experience might be appropriate.

11.2.4 Drives, paths, patios and steps

The formal drives on a property can have a considerable impact on the usability of the dwelling. If there is a long distance from the building to the entrance then it is important that the drive is safe to use. This is important not only for the occupants but also for delivery and service vehicles that are usually much bigger than normal cars. Paths can allow good access to the garden but can also be dangerous if they are uneven. Patios can be delightful features that add to the quality of the accommodation if well designed and built. Steps must be stable and safe to use. The range of materials used for paths and drives include (Noy 1995, p.290):

- flexible paving such as tarmacadam and other bitumen-based materials;
- concrete paving;
- tiles and setts;
- gravel and hoggin.

In all cases these features should be described and their condition assessed.

Key inspection indicators

- Are the drives/paths in good condition with a sound, even and durable finish? Are there excessive numbers of weeds pushing up through the path material? This could indicate poor workmanship and inadequate thickness of the path itself.

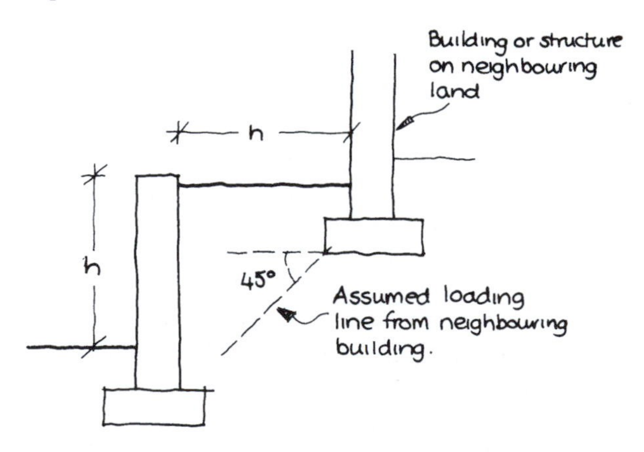

Figure 11.2 Relationship between imposed loads and retaining walls.

- Are there any signs to suggest the drives or paths are shared by neighbours? It could be an unadopted road that has significant maintenance liabilities.
- Is there any root damage from nearby trees or shrubs?
- Is there any evidence of cracks or areas of settlement to the surfaces? Where this occurs around or between inspection chambers it could point to drainage problems.
- Check to make sure that the drives, paths, patios, etc. next to the dwelling have not been built up above the dpc level or are blocking any air bricks, etc.
- Are there areas of ponding and poor drainage? Are there sufficient drainage points/ gullies?
- Are the surfaces level, stable and free of tripping hazards?

11.2.5 Trees and other vegetation

The effect of trees on buildings has been covered in section 4.6 and will not be repeated here. In respect of large trees that are in the grounds and are not within influencing distance of the building, they should be reported and the variety identified if possible. Although they might not undermine the foundations, trees can still be a liability.

Key inspection indicators

- Are the trees mature and in a poor condition? Are there any overhanging branches in danger of collapsing? These may require expensive maintenance.
- Are the trees close to the boundary wall. If so:
  - Are they likely to undermine the wall itself?
  - Are they within influencing distance of buildings on the neighbouring property?
- Are the trees blocking light from the dwelling or neighbouring buildings?
- Are the trees close to drain runs where the roots might cause damage?

Any recently planted ornamental hedges should also be noted. A row of 30 leylandii may seem to solve a privacy problem when they are 1.75 metres tall. Because they have a mature height of some 20 metres they can soon become a maintenance issue for the owner.

11.2.6 Boundaries

Establishing the ownership of the boundaries of the property is not the duty of the surveyor. The client's solicitor should be able to establish that through investigation of the deeds. The important role that the surveyor has is to report on the nature and condition of the boundaries. This can help:

- identify any encroachments on the land being inspected;
- help the client assess the financial implications that the boundaries may impose;
- highlight any future neighbour disputes.

The last point may seem frivolous but as Anstey (RICS 1990, p.22) states '... there is an awful lot of trouble to be found in the suburban back garden'. Disputes with neighbours over the position of fences, the condition of walls, the trespass of dogs, cats, chickens, sheep and even children because of inadequate fences have all been well documented. In short the surveyor can be the eyes and ears for the client's legal advisor, highlighting specific concerns that need to be checked further.

Ownership

Any surveyor experienced in boundary disputes will testify that there are no clear and fast rules in deciding the ownership of a boundary from visual signs. Over the years pragmatic repairs by a frustrated owner may contradict the information shown in the deeds. Despite this, Noy (1995, p.294) has attempted to identify some general rules (see figure 11.3):

- Garden walls
    - buttress or pier on one side only – belongs to the owner on whose side the buttresses are;
    - symmetrical piers – boundary runs down the middle – shared ownership;
    - plain brick boundary walls – boundary runs down the middle – shared ownership.

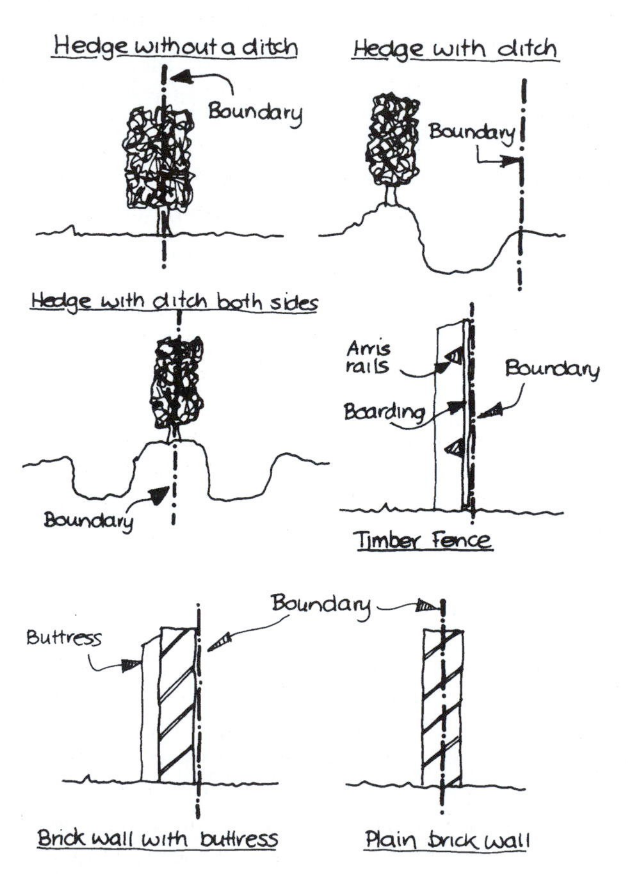

Figure 11.3 Typical boundary positions.

- Fences
  - post, arris rails and boarding – usually on the side of the person that owns the fence;
  - post and wire fence – usually constructed on the boundary itself so no presumptions can be made about ownership, This can also apply to modern ranch fencing, concrete panels dropped into concrete posts, etc.;
- Hedges and ditches
  - ditch with bank and possibly hedge on the top – boundary on the far edge of the ditch;
  - ditch without hedge – no presumption can be made but likely that boundary is in the middle although look for remnants of bank which could indicate ownership;
  - hedge – usual for ordnance survey to assume ownership runs down the middle.

Position of the fences

Most surveyors unless specifically instructed, will never see a copy of the deed plans. They will not have any idea of where the boundaries should be. But it may be possible to detect where the boundary has been repositioned and where this has happened the amount of this movement is important. If a fence has strayed a few millimetres from a true line then the courts are likely to decide that it is not worth considering the issue. In one case (Anstey 1990) it was found that a boundary was 175 mm out of line in a suburban garden that was 9 metres wide. The courts found that this was actionable. The following signs may indicate that a boundary has been moved:

- straight boundaries that have unusual 'kinks' in them;
- strangely shaped parcels of land that seem to encroach on the client's property;
- lengths of walls or fencing that are of a different constructional method than the rest of the boundary fences and follow a different line.

This is important because in certain circumstances the land that had been enclosed might actually become the neighbour's after the passage of time. 'Adverse possession' has strict requirements in the eyes of the law but the matter should be highlighted in the report and the client advised to seek clarification from their legal advisors.

Condition of the boundaries

If the boundaries are at the end of their useful life they can be very costly to replace. The main disadvantage is that they tend to be very long! If they have holes and gaps then a variety of animal life can get between properties and if they are unstable they might fall on people or property. This could expose the owner to legal claims especially if the boundary faces a public right of way. Therefore clients will be very interested in their condition. An excellently illustrated guide to the condition of boundary walls is included in *Good Building Guide 13* 'Surveying brick or blockwork freestanding walls' (BRE 1992).

KEY INSPECTION INDICATORS

- Brick and stone walls
  - Are the walls vertical or leaning? Is the wall sufficiently thick for its height?
  - Is there any evidence of cracking to the wall? Boundary walls often have the shallowest of foundations and are susceptible to subsidence. Many do not have expansion joints so thermal and moisture movement can be a problem.
  - Are the copings loose and liable to become dislodged?
  - What condition is the pointing and bedding mortar? Over the years these can be eroded by the elements leading to further accelerated deterioration. Sulfate attack can be common.
- Timber fences
  - What type of posts support the fence? Concrete posts generally last much longer than timber equivalents that can rot. Are the posts stable? Is there any evidence of remedial supports being fixed at the bottom of the posts (i.e. iron angles, spur posts etc.) that would suggest past failures;
  - Is the fencing of a good quality and properly fixed? Some fences can be a 'DIY' hotch potch of panels and timber not meant for the job.
- Other fences
  - Concrete fence posts and panels – still a common fencing method. Panels or planks slot into rebates in the side of the posts. These can be very robust but can become very unstable if the posts have not been set deep enough into the ground. Older installations can suffer from spalling of the concrete cover caused by corrosion of the reinforcement.
  - Post and wire fence – this can include steel angle or concrete posts with straining wires stretched between, usually clad with chain link fence. If the straining wires corrode or are broken, the fencing can easily distort and be vandalised.

11.2.7 Rights of way

Similarly to adverse possession, a right of way can be acquired over someone else's land through regular and continued use. This tends to be a problem associated with larger properties but rights of way issues have been known to affect the sale of smaller, inner-city sites.

Key inspection indicators

- A well trodden path or drive across the property not leading to the property.
- Stiles or even gates in the boundary walls/fences.
- Evidence of a footpath on an ordnance survey map of the area.

If another party appears to be using or passing through the property, it is vital that the client is notified so the deeds can be checked. A right of way may affect the enjoyment of the dwelling, in some cases affect its value and be very difficult to 'extinguish'.

11.3 Contaminated land

11.3.1 Introduction

Dwellings built on contaminated land can present two problems:

- a hazard to human health;
- a significant and detrimental effect on property value.

So far in the UK there have been few major pollution incidents on contaminated land sites that have lead to loss of life. However, incidents of properties exploding owing to the presence of landfill-derived methane gas are causes of major concern (Viney and Rees 1990). Therefore, it is very important for the survey to pick up this possibility.

11.3.2 Definition

Contamination is defined in the Building Regulations (HMSO 1991, p.8) as:

> Any material in or on the ground to be covered by the building (including faecal or animal matter) and any substance that is, or could become: toxic, corrosive, explosive, flammable or radioactive and so likely to be a danger to heath and safety.

The implication of building on contaminated land has recently become far more critical. In an effort to preserve the green belt, the government are encouraging the development of brown field sites within urban centres. Therefore it could affect a higher proportion of dwellings in the future. Also more forms of contamination have been identified. For example, the creation of methane gas from waste materials and its dispersal to adjacent areas has only been appreciated as a problem relatively recently. Therefore, both new and more established properties can be subject to contamination.

11.3.3 Recognising contaminated sites

Many contaminated sites can be identified through planning records or local knowledge of previous uses. The problems begin when the surveyor does not know the area or the site was developed a number of years ago. Spotting a contaminated site is not easy. Many liquid and gaseous contaminants are mobile and can affect neighbouring land. The building regulations state that gaseous substances can migrate by as much as 250 m. So surveyors need to take a broader view of the neighbourhood. Sites that are likely to contain contaminants include:

- asbestos and chemical works;
- gas, coal and coal by-product plants;
- landfill and other waste deposit sites;
- metal mines, steel-making plants and metal-finishing works;
- oil storage and distribution sites;
- railway sidings and depots;
- nuclear installations;

- scrap yards;
- former sewerage works, etc.

The environmental health officer of the local council might be able to help with identifying such sites in their area. Another good source of information is old ordnance survey maps. These might show the sites of old brickworks, railway sidings and sawmills – all clues that toxic substances might be buried on the site. Caution must be exercised when interpreting these clues. They rely on the surveyor having a good knowledge of what materials were used in the former process and where they were stored. This could mean the difference between isolated pockets of contamination and the need for whole-site treatments. In these cases employing a specialist may be worthwhile.

Key inspection indicators

If desk-top studies fail to turn up any information about a particular area, visual clues should be checked. These includes:

- large open sites that appear to have an uneven or unnatural topography;
- surface features of open ground that have the following characteristics:
    - an absence of vegetation or growth that is poor or unnatural;
    - unusual surface materials that may be strangely coloured and contours that suggest wastes and residues;
    - fumes and odours suggesting organic chemicals at very low concentrations.
- road and street names can give important clues, e.g. Pit Street, Cemetery Road, Pond Street, Tannery Row, Gas Street Run and Railway Terrace are just a few taken from the A–Z map of a typical town. In all of these cases the use suggested by the name was no longer visually evident.

Where evidence is noted, further desk-top studies or enquiries by the surveyor may be required. If contamination is confirmed or suspected then a referral to a specialist company would be appropriate.

11.3.4 Implications of contamination

If a site has been contaminated, making safe the subsoil can be expensive. There are three broad options:

- removal of the contaminated ground itself;
- filling of the ground by at least 1 metre thickness of inert material;
- sealing the site with an imperforated barrier. In these cases cultivation in the ground above the barrier is usually restricted.

Not only are these options expensive but in some cases it is impossible to identify the polluter let alone get them to pay. Therefore contaminated land can seriously affect the value of dwellings.

11.4 Radon

11.4.1 Introduction

Not all dangerous contaminants are unnatural. Radon is a naturally occurring, radioactive, colourless and odourless gas. It is formed in small quantities by the radioactive decay of naturally occurring uranium and radium in the ground. Radon can move through the subsoil and in some cases collect within the voids of buildings. Human exposure to high levels for long periods can increase the risk of cancer (Richardson 1991, p.179). The National Radiological Protection Board (NRPB) have identified the following counties as areas where the level of radon is potentially high:

> Cambridgeshire, Derbyshire, Leicestershire, Lincolnshire, Northamptonshire, Shropshire, Staffordshire, Warwickshire, Cornwall and Devon, Dorset, Gloucestershire, Hampshire, Hereford and Worcester, Oxfordshire, Somerset, Wiltshire, Cumbria, Lancashire, Northumberland and North Yorkshire.

For more precise location of these risk areas the NRPB should be contacted directly. Another source of information would be the local council for the area especially the Environmental Health Department.

Actual radon levels in homes can be measured by small plastic detectors that can be supplied the NRPB for a small charge. These should be left in place for three months. Because of this period of monitoring, the NRPB advises that householders should not let uncertainty about radon levels affect the purchase of their homes. Instead they should begin to monitor levels when they move in to the dwelling. If the levels of radon are higher than the 'Action Level' they can usually be reduced without difficulty.

11.4.2 Typical advice

Following consultation between the NRPB and the RICS it was decided that it was only necessary to make specific reference to radon gas where the dwelling is in an area that has been identified as a Radon Active Area. The following statements have been suggested.

For Mortgage Valuation reports:

> The National Radiological Protection Board has identified the area in which the property is situated as one in which, in more than one per cent of dwellings, the level of radon gas entering the property is such that remedial measures are recommended. The applicant should seek further advice on this.

For Homebuyers surveys:

> It is not possible in the course of a survey to determine whether radon gas is present in any given building as the gas is colourless and odourless. Tests can be carried out to assess the level of radon in a building. These are available by post from the National Radiological Protection Board and other approved laboratories. The minimum testing period is three months. The NRPB strongly advise against using shorter term tests as they can give misleading results.

Table 11.1

Remedial measures	*Average cost (£)*	*Radon reduction (%)*
Underfloor extraction (sump)	750	90
Positive whole-house ventilation	450	60
Increased underfloor ventilation	350	50–60

If tests have not been carried out it is recommended that they are. Where radon is discovered, it has been the experience of the NRPB that it is not expensive in proportion to the value of the property to effect the recommended remedial measures.

11.4.3 Remedial measures

Remedial measures are generally simple and effective depending on the level of the gas. It may include sealing around suspended and solid floors, ducts and openings but the most effective options usually include the provision of additional ventilation. The NRPB give guideline costings for different types of measures along with typical reduction factors (see table 11.1).

More advice can be obtained from:

- National Radiological Protection Board, Radon Survey, Chilton, Didcot, Oxon OX11 0RQ.
- Building Research Establishment, Garston, Watford WD2 7DR.
- The Radon Council Ltd., PO Box 39, Shepperton, Middlesex TW1 7AD (List of builders with radon remediation experience).

11.5 Deleterious materials

11.5.1 Introduction

The dictionary definition of 'deleterious' is 'harmful or injurious to the health'. The presence of these materials in a dwelling could in severe cases affect the health of the occupants. The value of the property could also be affected. The full extent of this is unlikely to be appreciated because often the materials appear to be performing an adequate function. For example, asbestos lagging to pipes may still be as it was originally designed and will not present an immediate danger. The problem is the hidden danger of future deterioration of the material. The other complicating factor is the attitude of the occupiers. Some will not want to share their home with a potentially dangerous material and will want it completely removed. Others will be prepared to undertake appropriate maintenance work to keep the material in a safe condition. The latter approach leaves an uncertainty as ever-increasing standards and research may result in the need for further work in the future. If this doubt is accepted a minimalist approach can give a base value reduction.

The typical substances included under this heading include:

- asbestos
- lead
- urea formaldehyde foam.

Some aspects of these have been covered under other sections and references are made to the relevant text.

11.5.2 Asbestos

Breathing in asbestos dust can lead to the development of asbestos-related diseases such as cancer of the chest and lungs (Addison 1990, p.87). The vast majority of the people currently dying from these diseases were exposed to the substance during the 1950s and 1960s when the material was more widely used. Surveyors should be able to:

- identify the presence of asbestos;
- give broad advice to the client on the implications of having it in their future home; and
- give guidance on what should be done about it.

Identifying asbestos

The risk from asbestos has to be kept in balance. Materials that contain a high percentage of asbestos are more easily damaged. Sprayed coatings, laggings and insulating boards are more likely to contain the more harmful blue and brown asbestos and can be made up of 85% asbestos fibres. These materials pose the greatest risk. Asbestos cement products on the other hand can contain only 10–15% asbestos often of the white variety. The fibres are tightly bound into the material and will only give off dust if damaged or broken.

KEY INSPECTION INDICATORS

- Asbestos is usually present in dwellings built between 1950–1980 especially in non-traditional steel-framed dwellings.
- High-risk locations include:
  - sprayed coatings or laggings as thermal insulation to boilers and flues;
  - insulating boards used for thermal insulation, fire protection to floors and service ducts, etc.
- Low-risk locations:
  - asbestos cement products including corrugated roof sheets; wall claddings; linings to the soffits of external features such as roof eaves, balconies, porches, covered ways, etc.
  - ceiling lining in garages;
  - gutters, rainwater pipes and water tanks.

Assessing the condition of the asbestos

The next stage is to identify the risk of asbestos fibres being released into the air. The risk is high if:

- the material is being disturbed. For example it is in a position where it is likely to be knocked and scraped by normal usage;
- the surface of the material is damaged, frayed or scratched;
- surface sealants (paint, etc.) are pealing or breaking off;
- the material is becoming detached from its base, e.g. sectional insulation is falling away from a flue, etc;
- there is dust or debris in the immediate area.

The risk is low if the asbestos component:

- is in good condition;
- is not likely to be damaged; and
- is not likely to be worked on.

Advising the client

If the asbestos is in good condition it is likely to be safest to leave it in place and introduce a management system to keep it safe. This could include labelling the asbestos product and keeping a record of where the asbestos is. This should be left with the deeds so it can be passed on with the sale of the dwelling to protect future owners and users. Slightly damaged asbestos can be made safe by repairing it and sealing it with a special coating.

If the asbestos is in poor condition and/or is likely to be disturbed during routine maintenance work it must be removed completely.

Because of the special health risks, whether the asbestos is being sealed in position or removed, the Health and Safety Executive (HSE) advise that a licensed contractor is used. A list of these organisations will be available from the local HSE office.

This work is of a specialist nature and so can be expensive. Depending on the extent this could affect the value of the property and certainly endanger the client's health if the work is not done properly. Therefore further specialist advice should be sought if asbestos products are identified.

11.5.3 Lead

The problem of lead water pipes has been fully discussed in chapter 10. The other way lead can pose a health risk is through lead in older paint coatings. There is a risk of inhalation of lead dust during DIY activities such as paint stripping and sanding. It poses a particular danger to young children. Studies have shown that household dust is contaminated with deteriorated lead paint and is one of the main sources of lead exposure. As much as 50% of the lead intake of two-year-old children is derived from dust ingested by hand to mouth activity (Osborn and Meeran 1998).

The use of lead in paint has been gradually phased out since the 1950s but enough exists to present a potential hazard in as much as 50% of the UK housing stock. Like asbestos the potential risk has to be kept in perspective. The main triggers for concern are:

- pre-1960 properties. Generally speaking the older the properties the greater the risk;
- large areas of loose and flaking paint that are creating residues of dust and smaller particles.

Lead-based paint be treated by overcoating sound material with modern lead-free paint. Where the original coating is unstable then removal may be the only option. Because of the health risks, stripping of lead-based paint should involve taking extra precautions including:

- increased levels of ventilation
- high efficiency vacuum cleaners
- precautions taken to limit dust from spreading around the dwelling, e.g. plastic curtains to doorways, etc.

Although the safety measures are not as rigorous as with asbestos, they will add to both the cost and inconvenience of the work.

Where a surveyor is concerned, Osborn (1998) suggests that they should advise the client about the potential risk of lead-based paints and suggest more advice should be taken. Additional information is available from the Paintmakers Association and the Department of Environment, Transport and the Regions.

11.5.4 Urea formaldehyde (UF) foam

The use of urea formaldehyde (UF) foam to insulate cavity walls of existing buildings was very common 10–20 years ago. Well-documented cases resulting in irritation to occupiers' eyes and throat (Greenberg 1990) led to a steep decline in the use of this material. During a typical survey of a house it will be difficult to establish whether the cavity wall has been insulated. Clues could include:

- evidence of 'pointed-up' holes to the external elevations. These are usually 12–15 mm in diameter and are usually located at the junction of the vertical and horizontal mortar joints;
- excess foam can sometimes be seen in the roof space where it has squeezed out through gaps in gable walls, wall plates, etc.;
- information from the vendor.

If it is suspected that UF foam has been used then the following assurances should be requested;

- the inner leaf of the wall should be built of either bricks or concrete blocks;
- the contractor that installed the foam should have been certified under the appropriate British Standard Institute (BSI) Registration scheme;
- the material and installation should have been carried in accordance with the appropriate standards (BS 5617:1985 and BS 5618:1985).

If these assurances cannot be given, clients should be notified of the potential irritant effects.

References

Addison, J. (1990). 'Asbestos'. In: *Buildings and Health. The Rosehaugh Guide to the Design, Construction, Use and Management of Buildings*, pp.85–104. RIBA Publications, London.

Allen v. *Ellis & Co* (1990). 1 EGLR 170.

Anstey, J. (1990). *Boundary Disputes and How to Resolve Them*. RICS Books, London.

BRE (1992). 'Surveying brick or blockwork freestanding walls'. *Good Building Guide 13*. Building Research Establishment, Garston, Watford.

Greenburg, M. (1990). 'Plastics, resins and rubbers'. In: *Buildings and Health. The Rosehaugh Guide to the Design, Construction, Use and Management of Buildings*, pp.133–54. RIBA Publications, London.

HMSO (1991). Building Regulations 1991. Part C. *Site Preparation and Resistance to Moisture*. HMSO, London.

Noy, E.A. (1995). *Building Surveys and Reports*. Blackwell Science, Oxford.

Osborn and Meeran (1998). 'Warn home-buyers of the dangers of lead paint'. *Chartered Surveyors Monthly*, April, pp.36–37.

Richardson, B.A. (1991). *Defects and Deterioration in Buildings*. E&FN Spon, London.

Rylands v. Fletcher (1868). L.R. 3HL.

Viney, I.F., Rees J.F. (1990). 'Contaminated land: risks to health and building integrity'. In: *Buildings and Health. The Rosehaugh Guide to the Design, Construction, Use and Management of Buildings*. RIBA Publications, London.

Chapter 12

Non-traditional housing

12.1 Introduction

In the early 1980s the increasing popularity of the 'right to buy' policy resulted in thousands of non-traditional dwellings entering the private housing sector. As these types of properties have their own unique problems, surveyors should have an insight into:

- why and how they were built;
- how to recognise a non-traditional dwelling;
- what common defects can be expected;
- the implications of inspecting one;
- the impact upon price relative to traditional construction.

12.1.1 Historical development of non-traditional housing

The First World War had slowed down house building and a shortage of skilled labour had resulted in an urgent need for new houses. The Tudor Walters Report in 1918 tried to co-ordinate a centralised view of the industry but largely failed and only a few industrialised buildings were developed by contractors and manufacturers. These included:

- Waller System – precast storey height slabs;
- Dorlornco System – promoted by Dorman Long, used a steel frame with rendered ribbed metal lathing externally and 50 mm clinker blocks internally;
- Atholl House – steel plate externally supported by steel 'T' frame;
- Boswell House – precast reinforced concrete columns at corners with cast in situ concrete clinker cavity walls.

The Second World War led to a similar dislocation of the construction industry. Special legislation was enacted that aimed to produce large numbers of houses quickly. The 'prefab' was the first to be built followed by the Aluminum Bungalow, the Arcon, Uni-seco and the Tarran types of prefabricated dwellings. The politicians encouraged the use of non-traditional systems to speed up the post-war re-construction and initially approved the BISF and the Airey House types. Many different varieties followed with national distributions. Others were limited to specific regions with a significant number confined to cities and towns, often in quite small numbers.

As the 1950s progressed these non-traditional houses still could not satisfy the demand for house building. Local authorities were encouraged to build upwards through generous government subsidies. Prefabricated high-rise blocks soon overtook low-rise development. As a result hundreds of different types of non-traditional dwellings were developed and fall into the following categories:

- precast reinforced concrete (PRC);
- steel framed;
- steel frame with brick/concrete/block cladding;
- timber framed;
- Cast in situ concrete types.

12.2 Non-traditional housing and the market

12.2.1 Background

In the early 1980s, the BRE investigated a fire in a precast reinforced concrete (PRC) house type. During this study they discovered that the concrete had deteriorated through a process called carbonation. The extent was so surprising that further investigations were carried out in other similar dwellings. This revealed that a large number of PRC properties either suffered or had the potential to suffer from the same defect. By this time, thousands of people had bought their homes, unaware of the extent of the physical problems with them.

After early attempts to award grants to homeowners, the Government brought in the special provisions under the Housing Defects Act 1984 (consolidated into the Housing Act 1985). This allowed the Secretary of State or the local authority to designate a particular type of dwelling as 'defective by reason of their design or construction'. Once designated, the owner of the dwelling could apply for a grant to fund either remedial work or repurchase by the original landlord depending on the financial viability of each option. Each type has a cut-off date before which the owner must have acquired the interest.

A total of 24 systems were designated which included:

> Airey, Boot, Boswell, Butterly, Cornish, Dorran, Dyke, Gregory, Hawksley SGS, Myton, Newland, Orlitt, Parkinson, Reema, Schindler, Smith, Stent, Stonecrete, Tarran, Underdown, Unity, Waller, Wates, Wessex, Winget, Woolaway.

This list contains 26 types but often 'Schindler and Hawksley SGS' and 'Unity and Butterly' are combined to give a total of 24. Some lenders have a longer list as they include those dwelling types that were designated in Scotland and may include:

> Lindsay, Blackburn-Orlitt, Boot, Dorran, Myton-Clyde, Orlitt, Tarran, Tarran Clyde, Tee Beam, Unitroy, Whitson-Fairhurst, Winget.
>
> Only Orlitt was designated in Northern Ireland.

The Smith and Boswell Houses were the last to be designated in 1986 and 1987 respectively. This left a large number of other house systems that were not designated including all of the steel framed types. Figures 12.1 to 12.5 show just a few of the most

Figure 12.1 An Airey house. The shiplap concrete panels are the distinctive feature of this house type.

Figure 12.2 Unity house. This type is faced externally with concrete panels.

Figure 12.3 Smith house. This house is clad with clay tiled panels that resemble traditional brick and so it easy to mistake for a traditional dwelling.

Figure 12.4 Boot house. The externally-rendered exterior can make them difficult to identify. Vertical cracking at the corners is a very common feature.

Figure 12.5 Boswell house. The Boot and Boswell houses can appear very similar. When they are well maintained and painted they can be mistaken for a 'Wimpey no fines' house which is not designated.

common types. The Building Research Establishment (BRE) have produced a large number of Information Papers and reports on prefabricated houses. A few publications have been identified in the reference section of this chapter under 'Further reading'.

Home owners had ten years to claim assistance under the scheme. Although local authorities have the discretion to extend the period by a further 12 months, no further grants have been available since March 1998.

12.2.2 Administering the scheme

The National House Building Council (NHBC) set up PRC Homes Ltd. to administer the licensing and insurance of repair systems associated with the passing of the Housing Defects Act 1984. Various structural engineering firms and other organisations applied and obtained licences for the repair of the different standard housing types. There were three parties involved with the design and execution of the repair work carried out under the PRC Homes Ltd. scheme:

- the Designer – a professionally qualified architect, engineer or surveyor who has designed the repair system;
- the Repairer – the builder who carried out the repair;
- the Inspector – a professionally qualified architect, engineer or surveyor approved to check that the work is in accordance with the original repair system.

If the repair was completed in accordance with the scheme then PRC Homes provided Repair Insurance which covered:

- loss due to repairer bankruptcy;
- all defects that occur during the first two years;
- structural defects that occur between the 3rd and 10th years after completion.

It is important to distinguish between two terms:

- licensed systems – this is the technical repair scheme of the particular housing system;
- licensed scheme – this is the framework that the repair is carried out within, i.e. the design, supervision and insuring of the repair project.

In September 1996, PRC Homes officially ceased approving new schemes and licences and will only administer 'run-on' insurance until September 2006. The ownership of the original licences still remains with the agencies that obtained them so they can still be offered to dwelling owners requiring the repair but it will not:

- be carried out by a registered builder (the repairer);
- be checked by a registered inspector; and
- will not have a 10-year insurance cover.

Consequently, repair projects carried out using licensed systems since September 1996 must be viewed with great caution. They will not necessarily have the overall assurance of projects carried out under the original licensed scheme.

12.2.3 Other repair schemes

In addition to the official licensed schemes, non-traditional houses may have been repaired by other means:

- PRC types repaired before PRC Homes Ltd. came into operation (1984);
- local authority designed and supervised schemes;
- privately organised contracts;
- DIY repairs.

All of these approaches must be judged on their own merits but many will prove to be inadequate if judged by the PRC Homes criteria. Even where local authorities have used licensed systems, they may not have followed the procedures fully. For example, financial restrictions may have led to a reduction in the scope of the specification resulting in a different standard of repair. Consequently when assessing non-licensed schemes, the following criteria should be taken into account:

- Who designed and supervised the scheme? Were they recognised professional people with experience of repairing non-traditional buildings?
- Who carried out the building work? Did they have experience of this sort of work? Were they appropriately supervised?
- Does any certificate of final completion/structural worthiness exist? Is there any form of insurance-backed guarantee?

If these inquiries fail to confirm the quality of the repair work then the suitability of the property must be seriously questioned.

12.2.4 Attitudes of the lending institutions

Faced with this very complex situation, lending institutions have issued a wide variety of guidance ranging from wholesale refusal of certain types through to judging each case on its own merits. The list below is a summary of the advice recently given by leading lending authorities.

Dwellings not suitable for mortgage

- Unrepaired PRC homes and repaired PRC houses without scheme repair.
- Steel framed and steel clad houses without a structural survey.
- Older style system-built timber framed houses (especially pre-1970).
- In situ/poured mass concrete houses that are badly cracked.
- Most large-panel system-built properties (especially multi-storey).
- Buildings constructed entirely of timber apart from those with appropriate Agreement Certificates.
- Timber-frame properties with fully insulated cavity.
- Certain properties containing Mundic concrete (mainly Cornwall area).
- Underground properties.

The precise requirements of each lending institution should be sought and carefully followed.

12.3 Inspecting and advising on non-traditional properties

Particular care is needed when inspecting non-traditional houses as many may have been altered or improved. New roofs, rendered finishes and windows can easily mask the main characteristics of a non traditional dwellings. Surveyors will be expected to be familiar with the published material of the main systems and have some knowledge of local types and their locations. Many local councils publish useful guides that help identification.

12.3.1 Inspection strategy

The inspection strategy can be broken into a number of stages:

- Identify that it is a non-traditional dwelling. Look for the following clues:
    - chimney stack above and below roof covering. Often constructed of mass concrete or precast concrete blocks;
    - cracks to joints of panels;
    - loft space inspection of party walls. Even if the external elevations have been covered over, concrete sections can still be seen in the loft;
    - size of windows – often quite small in comparison to modern standards because they need to fit between structural members;
    - contained within an area of well-established housing (40–50 years old);
    - neighbouring properties. Non-traditional dwellings were always built in groups. Unless all neighbouring properties have been demolished or improved then a simple walk around the local area might reveal a few dwellings in their original state.
- Identify the particular system. If necessary take photographs and identify later.
- Establish whether the system is designated or not.
- Make a broad assessment of condition by following the trail of suspicion. For example are there any defects that may cause concern? For example:
    - spalling, cracking, splitting of concrete cover revealing the reinforcement below;
    - bulging or leaning of walls;
    - corrosion of the cladding;
    - poor seals at panel junctions, etc.
- Make a note of any repairs, renovations or extensions that have recently been completed. Ask for full details of the schemes including planning and building regulations, etc.
- Obtain any guarantees, reports, specifications that have been produced on the dwelling.

Disguised non-traditional housing

Local authority repair schemes, unlicensed contractor packages and DIY alterations may result in the masking of the essential visual characteristics of many non-traditional dwellings. Figures 12.6 and 12.7 show typical examples. Surveyors have been known to mistake this type completely. Whatever the extent of the improvement work, the best indicators are still the chimney, the loft space inspection and the presence of non-disguised dwellings in the neighbourhood.

Figure 12.6 A reclad Unity house. The new cladding is easy to spot when the other semi-detached house is unimproved but what if the whole street has been altered?

Figure 12.7 This Orlitt house has been completely refurbished. Not only has it been reclad but extended as well. This changes its proportions completely. The give-away is the precast concrete chimney stack to the left-hand side of the photograph.

12.3.2 Giving advice

The precise guidance of every lending institution should be followed in each case. In the absence of this, the following advice should provide a basic framework. It is an amalgamation of the attitudes of several lending institutions.

If the dwelling is a PRC designated type and it has been repaired:

- It will be acceptable if the appropriate PRC Homes documents are available in full.
- If the repair is not covered under a PRC Homes scheme then request certification that the repairs are of a standard equal to the PRC approach. Even if this is obtained it must be pointed out that this may not be acceptable to some lenders.
- If no certification is available then it should be considered unacceptable for mortgage purposes until a structural survey and the appropriate repairs have been completed.

If the dwelling is a PRC designated type but is unrepaired:

- A repair scheme equal to the former PRC Homes scheme should be organised. An appropriate certificate should be kept with the title deeds. Point out that this repair may not be acceptable by other lenders and could restrict future saleability.

If the dwelling is an in situ concrete type of dwelling:

- It will be generally acceptable if it does not suffer from obvious defects.
- If cracking is present or it was built before 1950 then some lenders require a more detailed assessment of the property.

12.3.3 Other PRC systems

Dwellings should be assessed taking account their merits including:

- resaleability;
- location – from a value point of view as well as exposure. Some systems are particularly vulnerable to rain penetration for example;
- condition of repair.

In most cases a full building survey is almost always requested and warnings given about possible future restrictions on marketability and saleability.

12.3.4 The special case of steel-framed dwellings

Despite representations to the Government by groups of owners, no steel-framed dwellings were designated. One of the main reasons given was that even if a steel-framed dwelling does begin to corrode, the repair is fairly straightforward. The same could not be said of concrete buildings where the only way of properly resolving carbonation defects is by removing the concrete components completely.

Therefore, most steel-framed houses have to be taken on their own merits. Some lenders automatically call for a building survey while others ask for this only if the original inspection identifies defects that give cause for concern. The main type of steel-frame dwelling is the British Iron and Steel Federation (BISF) type. Assuming a typical example, worrying visual signs might include (see figure 12.8):

- cracking to any rendered finishes in the position of underlying steel stanchions. This could suggest corrosion and expansion of the structural frame;
- excessive corrosion to the steel sheeting to the walls.

Having a building survey is good advice for any prospective owner but with a steel-framed dwelling the extent of the building survey would have to be more extensive. The BRE (1987) state that the surfaces of the steel frame should be exposed and inspected in vulnerable locations. In most cases this could be done by an inspection via a boroscope with only minimal disruption to the internal surfaces. In others, the removal of small areas of internal linings may be required. At the very least this will be expensive and

Figure 12.8 The front elevation of a typical BISF house.

disruptive. It can frustrate the sale process and so clients should be made aware of the impact this might have on marketability and the value of the house.

12.4 Timber-framed dwellings

12.4.1 Introduction

In this country, houses have always been constructed using timber framing in one form or another. The modern form was developed after World War Two as the interest in industrialised buildings increased. A timber frame was of particular appeal as it was promoted as a 'warm' building. Essentially the different forms were based on timber stud walls covered by a variety of directly applied claddings including:

- timber
- plywood
- tiles and
- brick.

Insulation in earlier types was often minimal consisting only of a 25 mm layer of fibreboard. A layer of breather paper was fixed to the external face of the studs behind the cladding with foil-backed plasterboard internally to provide a vapour check (see figure 12.9). Changes in the building regulations and development of the various systems have resulted in most modern timber framed construction consisting of the following:

- cladding – this is usually brick but can be any masonry material including stone;
- cavity – usually 50 mm;
- breather membrane – a micro-porous layer that will allow water vapour out but will not let water in;
- sheathing plywood – usually about 13 mm thick nailed or stapled to the timber studs. Very important for the stability of the whole structure;
- timber-frame member – usually ex 50 × 100 mm section at 450–600 mm centres;

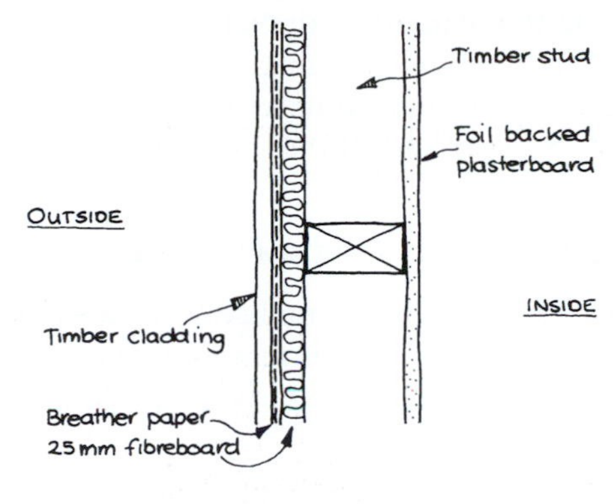

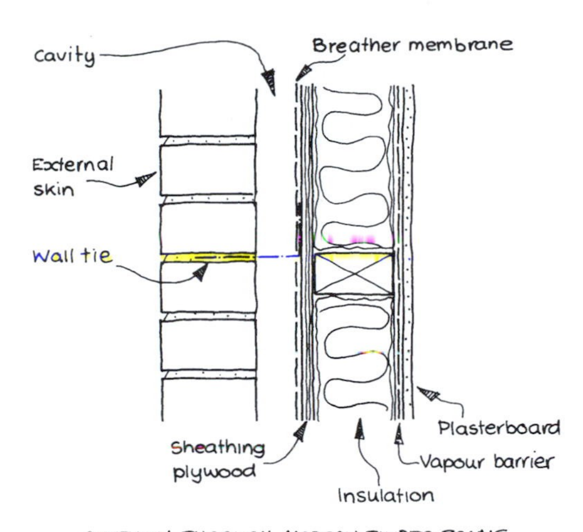

Figure 12.9 Sketch section through two different types of timber framed walls.

- insulation – this usually fully fills the cavity within the timber frame. Commonly mineral fibre quilt;
- vapour control layer – must be at least 500 gauge polythene sheet. Because it is punctured by nails and other fixings, the name 'vapour barrier' has been dropped;
- interior lining – usually plasterboard.

Although timber frame construction is increasing in popularity and modern forms are fully approved by local building control departments and the NHBC, they have to be designed and constructed correctly. They are dependent on good construction practice because of the vulnerability of the timber components. When assessing these dwellings there are two important stages:

- identifying that the dwelling is timber framed. If one is mistaken for a traditionally built property then important problems may be missed;
- once properly identified assessing the condition of the actual dwelling.

One additional issue that can affect timber-frame dwellings in 'holiday' areas are conditions that may imposed by the original planning permission. The local planning authority may have limited the number of weeks per year that the 'holiday' homes can be occupied. This may be a problem if the owner wishes to live there all year round.

12.4.2 Identifying timber-framed dwellings

It is very easy to miss that a dwelling is timber framed especially when carrying out a valuation inspection. There is not just one feature that distinguishes timber frame from a traditional dwelling but a combination of features (Marshall 1998):

Externally

- If the wall has a claddding directly fixed to the timber frame then the overall thickness will be much less than a conventional cavity wall.
- Windows tend to be fixed on the internal skin and give rise to deep reveals.
- Movement joints with mastic infills are often provided around all the windows, beneath the eaves and sometimes at the party wall positions. On three-storey properties these gaps can be up to 15–18mm wide.
- Weep-holes at all the normal locations (i.e. above the openings, just above damp-proof course, etc.) but also at first and other intermediate floor levels. This may indicate a cavity tray dpc over a horizontal fire stop.

Internally

- The party walls in the roof space may have a plasterboard or mineral board finish. The internal face of the gable wall may also be similarly lined. In older timber-framed properties this boarding could be asbestos based. Some older dwellings may also have 'cross-walls' of brick or block with only the front and rear walls consisting of timber frame.
- If it can be seen, the wall plate of a timber-framed dwelling that supports trussed rafters will be in 'planed' or finished timber. This is because it is usually made in the factory. Wall plates of traditionally built properties tend to be in 'sawn' or rough timber.
- When knocked, the internal linings of the external walls will give a hollow sound. Care must be taken as dry-lined solid walls can also give a similar response.

12.4.3 Assessing timber-framed dwellings

There are a number of publications that advise on inspection routines that should be followed when assessing the condition of timber-framed dwellings (BRE 1995). Many of these involve a variety of destructive tests including drilling into sole plates, removing plasterboard and skirtings and inspecting fire barriers in cavities. All of these are beyond the scope of most standard surveys. Therefore, once a dwelling has been identified as timber framed, the surveyor should look for a number of key indicators that may trigger the need for further investigations.

Key inspection indicators

EXTERNALLY

- Establish whether the cladding is fixed directly to the frame. These can be more vulnerable than those that have a cavity between the cladding and the frame.
- Determine the age of the dwelling and if possible the system used. Older timber-framed dwellings will be more likely to have problems (say before 1985).
- Make a note of the orientation of the dwelling and the exposure of the surrounding terrain. Fully exposed sites warrant greater levels of attention during the survey.
- Look out for evidence that the cavities of the external walls have been filled with insulation. Large 10–12 mm circular drill holes at the junction of the vertical and horizontal mortar joints are the typical signs. Insulated cavities will dramatically affect the build-up of moisture within the wall creating ideal conditions for rot to develop.
- Check the weathertightness of the building envelope. Critical areas include:
    - chimney to roof details – especially important as frame shrinkage may cause maximum movement here. Check inside the loft as well;
    - junction of different claddings;
    - deteriorated claddings – any damaged, decayed or displaced cladding panels or boards. Prod for decay especially on the most exposed elevation at corners or near foot of wall;
    - damp patches, green staining or staining on cladding – this suggests excess moisture and should be investigated further;
    - weather seals around all openings – check that the mastic seals are in good condition and still effective.
- Structural integrity – most problems are associated with brick-clad housing and are attributable to inadequate allowance for movement between the frame and cladding. This can result in cracking, bowing and leaning in the cladding especially around openings. Other critical locations include:
    - eaves, verges, porches and balconies;
    - extensions and new additions – check that these have the correct movement joints between the new building and main house;
    - roof – check for any roof-tile displacement that may suggest roof sagging.

INTERNALLY

Linked to the external inspection, visible internal signs can help to confirm a potential problem and a need for further investigation. There are three main defects to look for:

- **Dampness related** – any evidence of water staining, mould growth, breakdown of plaster surfaces and decorations, etc.
- **Structural problems** – cracking or bowing of the floors and ceilings, splitting along plasterboard joints and popping of nail heads.

- **Evidence of alterations or repair** – timber-framed dwellings are sensitive to inappropriate repair and alteration work. If a vapour control layer is punctured or a plywood stressed skin is cut through, the integrity of the whole may be affected. Look for the following indicators:
  - plumbing leaks and inferior plumbing installations;
  - changes to services that may have damaged the vapour control layer or affected the fire resistance of the party wall. Examples would include new electrical sockets, TV aerials, etc.;
  - installation of new windows that may not be properly sealed with dpcs or have appropriate movement joints, etc.;
  - structural alterations that may have weakened internal load-bearing partitions, e.g. 'through' lounges, new door openings, serving hatches, enlarged bedrooms;
  - sagging or springy floors;
  - any gaps or spaces through or over the separating walls in the loft that may compromise the fire resistance.

12.4.4 Giving advice

Like all assessments, giving advice about the condition of a timber-framed house is a matter of judgement. No one sign or symptom should automatically result in a recommendation for further investigation. It will often be a matter of balancing the inadequate features against the damage that is likely to arise.

References

BRE (1987). *Steel-Framed and Steel Clad Houses: Inspection and Assessment*. Report no. BR 113/1987. Building Research Establishment, Garston, Watford.

BRE (1995). 'Supplementary guidance for assessment of timber framed houses. Part 1: examination'. *Good Building Guide 11*. Building Research Establishment, Garston, Watford.

Marshall, D., Worthing, D., Heath, R. (1998). *Understanding Housing Defects*. Estates Gazette, London.

Further Reading

BRE Reports on individual house types:

no. 34 (1983). 'Structural condition of Boot pier and panel cavity houses'.

no. 35 (1983). 'Structural condition of Cornish unit houses'.

no. 36 (1983). 'Structural condition of Orlitt houses'.

no. 38 (1983). 'Structural condition of Unity houses'.

no. 39 (1983). 'Structural condition of Wates houses'.

no. 40 (1983). 'Structural condition of Woolaway houses'.

no. 52 (1984). 'Structural condition of Parkinson framed houses'.

INFORMATION PAPERS

10/84 (1984). 'The structural condition of some prefabricated reinforced concrete houses'.

14/87 (1983). 'Inspecting steel houses'.

15/87 (1983). 'Maintaining and improving steel houses'.

Chapter 13

Other issues

13.1 Introduction

When advising a client on the condition of a property, the surveyor has to be aware of more than the state of the building. For example if the dwelling has been altered, extended or in some cases just repaired, permission may have been required from the local authority. Without this, the work may be sub-standard and in breach of local by-laws and vulnerable to enforcement action. This section will look at three factors that may affect the appraisal process:

- Building Regulations
- planning permission
- listed building consent.

In all of these cases the role of the surveyor is to:

- identify work that has been carried out which may require permission from the local authority; and
- pass on appropriate concerns to the client so their legal representatives can make further inquiries.

13.2 Building regulations

The 1991 Building Regulations are complex and apply to all areas of England and Wales. Although they are national standards the way they are interpreted and applied can vary between different local authorities. The different parts of the Building Regulations will not be described here but it is worth noting the latest edition dates of each one:

Part A – Structure 1992
Part B – Fire Safety 1992
Part C – Weather Protection and Ground Moisture 1992
Part D – Cavity Insulation 1992
Part E – Sound Insulation 1992
Part F – Ventilation 1995
Part G – Hygiene 1992

Part H – Drainage 1992
Part J – Heat-producing Appliances 1992
Part K – Stairs and Guarding 1998
Part L – Power Conservation 1995
Part M – Disabled Access 1999
Part N – Safety Glazing 1998

There is some suggestion that Part B is about to be revised. Part M has been adjusted in 1999 to include provision for the disabled in housing.

13.2.1 Exempted work

The following work does not require building regulation approval (see figure 13.1):

- Conservatory extension at ground level less than 30 m^2 in floor area. Although it is not stated in the regulations, most local authorities use the following 'rule of thumb' that at least ¾ of the roof and ½ of the wall area should be of translucent material. Ideally a conservatory should not be built so it restricts ladder access to higher windows serving a room. This may restrict escape in the case of fire but is only mandatory where the window is a designated means of escape from the dwelling.
- Carports, covered ways and covered yard extensions that are fully open on two sides. They must be single storey and less than 30 m^2 in floor area.
- Entrance porch extension at ground level that is less than 30 m^2 in floor area. Functionally it must act only as an entrance porch. For example, if a sink or washing machine had been installed then it would become a utility room which is not a permitted development. Any glazing must meet the requirements of Part N of the regulations (see below).

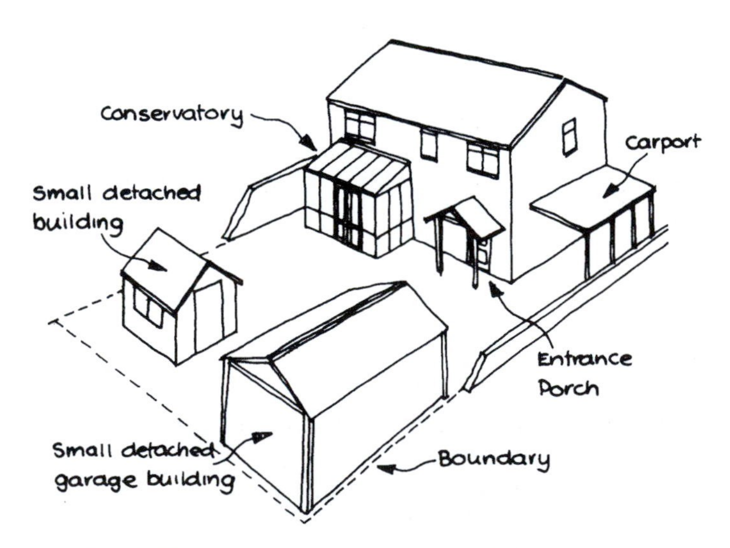

Figure 13.1 Exempted work. The illustrated projects do not usually require building regulation approval.

- Small detached single-storey garage-type building less than 30 m² in floor area. It must not contain any sleeping accommodation. If it is built substantially of combustible material (i.e. timber) then it must be at least 1 m away from any boundary.
- Small detached buildings (e.g. garden shed, etc.) less than 15 m² with no sleeping accommodation.
- Any boundary walls unless they are retaining the building.
- Replacement windows as long as the window is not enlarged or the original windows did not have load-bearing frames. Where the windows are required for means of escape purposes they should maintain the same opening space.

Other work that may have exemption includes:

- greenhouses;
- buildings not frequented by people e.g. plant rooms, etc.;
- agricultural buildings or buildings used mainly for keeping animals as long as they contain no sleeping accommodation, no part is used as part of a dwelling and it is more than 1.5 times its height away from a building containing sleeping accommodation;
- temporary buildings that are erected for less than 28 days;
- shelters from nuclear, chemical or conventional weapons.

13.2.2 Work requiring Building Regulation approval

All other types of work may require Building Regulation approval. Technically there are two categories of this:

- building work including the erection or extension of a building; and
- material alterations where the work would at any stage result in a building or a controlled service or fitting not complying with the Building Regulations where it previously did. This will also include work that did not originally comply but it is made worse by the carrying out of the work.

An example of the latter category would be the removal of a load-bearing wall. Although the finished work may comply with the Regulations, when the original wall is removed the building will be temporarily worse than it was before the work started. This allows Building Control departments wide powers to look at a broad range of work if they wish.

Glazing in domestic dwellings

Many homeowners build their own porches, conservatories and carports; some of them even make a good job of it! Even where these features do not require regulation approval, the often large amounts of glazing have to meet the requirements of Part N. This will not be described in detail and the main features are illustrated in figure 13.2. If the construction falls well below these basic standards then a request for further investigations (see section 13.2.3 below) might be appropriate.

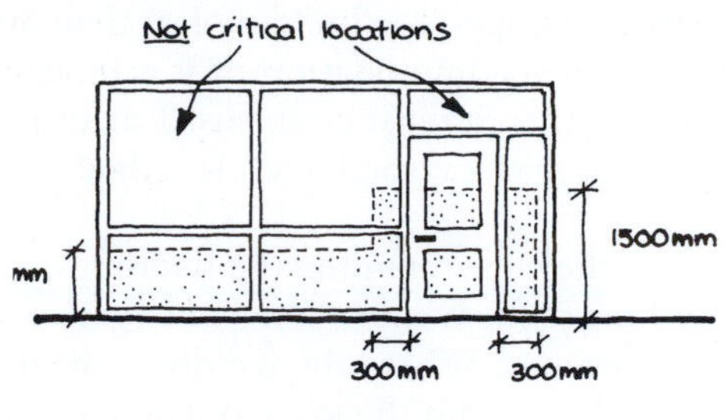

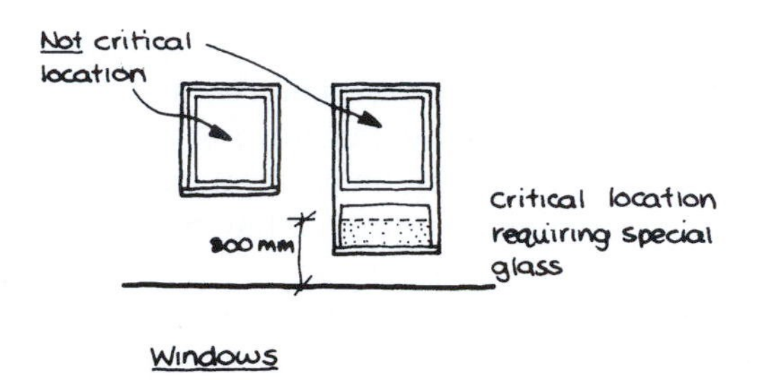

Areas fully or partially shaded required to be laminated or toughened safety glass.

Glass manufactured to BS 6206: 1981. Kite marked A, B, or C.

Figure 13.2 Safety glass required in domestic dwellings (Crown copyright. Reproduced with the permission of the Controller of Her Majesty's Stationery Office).

Many owners are simply not aware that they have to apply for Building Regulations for some of the work they have done. For example, the following work all needs Building Regulation approval:

- insulating the cavity walls;
- installing or altering the position of some gas, solid fuel and oil heating appliances;
- altering the position of a WC, bath, basin or sink that involves new or an extension of the drains;
- all loft conversions;
- works involving underpinning;
- roof re-coverings except like-for-like replacement.

The surveyor should seek to identify alterations of this type during a standard survey.

13.2.3 Approvals

Not only is it necessary to get initial Building Regulation approval but it is also important for local authority building inspectors to approve the work at various stages during the construction phase. This involves:

- The owner or builder must request the Building Control authority to carry out staged inspections to ensure the work is being carried to the appropriate standard.
- These are called 'statutory inspections' and must occur at the following stages:
  - excavations – when the foundation trenches have been dug out;
  - foundations – when the concrete has been placed into the excavations;
  - drainage – after the excavation and before the drains are covered over;
  - damp-proof course – after this has been laid;
  - concrete oversite to floors, etc.;
  - notice of occupation;
  - notice of completion.

Apart from these stages, Building Control departments are under no obligation to visit the work at any other time although many do. This means that a considerable amount of construction work between the oversite concrete stage and completion of the project may not be inspected. A wide range of defects can be hidden away during this time.

Another gap in the system exists where an owner never requests an inspection on completion of the project and so a Completion Certificate is never issued. Building Control departments rarely have the resources to chase this up in a pro-active way. The first time that many of the outstanding Completion Certificates are detected is when the house is being sold and a search request is received by a purchaser's solicitor. It is at this point that many house sales are frustrated because these formalities have not been properly completed.

13.2.4 Unauthorised work

If unauthorised work is carried out (i.e. no application for approval was made) the building control authority has two main options:

- If the work is reported to the authority within six months of the work being completed, prosecutions can be taken out against the person(s) responsible for the breach. If the unauthorised work is reported to the authority within 12 months, no personal prosecutions can be taken but action to ensure corrective work is done is an option.

Searches and building regulations

The standard search carried out by most conveyancers asks whether any notifiable work has received building regulation approval. The answer 'yes' can often refer to the original plan application. It can have little to do with how the work was done; whether the statutory inspections were carried out and whether the final completion certificate was issued. Most conveyancers offer the purchaser the opportunity to commission a more detailed search but many decline this service because it usually costs more. A detailed search would reveal the status of the necessary certificates.

- Encourage the use of regularisation procedures. This allows an owner to make an application to have the unapproved work 'regularised' which is equivalent to having the building work approved at the time it was constructed. There is a fee charged for this work and the authority may require drawings. Key areas of the building may have to be exposed for easier inspection (see chapter 7 on loft conversions). This level of inspection may require further remedial building work to bring the dwelling up to standard. Therefore, the process may have a cost implication for the owners.

In one case reviewed by the authors, a search elicited that an extension on a semi-detached house had received Building Regulation approval. It had been inspected at foundation, dpc and floor joist stage but no final completion certificate had been issued. In a letter, the local authority admitted that they would be taking no further action on the matter. What should the surveyor recommend? The Building Control department may not be taking action because of resources and not necessarily because the work is satisfactory. There are two main options:

- suggest that the owner applies for and obtains a regularisation certificate; or
- recommend that the work is inspected by a building surveyor or other approved professional who can carry out an independent assessment and make appropriate recommendations.

13.2.5 Key inspection indicators

When carrying out the survey, alarm bells should sound if building work that has been carried out relatively recently:

- has been carried out since 1985. Work before that time cannot be regularised;
- falls outside the exempted building work classification as described above;
- is poor quality with many DIY-type features that suggest Building Control approval would not have been granted even if it had been obtained. The 'rogues' gallery typically includes:
  - informal loft conversions that owners say are for storage only but contain the children's bunkbeds and toys. One person's storeroom is another occupier's spare bedroom;
  - removal of internal walls to form 'through lounges';
  - partial removal of a chimney breast;
  - connection of a shower or washhand basin into a rainwater pipe;
  - installation of a ground floor WC (often into the conservatory!);
  - the removal of the French doors between the main house and a conservatory. This would make the conservatory part of the room and lead to high levels of heat loss.

All of these features should trigger the need for further consideration by the surveyor.

13.3 Planning permission

13.3.1 Introduction

Planning permission is the other type of control local authorities have over building activities in their areas. If this is not obtained and the work is carried out, it may lead to court action, costly alterations to the property and in some cases the complete removal of unauthorised development.

13.3.2 Works not requiring planning permission

The following work does not normally require planning permission and is called 'permitted development':

- a house extension that:
  - does not bring the house nearer to the highway;
  - is less than 4 metres high or is further that 2 m from the boundary;
  - has a combined volume of less than 50 m^3 for a terrace house or 70 m^3 for a semi-detached dwelling;
  - does not project higher than the existing house.
- porches can be built nearer a highway if they are less than 3 m^2 in floor area; less than 3 m high and more than 2 m away from the highway.

13.3.3 Works requiring planning permission

Planning permission is normally required for the following work:

- most building work involving the erection of new buildings or extensions (not covered by section 13.3.2);
- changes in the use of land and/or buildings;
- large adverts fixed to the side of buildings;
- certain demolition works.

13.4 Listed buildings

13.4.1 Introduction

Appraising properties that have a significant historic nature poses special challenges. Since the Town and Country Planning Acts of 1944 there has been a system of listing buildings that are of 'special architectural or historic interest'. Before any work can be carried out to a listed building the owner may have to obtain special 'listed building consent'(LBC) from the local planning authority. When assessing buildings of this type there are a number of key questions:

- Is the property a listed building or in a conservation area?

- If it is, what type of restrictions on future development, repair and maintenance might there be?
- What are the implications of repairing and maintaining the property?

These are discussed in more detail below.

13.4.2 Is the dwelling protected?

Listed buildings have three different gradings:

- **Grade I:** Buildings considered to be of exceptional interest;
- **Grade II*:** Important buildings of more than special interest;
- **Grade II:** Buildings of special interest that warrant every effort to preserve them.

The higher the grade of listing, the greater the control over what can be done to the building. In terms of surveyors advising clients, because all three categories will have a similar impact on ownership then distinguishing between them is not as important as knowing they are listed in the first place. Recognising listed buildings is not a straightforward process as they do not have a convenient label attached. Key indicators include:

- **The age of the building** – the older a building is the more likely it is to be of historic interest apart from the exceptions! For example a number of post-1939 buildings have recently been listed.
- **Unique architectural or contructional styles and methods** – unusual or fine examples of design or construction such as a classic Georgian terrace, a 17th-century timber-framed house with white daubed infill panels; traditional 'Cruck' timber structure, etc. The list is extensive. Figure 13.3 for example is a dwelling built in the 1930s that incorporated the reconstructed timber façade from a much older building. The principle here is to be alert to the unusual.

Figure 13.3 Photograph of a listed building that was built in the 1930s but using a reconstructed façade that is over 200 years old!

- **Location** – if the dwelling is in a conservation area or a historic location then it is more likely to be listed or protected. For example if the building is in the middle of a fine terrace then it could be protected because of its group identity.
- **Associated with well-known characters or events** – without local knowledge this may be difficult to determine – a blue plaque on the front of the building is a clear clue!
- **Information from the vendor** – if a building has special protection the owner is usually only too pleased to let others know. Information given in this way is not always correct and will need checking.
- **Information from the local authority** – copies of the statutory list of listed buildings are made available for the public by English Heritage. Copies are usually available from the local planning authority. Many councils have edited the national list to produce booklets that locate listed buildings in their own areas. A copy of this booklet would be a vital addition to a surveyor's information file.

13.4.3 Restrictions on development

Work on historic buildings may be restricted through two different mechanisms:

- the building itself being listed; and
- the building being in a conservation area.

Listed buildings

The owner of a listed building is restricted in what can be done to the building. Listed building consent (LBC) will have to be obtained for 'works for the demolition of a listed building or for its alteration or extension in any manner that would affect its character' (Planning (Listed Buildings and Conservation Areas) Act 1990). Although there is no clear definition of some of these terms, Pickard (1996) suggest the following work may required LBC:

- **The demolition of all or part of the building** – in some cases this could even include part of a garden wall or an internal brick partition.
- **Alteration or extension of the building** – extension of a building is probably easier to define whereas alterations are more difficult. The removal or replacement of windows, doors, small sections of plasterwork and internal panelling could be seen as alterations. According to Pickard, each case has to be taken on its own merits. He goes on to say that 'without doubt the national amenity societies and other bodies with a conservation interest ... will stringently argue the case for greater scrutiny where a major change is proposed'.
- **The painting of a listed building** – not only would this limit what colour the whole façade could be painted but the woodwork around window openings might also be controlled.
- **Replacement windows** – several court decisions have ruled that PVCu double-glazed windows are not acceptable for listed buildings.

- **Change of use** – if a client intends to change the use of any dwelling, she will almost certainly have to apply for planning permission in the normal way. If the building is listed, then the requirements of the planning authority will be a more prescriptive.

Conservation areas

The Planning Act 1990 placed a duty on local planning authorities to 'determine which parts of their area are areas of special architectural or historic interest' and to designate them as conservation areas. There is no typical conservation area, all can vary in size, shape and nature. Some local authorities have designated only one area while others (City of Westminster in London) have designated over 70% of their area. Pickard (1996) gives a few examples:

- Hampstead garden suburb in north London;
- inter-war housing estates in London Boroughs of Havering and Harrow;
- the Jewellery Quarter in Birmingham;
- a cemetery in Bradford;
- a workers' village in Essex.

This shows the range of areas that could be included and so surveyors should keep an open mind. Each local planning authority will keep a record of the conservation areas in its area.

Controls over development in a conservation area are similar to listed buildings but less rigid and less well defined. They usually relate to the external features of the building. The main thrust of control is to bring items that would otherwise be considered 'permitted development' (i.e. not require planning permission if it was outside the conservation area) under planning control. Although there is currently some debate over the effectiveness of this process, Pickard identifies the following works which may require special approval:

- extension to a single-family dwelling of up to 50 m^3;
- cladding of any part of the exterior of the property with stone, artificial stone, timber, plastic or tiles;
- alteration or addition to the roof;
- changing of the roof covering and the insertion of roof lights;
- erection of a porch to the front door;
- provision of hardstanding to the front of the property for parking a car;
- fixing of a satellite antenna;
- new fence, gate or wall to a property;
- painting of the exterior of a building.

This is a considerable list and includes works for which home owners outside the conservation area would not need special approval. Therefore, clear advice from the surveyor is vital.

13.4.4 Implications of owning an historic property

The special problems associated with the ownership of a listed building or one in a conservation area are not restricted to whether permission to carry out the work is granted or not. Once work does go ahead the special nature of the building may require additional professional advice and the appointment of expensive specialist contractors to make sure the work conforms to the local panning authorities' requirements. This is described in more depth in the Guidance Notes Appendix 2 of the *Appraisal and Valuation Manual* (RICS 1996). It points out that many historic buildings have been harmed by standard repairs that are common in more conventional buildings. Examples include:

- **Injected chemical damp-proof courses** – ineffective in stone rubble walls and can cause damage to cob, pise and clay lump walls. Dense waterproof renders internally can disturb the moisture balance in thick walls.
- **Penetrating dampness** – old walls are usually porous and need to lose moisture by evaporation. Application of dense renders, impervious coatings and silicone treatments can all exacerbate the problems.
- **Structural movement** – the movement and consequent cracking of conventionally built dwellings are relatively easy to identify and interpret. Historic buildings are far more complex. Sometimes it is better to live with minor structural deformations that occurred centuries ago rather than embark on underpinning that must be viewed as a last resort.
- **Timber treatment** – some of the preparatory works required by timber treatment specialists may damage the character of a listed building. Exposing the full extent of a rot attack or blanket treatments for wood boring beetle may be totally inappropriate. Specialist in situ treatments are increasingly used including epoxy resin strengthening technologies and paste 'mayonnaise' treatments for beetle attack.

13.4.5 Giving advice

Dwellings that have a significant historic interest impose special duties on the surveyor. These are summerised below:

- Every effort should be made to find out whether a dwelling is listed or in a conservation area. This may have a financial implication for any owner or at least restrict how that dwelling can be used.
- Once it is known that the building is protected the client should be advised of the implications of owning such a dwelling. Getting listed building consent may result in more expensive work being carried out to a standard that the client would not normally choose.
- During an inspection or a survey, any repair or alteration work that does not seem to be in keeping with the character of the building should be reported. The client should be advised to check whether the appropriate approvals have been properly granted. This is because some local planning authorities keep a very watchful eye for unauthorised developments. In the worse cases the new owners could be served

with a listed building enforcement notice requiring them to restore the original work even if it was done before they owned the property.
- Where alteration or remedial work will be required on a historic building then the client should be clearly advised to get further professional advice. This should be from consultants and contractors who have knowledge and experience of dealing with older buildings.

13.5 Surveying new properties

The technical content of this book supports the assessment of the condition of all residential properties whatever their age. Although a lot of the information does address defects with traditional construction techniques, the problems associated with new dwellings has been included and integrated within the appropriate sections. Furthermore, the diagnostic approach proposed in this book is as relevant to houses built last month as it is to those built 150 years ago.

Despite this, the issues presented by a newer property are unique in two main respects:

- the vast majority of newer properties are covered by warranty schemes;
- newer properties tend to be in a better condition (at least initially) where the signs of any deficiencies in the construction have yet to manifest themselves. This often results in a report that may have to speculate about what defects may occur in the future rather than focus on what defects exist at the time of inspection.

This section will highlight some of these issues in more detail.

13.5.1 Warranty schemes for new dwellings

The vast majority of dwellings are now built under a variety of warranty schemes offered by either the National House-Building Council (NHBC) or Zurich Municipal. The practice is so widespread that many buyers would find it almost impossible to get a commercial loan from many lenders without the dwellings being covered by a recognised scheme. But the authors are aware that a number of smaller developments in many parts of the country are not covered. Because the houses are being sold, someone must be lending the money to buy them. Therefore surveyors must always be vigilant with new property and clearly establish whether it is covered by a scheme or not.

Lenders will have their own prescribed requirements for these situations but it will be useful to review the main features of the schemes.

National House-Building Council (NHBC) Buildmark scheme

The NHBC claim that they provide about 85% of the market through their Buildmark products. The main provisions are summarised below:

- **The Buildmark scheme** – this covers new home owners during three time periods:
    - before completion of the house if the builder becomes insolvent;

- during the first two years – the builder is responsible for rectifying any defects that occur. The definition of a 'defect' is wide and includes any 'breaches of NHBC's extensive building standards'. Exemptions include normal shrinkage, condensation due to drying out, general wear and tear and damage that arises from a failure to maintain the property;
- during years three to ten – the NHBC will cover defects costing more than £500 to put right in the following elements:
  - foundations
  - load-bearing walls
  - external render and tile hanging
  - load-bearing parts of the roof
  - tile and slate coverings to pitched roofs
  - load-bearing parts of the floors
  - floor decking and screeds where they fail to support normal loads
  - retaining walls where they are necessary for the stability of the building
  - multiple glazing panes to external windows and doors
  - below-ground drainage
  - defects to a chimney or flue which causes a present or imminent danger to the physical health and safety of anyone normally living in the building
  - special cover is also available for properties built on contaminated land.

- **Buildmark Choice** – this is specially designed for housing associations and is very similar to the 'Buildmark' scheme but can be extended to include physical damage.
- **Conversions and renovations** – 10-year cover is offered for newly converted and renovated homes as long as they are carried out by an NHBC registered builder.
- **'Solo' for self-build** – this is for people who build their own homes and offers protection as work proceeds and for 10 years after completion.

Like any warranty scheme their are a number of exceptions and qualifications. For the precise nature of these, refer to NHBC's literature.

Zürich Municipal 'Building Guarantees'

Zürich claims to be the world's leading financial risk manager across 90 countries. They market a range of Building Guarantees that are recognised by most mortgage lenders. In many ways they are similar to the NHBC schemes and include the following products:

- **Newbuild solutions** – this offers new home buyers a full 10-year structural insurance against latent defects. This can be extended to 15 years and the scheme includes a variety of optional defect packages that go beyond the normal 'structural' items.
- **Newstyle solutions** – this is a latent defect warranty for housing associations.
- **Restyle solutions** – similar to the Newstyle scheme but covers buildings where existing structures are retained as part of the scheme.
- **Rebuild solutions** – offers cover to buyers of newly converted homes. This is similar to the Newbuild solutions.
- **Custombuild and Custombuild conversions solutions** – Zürich's scheme for self-builders.

Zürich has a range of other products and the full details are available directly from them.

13.5.2 Inspecting new properties

Defining 'newer' properties

Firstly it may be useful to draw a distinction between different types of 'new' properties. Adapting a framework suggested by Melville and Gordon (1992, p.374):

- 'New' properties are still being built at the time of inspection, have been very recently completed (say still within the six-month defects period) or still within the 2-year builder's warranty period.
- 'Near new' houses that are between two and 10 years old.

Anything older than 10 years will be outside a warranty scheme and so must be approached in the same way as older houses.

Surveying techniques for newer properties

As discussed above, new and nearly new properties usually offer fewer clues of hidden or developing defects. Therefore the survey of these properties should focus on projecting possible defects (Melville and Gordon 1992, p.374). This involves critically analysing the design and construction and speculating '... on the ability of the structure, services and finishings to fulfil the client's expectations for the future'. Although a warranty scheme is a clear indication of a certain level of quality, mistakes and problems can still occur. If the surveyor carries out a superficial inspection limiting comments to minor blemishes and omissions the client would be justified in being unhappy. Even if the house is insured against major defects, most people would rather not buy a dwelling that will soon be entangled in a lengthy and complex claim against the original builder. The surveyor can offer 'added value' by spotting possible future problems early.

This 'prognosis for the future' (Melville and Gordon 1992, p.376) can be a very challenging task that will involve closer than normal comparison to current Building Regulations, British Standards and even the NHBC's and Zürich's own standards. The process is further complicated by the fact that most new properties will show signs that could be mistaken for more serious deficiencies. Consider two examples:

- **Cracking** – all new dwellings will suffer from initial drying shrinkage that may result in hairline cracking at the junctions of wall and ceilings, over window openings, etc. The skill is telling the difference between this and early signs of more serious structural movement.
- **Dampness** – the massive amount of water used in the construction of a house can take a long time to dry out. Because most builders are keen to sell as quickly as possible, much of this 'drying out' will occur after the first owner has moved in. The extra moisture in the atmosphere can lead to excessive condensation and in some cases mould growth – signs that can easily be confused with more permanent dampness defects.

The surveyors only guide through this maze is a careful examination of all the signs and symptoms and the formulation of possible causes. To do this the survey needs to be as thorough and as extensive as for older properties. Any surveyor who thinks time can be saved on inspecting a new property will not be providing a good service to the client.

Another difficult area is the need for specialist testing. In all cases, the only way a client can be sure that a service system will perform adequately is to have it tested by the appropriate specialist (see section 10.1.2). But with new properties the probability of major faults existing must be more unlikely than with nearly new dwellings. Therefore the following advice can be given:

- New properties – tests are usually unnecessary at this stage. If any faults do develop then they can usually be covered by the warranty scheme. The exception to this would be where there is evidence that the service system has been either altered (DIY sockets, etc.) or damaged in some way (fire or flood, etc.). In these cases a test is fully justified.
- Nearly new properties – these should be assessed against the criteria outlined in figures 10.11 and 10.14 and where there are visual signs of deficiencies then a specialist test should be advised.

13.5.3 Giving advice

When asked to carry out a survey of a new property, in addition to the normal reporting of the condition of the property a surveyor could (or should!) comment on the following:

- **Establish the precise nature of the warranty scheme** – this would include the nature of the cover provided, the time period left under the scheme, etc. One important aspect for buyers who are not the first owners of the property is to be clear whether any faults were known about at the time of previous sales that led to a reduction in the purchase price of the property. In these cases, the warranty scheme will not apply.
- **Snagging advice** – if the survey is carried out prior to practical completion or the end of the defects liability period then it will be useful to provide the client with a list of 'snags' that are the responsibility of the builder. This can help a new owner resolve minor irritations that are common in new properties. In the case of nearly new properties, opinion can be given about whether any noted defects should be covered by the warranty scheme.
- **Whether defects are covered by warranty schemes** – although properties are covered by warranty schemes it is useful if the surveyor gives broad advice of whether the defects or deficiencies noted during the survey will be covered by the warranty scheme. Although this sort of advice should always be measured, clients will find this initial guidance useful as it will help them develop a realistic approach to negotiating with the appropriate organisation if a claim is made.

References

Melville, I.A., Gordon, I.A. (1992). *Structural Survey of Dwelling Houses*. Estates Gazette, London.

Pickard, R.D. (1996). *Conservation in the Built Environment*. Addison Wesley Longman, Harlow.

RICS (1996). *Appraisal and Valuation Manual*. RICS, London.

Part III

The report and future opportunities

Chapter 14

Writing the report

14.1 Introduction

There are many publications advising on how to produce good reports and this guidance will not be repeated here. Of particular importance is the 'Red Book' (RICS 1995, p.57) which gives targeted advice on completing professional survey reports. To complement the standard guidance this section identifies other key features that the surveyor should consider when producing a report.

A surveyor is only as good as the person reading the report.

This applies to any profession where the person with the special knowledge has to relay information to the client. If the specialist cannot convey the information in an understandable form then it will be of little value to those who have paid for it. For example a comment such as:

> the brickwork is affected by efflorescence

has little meaning on its own. It would be more helpful if replaced by the following:

> The brickwork to the left elevation was affected by white salts (efflorescence), this indicates that it is affected by dampness. This appears to be caused by a failure of the damp-proof course and you should seek advice from a damp-proofing specialist to establish the costs of remedial works.

This example is more likely to be used in a Homebuyers report than in a mortgage valuation. Throughout this section other examples are taken from genuine reports and related correspondence with slight amendments to ensure confidentiality is maintained. They are not the work of incompetent surveyors operating at the margins of the profession. They are errors that all surveyors either have or could make at some time in their careers. They are used here to emphasise key points.

14.1.1 Focus on the reader

Any message should be simple and clear with the principle objective of satisfying the needs of the reader.

Think of the reader.

To put this in context consider some popular forms of the media such as a newspaper. The key to an article is the headline. This has to both attract the reader and encapsulate in a few words the message of the article. A similar comparison is the use of comic strips to tell a story; the combination of a picture and key words allows the reader to quickly assimilate a lot of information.

Currently there is an increasing use of 'sound bites' where a paragraph or phrase, usually longer than a headline, puts across a key message. When faced with a long report, how many people succumb to reading just the introduction and the conclusion? Many busy managers will only read the executive summary of a report. The development of these techniques has arisen because of time pressures on individuals. The greater detail in the remainder of the document often satisfies operational or legal requirements and has little to do with informing the client.

Thinking of the reader can be approached from another angle in respect of what is important to them. The following was contained in a letter from a surveying practice to a complainant:

> The mortgage valuation clearly states under the Gas section 'Yes' to whether the service is available. We understand from you, however, that there is not a gas supply to the property. Clearly, this is incorrect. We are obviously sorry that this error has occurred in completing the form.

Was it incorrect that there is not a gas supply to the property or that the form was completed incorrectly? What is clear from this example is that an aggrieved customer or even a judge trying to arbitrate in such a matter would not be impressed by this choice of wording. The credibility of the organisation would be seriously questioned.

14.2 The mortgage valuation

14.2.1 Standard forms

The mortgage valuation is usually specified by the client lender in respect of format. This places constraints on the style that can be used. The key reason is that the lender needs specific information that can be used to assess the degree of risk in undertaking the lending. The information has to be passed on to other advisors involved in the transaction process including the mortgage applicant. It is therefore essential that the surveyor provides the information in a way that cannot be misinterpreted. To minimise on the variation in reports, lenders tend to use tick boxes or a limited form of multiple choice. In many cases where 'free-format' boxes are used they are kept very small indicating the size of message that is expected. Many surveyors find it difficult to limit their comments especially if there is a problem with the property. This is mainly because of their duty of care not only to the lender but also to the applicant.

The use of standardised paragraphs means that the underwriter to the mortgage becomes accustomed to certain situations and those that are exceptional and stick out like the proverbial sore thumb. It also means that the surveyor's comments can be easily lifted and used in other forms of communication, for example to the solicitor, intermediary or the customer where a condition is imposed on the mortgage offer. Free-format boxes take on a standard approach. Adopting this practice is the first step to developing a system that can be incorporated into a computer programe that can enhance efficiency (more of this in section 15.4).

14.2.2 Use of standard paragraphs

There are many examples of standard paragraphs and the best should adopt a simple pattern:

- identify the problem;
- state the action necessary to resolve the problem.

Consider the following example:

> Gutters and downpipes appear to be of an asbestos material that is damaged and causing them to leak. As this material can be a health hazard specialist advice should be obtained before carrying out any work to these components.

This good example could have been further improved by stating the type of specialist that would be expected to undertake the work.

14.2.3 Identifying the problems

Many building defects may affect one or more related elements. For example, with penetrating dampness the surveyor needs to identify the source and the full extent of the damage. This could include a timber specialist to cover damp-affected timber and a general builder to do an external repair as well make good any internal damage to plaster work.

Paragraphs should not group items that are not connected as this can cause confusion. It could result in the lender trying to split the paragraph to follow up the action considered necessary. In essence this places the surveyor in the position of only being as good as the person rewriting the report. This reinforces the message of understanding what the reader wants and making sure it is delivered in the form and style required.

The following examples all would benefit from more clarity.

Example one

> General condition would appear to be reasonably satisfactory consistent with age, although there is some scope for improvements, redecoration, up-dating of fittings and services, etc. There are no essential repairs. Further works may become necessary when the property is opened up for improvement and redecoration. The cost of works could exceed the subsequent increase in value of the property.

What does this mean to the lender and the purchaser? Sympathy can be expressed for the surveyor who does not have the opportunity to explain to the purchaser what exactly this devastating statement means. Further work on this property might reveal the need for unforeseen repair works. However, on the face of it there is a suggestion that the lender could become unsecured, a suggestion that will set alarm bells ringing for both the purchaser and lender alike.

Example two

This included two sections:

> 1) The property is affected by structural movement, with distortion/fracturing being noted to the flank elevation. A full report identifying the cause and recommended remedial work necessary to ensure future stability should be obtained from a structural engineer (or Chartered Building Surveyor), together with an estimate for the cost. The recommended works should be carried out in full under the Engineer's supervision.
>
> 2) At the same time a check of the adjacent drainage is recommended.

A good start paragraph by identifying the extent of the problem and what should be done about it. The problem arises with the second part that refers to the drainage. Where is 'adjacent', could it be next to the affected wall or next door? Is the problem related to the structural movement or separate? Is there a reason to check the drainage as a separate item? What sort of 'check' needs to be carried out and by what sort of contractor? A customer would be very unsure on how to proceed.

Example three

> A number of other items were noted including flat felt roofs have a limited life and can fail unexpectedly, missing door furniture and the applicants are advised to obtain an electrician's report.

An attempt to save space that does not work. Are the applicants only concerned about the electrician's report and why do they need one? Is missing door furniture on the same level of importance as a failure in the electrical system? Is there a problem with the flat roof at the subject property or is it a statement that could apply to any property with a flat roof. Such a phrase is unlikely to provide a defence should the roof fail as it makes no reference to whether the roof was even inspected, the age of the covering or current condition.

Questions can be raised as to the level of detail necessary to report for a mortgage valuation but the point here is that the series of unconnected statements that have been made are meaningless. A judge reading this report would be given a picture of a surveyor who could not put together a reasonable sentence and would question his/her ability to undertake the job.

14.2.4 State the actions necessary to resolve the problem

Identifying what action is required to resolve the problem can take two forms:

- briefly specify the work necessary (in respect of repair); or
- state the course of action required to identify what is required to resolve the problem.

This could involve the nomination of a specialist or a contractor who can cost the works. This would allow the surveyor to establish the value of the security as the full implications of the problem will be clearer.

Some problems may not be easy to resolve. For example take planning restrictions. It may be that the surveyor knows exactly how to resolve the problem or considers it necessary to appoint a planning consultant in more complex cases. Changes of use of a building are just one example. This could lead to a situation where the time it could take to resolve a planning-related issue may not match the time scales of the transaction. In this case, the action necessary to resolve the problem is known but there is not enough time to solve it. Always try and create certainty.

Whatever the problem the principle needs to be adopted:

- identify the problem; and
- state the action necessary to resolve it.

This then helps to provide the basis for determining the value of a property by comparison to others that do not have similar constraints.

14.3 The Homebuyers survey

14.3.1 Introduction

The Homebuyers report is produced to a specific format allowing a considerable amount of free text within each section. Since the launch of the new Homebuyers survey there has been a large amount of advice, comment and opinion in the professional press. This will not be repeated here but given the nature of free text it would be sensible to follow a pattern using the techniques described above. The more consistent the approach the easier for the surveyor to maintain the routine and establish a format that is easily understood by the customer. It is important to remember that the customer is unlikely to have read many survey reports before, unlike the lenders, who will process many similar reports every day. Therefore use:

- summaries
- clear action points
- simple language.

The following examples of problematic reporting that highlights the need for clarity. This approach applies equally to the narrower scoped Mortgage Valuation.

14.3.2 Case study – dampness

> Readings were taken wherever possible and appropriate with an electronic damp meter. However, the scope of the investigation was severely restricted by part

> internal dry lining and particularly by the tiling in the shower room and panelling in the separate WC and lobby. No evidence was detected of rising damp, the vendor states that damp-proofing was previously carried out, backed by a guarantee. You should satisfy yourselves over the scope of the work previously carried out and of the effectiveness of the guarantee.

The surveyor has identified a situation and concluded that there is no damp. However, he has confused the situation by making a recommendation for action that would be difficult for the customers to follow. They were probably expecting the surveyor to do exactly that.

It is also interesting that the surveyor found it necessary to provide a caveat in the report that was also included in a similar fashion within the margin. 'Damp meter readings have been made where appropriate and possible to the external and internal walls, floors, etc, without moving heavy furniture fixtures and fittings.' This over-emphasis is not only unnecessary but would create a negative image in the mind of the customer.

This example suggests a property has no damp recorded and there is evidence to suggest work has been carried out to prevent rising damp. There are always restrictions on any inspection and the verbal evidence about the guarantee is insufficient on its own but what reason is there to be so negative? Compare it with a visit to a doctor. How would a patient feel if the doctor did the examination, advised she could find nothing wrong, then compounded this by suggesting that this did not mean there was nothing wrong and really you should satisfy yourself that you were actually all right!

A more positive response would be the following:

> No evidence of damp was detected from the readings taken at various points throughout the property. It is understood from the vendor that previous work was carried out to remedy rising dampness, which would have affected ground floor walls only. We have had sight of the documentation and this concludes ... **(the areas repaired)** ... and confirms that the guarantee passes to you as the new purchaser. We would advise that our investigation was limited by use of dry lining ... **(location)** ... tiling in the shower room and panelling in the separate WC and lobby. However, based on the documentation provided it is confirmed that these areas are covered by the guarantee.

It is not always this simple. The documentation cannot always be provided and some may say that the analysis of existing guarantees is beyond the scope of the Homebuyers Report and Valuation. If the surveyor is genuinely trying to provide a service to the client then such concerns about the nuances of a standard service may be irrelevant. Even if the documentation cannot be found the simple facts of this case are that:

- at the present time no damp can be found;
- the internal finishes are not affected by damp staining or showing any other signs of problems;
- there are no external features that suggest the damp-proof course has been bridged.

As a consequence on the balance of probability a surveyor would be safe to advise the client the property is currently free from rising dampness. There is always the potential that the dpc could breakdown but this has to be put in context and set against the individual circumstances at the time of the survey. This leads to a further point – state facts and include your client in your thought processes.

It is also important to relate to your client. The written report is only part of the story. How can a surveyor think of the reader if they have no idea of what the reader actually wants from the property inspected? It is rather like going out and doing someone else's shopping without knowing whether they are vegetarians or carnivores. In the vast majority of cases, surveyors will do the report and forward it to the client without ever making verbal contact. As many survey reports make recommendations to contact specialists, builders or other types of contractors it is those organisations who actually make the biggest impression on the customer because they are real people. To many customers the surveyor can remain a faceless stranger who charges lots of money for long reports full of jargon that they do not understand. It can often be the builder who actually explains what the report means. The following is an example of how this can go wrong.

14.3.3 Case study – garages and outbuildings

A Homebuyers report stated:

> A brick shed is situated in the rear garden but this is in fair condition only and requires some attention. The concrete floor slab to the shed at the rear has been undermined and additional support should be provided to the rear.

A builder looked at the property and gave his report:

> The block built shed is in poor condition and is beyond suitable and economic repair, I consider it should be demolished and rebuilt, as it is also an unsound structure and must be considered dangerous.

An outbuilding was clearly in need of some attention but had the surveyor anticipated a complete rebuild? His judgement was challenged by someone who had a financial interest in achieving more extensive building work. Most builders are skilled at what they do. Being at the operational end of the constructional process, they do not often become involved with analysing the condition of structures. Objectively recommending a course of action that is in the best interests of the client is not a role that is well developed. Yet in this case it is safe to assume that the client trusted the builder's opinion before the surveyor!

The surveyor's comments were not specific, open to interpretation and the surveyor had not met the customer. The builder on the other hand has acted as interpreter and given more positive advice.

Reviews of other Housebuyer surveys reveal a common problem with reporting on asbestos roofs of garages. In Allen v. Ellis & Co. (1989) a surveyor was found liable for not warning a prospective owner about the fragile nature of an asbestos roof. The owner later fell through it. For example, in a Homebuyers report on a detached house, one surveyor stated:

> There is a detached single garage that has a corrugated asbestos roof and rather ill-fitted doors. An internal inspection of the garage was not possible, but we have reservations as to whether the roof covering is entirely water tight at the present time.

This paragraph concentrates only on the roof's ability to exclude water and does not mention either the fragility of the sheet roofing or the health implications of its presence and removal. This extract from another report by a different surveyor is a little better:

> A detached single garage is provided ... walls are of brick and the roof is of asbestos cement. The roof is weathering with age and replacement in the short term will be required. Care is required with the disposal of the asbestos cement sheeting.

This is more helpful but merely suggests that 'care' should be taken when disposing of the sheeting. This caution is very different to what is required. In addition to warnings about the durability and fragility of the roof the surveyors should have made explicit reference to:

- the health risks associated with older asbestos components that are deteriorating;
- the problems associated with asbestos products that are worked on or removed;
- the additional costs involved in removing asbestos and the need to employ properly licensed contractors.

Without this guidance the client would be poorly advised.

14.3.4 Case study – services

As discussed in chapter 10, assessment of the services in dwellings is the one area where many surveyors lack confidence and knowledge. Homebuyer reports often reflect this fact. The following extracts are associated with foul water drainage. The first one relates to a semi-detached house built in the 1950s:

> This property is connected to the main sewer. It was not possible to locate an inspection chamber within the curtilage of the house but we have no reason to suppose that the underground drainage is defective.

It is to be hoped that the surveyor made this statement after a careful review of other key inspection indicators (see section 10.15.3) but somehow it is doubtful. Even if the drains were not defective the fact that the house has not got an inspection chamber is a problem in itself. There should be one at every change in direction and at locations that allow all parts of the drainage system to be cleaned and properly maintained. The owner of this property could face a big repair bill should the drains block, not only for the cost of clearing it but also for excavating and breaking into the drain itself. At the very least the client should have been warned about:

- the purpose and need for inspection chambers;
- the implications of not having one;
- clarification that the drains cannot be properly assessed without at least looking into an inspection chamber;
- advice that the matter should be investigated further, the drains traced and assessed, the possibility that a new inspection chamber might well be required.

The surveyor should also be alert to the likelihood that the chamber might have been concealed under a new path or flower border!

A similar problem was noted in a Homebuyers survey of another semi-detached house built in the 1930s. The surveyor stated:

> No manhole provided within the curtilage of the property for inspection. Drains are shared with neighbouring properties and solicitors should verify with the deeds your rights and liability concerning maintenance and repair.

Useful advice about the legal implications but of little practical value in relation to the drains themselves. In both these cases the drains will almost certainly be in rigid jointed clay pipes that are susceptible to damage and leaking.

Similar misconceptions about other services can easily be found in many recent reports. Lead piping attracts a broad variety of guidance. Where lead rising mains are observed, most surveyors make some mention of the possible health risks. They give a variety of advice ranging from no action because 'there are miles and miles of lead water main in the street anyway' through to total replacement back to the boundary 'especially where children are to occupy the property'.

More worrying is the following example. A house built in the 1930s was noted as having a lead rising main that the client was advised to have replaced. The paragraph went on to state:

> The lead main connects to copper that serves the appliances. The galvanised steel cold water tank feeds the indirect hot water cylinder. We would caution that this tank has a limited life and replacement in the short term is required. Again the lead service pipe should be replaced to both hot and cold water tanks.

Under the *hot water and heating* section the surveyor states that:

> Lead circulating pipes are provided to this (hot) water tank although they do not seem to be connected to any particular appliance.

This is obviously a property with a considerable amount of lead piping beyond the rising main stopcock. Setting aside the possible maintenance problems associated with lead, copper and galvanised steel on the same system, the presence of lead in the hot water system (even if semi-redundant) is dangerous! Lead will easily be taken into solution in hot water and so present a very great risk to the occupiers. The surveyor should have advised urgent investigation and warned about extensive replacement work. A warning not to use the system in its present condition might also have been appropriate.

14.3.5 Use of language

It is even more important to use simple language when dealing with clients who rarely purchase a property. Few members of the public have much technical knowledge and lose patience if presented with a barrage of technical jargon. The following example shows how comments can be misleading:

> Internally, one or two floors flexed substantially under load and these should be checked to ensure that adequate support is given.

Is this the load of furniture, humans or what? Who should check them and what should they look for? The fact that they flex means that adequate support is obviously not 'given' at the present time. Does this mean that the floors are going to collapse and pose a danger to the occupants of the building? All this may be obvious to a technical fraternity but to intelligent and proactive lay people who pay surveyors' fees it will just seem like a poor service. Also if additional support was required, it could be disruptive and expensive to the occupier. No provision was made for this in the valuation and so reveals another deficiency.

Another example of how clients can misinterpret technical advice occurred when a homeowner approached one of the authors about a leaking flat roof over their single-storey bathroom. In an effort to save the person some money the author told him to go to a DIY superstore and get a large tin of bitumen-based sealing compound. The owner was told to apply it '... over the area of the leak'. Although it was not a permanent repair it might keep the room below dry for a short time. A few weeks later at a chance meeting, the author asked the owner how the leak was. The owner replied that he did not reckon much to the advice. He had bought the sealant and applied over the area of the leak but:

> ... it had not cured the leak, water still poured through every time it rained. The black paint had also ruined the decorations to my bathroom ceiling! It was so bad we had to have the ceiling replaced as well as the roof recovered.

The author was horrified to learn that the homeowner had applied this black sticky bitumen sealant to the bathroom ceiling on the inside of the bathroom. After all, this was applied 'in the area of the leak'. There are many lessons to be learnt from this example. The principle one is never give casual advice!

14.4 Building surveys

This publication has concentrated on mortgage valuations and Homebuyer surveys. The more broad-ranging building surveys are beyond the scope of these pages. Yet there are many similarities between them all. For example Judge Phillips in Cross v. Mortimer (1989) stated that there is little difference in the skill level required:

> We are convinced that the same level of expertise is required from the surveyor in carrying out an HBRV as that for a structural survey.

Detailed guidance on building surveys is given in the RICS's *Building Surveys of Residential Property: A Guidance Note for Surveyors* (1996). The guidance states that in view of the great variety of building types, designs and construction methods, no simple pro-forma report can convey effective, balanced and complete advice. Therefore this gives the building survey report a very different format to a Homebuyers report. The structure and content will vary between surveyors as every report is produced to suit the property that was surveyed. Because the report includes advice on costs of actual maintenance and repair work, it is usually beyond the scope of those whose professional activities are focused in valuation activities. Therefore, this type of report is not considered any further.

14.5 Conclusions

When the work of surveyors is reviewed a standard phrase comes into mind:

> Hindsight is an exact science.

It is very easy to pick fault with phrases that busy surveyors may have inadvertently used in reports to their clients. But these deficiencies are so common and easy to locate that it must be indicative of a widespread problem. In the authors' view this is partly due to the surveyors not writing the report *for* the client. It is a matter of how the task is approached that must be changed.

In summary, when writing a report following any type of survey there are a number of key points that need to be kept in mind:

- Get to know your client if possible.
- Think of the reader of the report.
- Include the client in your thought processes.
- Identify the facts.
- Identify the action points and give positive advice.
- Use simple language or interpret and explain technical jargon.

If these principles are followed then the surveyor's clients will feel much better served and are much more likely to come back in the future.

References

Allen v. *Ellis & Co.* (1989). 1 EGLR 181.
Cross v. *Mortimer* (1989). 1 EGLR 154.
RICS (1995). *The Appraisal and Valuation Manual*. RICS Business Services, London.
RICS (1996). *Building Surveys of Residential Property: A Guidance Note for Surveyors*. RICS Business Services, London.

Chapter 15

Benchmark for change

15.1 Introduction

The now famous case of Yianni v. Edwin Evans (1982) triggered one of the most dynamic periods within the history of the surveying profession, particularly in respect of the debate of who is the surveyor's customer and how far does the liability extend to third parties.

The actual product has only changed marginally but the ownership structure of the business coincidentally has changed significantly. It is the combination of these factors that has set the scene for the business of undertaking residential property appraisals. Without a clear understanding of the market place, its customers, its products and the business environment in which they have to operate, surveying practices will find it hard to survive, especially where rapid change is increasingly evident.

The main reason for this section is that an understanding of how to assess the condition of a property or formulate a valuation is only part of the residential appraisal process. A sound knowledge of the product has always been and will continue to be a key element of the business. Using that skill and converting it into a package to suit the changing needs of changing customers is an ability that many surveying organisations have to come to terms with.

The intention of this section is to give an outline of some of the techniques that could be adopted to ensure that surveying practices are meeting the needs of their clients. The principle behind this is that a practice or business must:

- first establish the key drivers in that area; and
- then establish where the business fits within the market place and the industry.

This will then give a benchmark against which practices can periodically measure change to establish how they need to react to meet the needs of the customer.

Assessing market position and customer needs will not come without some form of catalyst. It is hoped that the examples given in this section will form the basis for further thought. It could be sufficient to draw up a brief for a facilitator or trainer who could lead a workshop to help develop an organisation's ideas.

15.2 Business planning

Residential surveying now finds itself positioned within the financial services industry and as such is required to match up to similar levels of return.

The change in ownership from small partnerships to corporate-style practices has resulted in a higher degree of importance being given to managing cost and producing a return for shareholders rather than the partnership. Whether this is achievable is debatable as the industries vary significantly. It is clear at the time of writing that the surveying industry is searching for products that the public truly appreciates will add value.

What can be done to keep costs down yet maintain a service or product value while at the same time having one eye on the changes demanded by regulation and product knowledge? A key element is to establish whether the customer base has or is liable to change. The initial step is to identify the drivers of change. Reviewing recent history the key drivers have been:

- the changing needs of the customer
- litigation
- changing ownership structures
- technology.

These factors also interact. For example technology drives change because it allows more to be achieved. It is the customer who drives the need for more to be achieved ultimately by their litigious activity. Many questions need to be asked. For example are these areas going to be the drivers of the future? Will competition become a key driver or is it implicit in the others?

One method of assessing any market position is to develop a benchmark of current activity in certain key areas. From time to time this should be reviewed and measured objectively. This should reveal whether there has been any change and if so what it could mean for that organisation. There are various models that can help establish where a business is and where it should be going. A few examples are described below.

15.2.1 PEST

PEST is one method described by Johnson and Scholes (1993). A possible interpretation of current-day activity is shown in figure 15.1. The method focuses on four key areas of activity:

- Political
- Economic
- Sociological
- Technological.

The use of these terms is best explained in an example that applies to the property profession. Setting a scenario in the summer of 1998, a major decline in an international market (the Far East) triggered an action within the home market. This resulted in the closure of a large local factory owned by a multi-national company. This resulted in significant job losses, which could lead to a corresponding reduction in the local economy, unless there are significant other job opportunities. The local surveyor should know whether this would ultimately affect the housing market. The effect on local businesses, including surveying practices would be potentially very serious.

Political/legal	– Increased regulation of estate agents – Property mis-descriptions influenced the skills base for the emerging revamped estate agency businesses – Yianni heralded a significant change in attitude to mortgage valuation reports – Introduction of ombudsmen and increased levels of consumer protection, including reference to conflicts of interest – Less subsidy on home purchase by the reductions in MIRAS – Continued release of local authority property, not originally designed for home ownership (planned maintenance issues etc.) – Change of government, more of the same or subtle differences
Socio/cultural	– Changing family values and structure, with more single parents and more parents in work – Levels of mobility, more car ownership and degrees of congestion. Could this lead to political intervention in an integrated transport policy? Questions then need to be raised on the cost implications to the individual, business and the environment – Higher standards of living and expectations – Need for increased levels of security
Economic	– A relatively buoyant economy, but with house price income ratios rising steadily – European considerations, differing policies and influences – Housing market conditions and attitudes, limited supply in some areas, but still some fears of over commitment and lacking confidence in home ownership – Diminishing unemployment, but increasing use of contract staff – Macro considerations need to be applied to the local market and an analysis done to interpret the differences and to judge whether the national picture predicts any changes for the local area
Technological	– Improvements in technology. How flexible are your systems to respond to those changes? – Business clients are demanding their suppliers change methods of communication – Method of reporting does it meet client expectations when compared to other businesses – New forms of equipment are available, damp meters, digital cameras. Can these reduce costs or help provide a better service?

Figure 15.1 Example of a PEST analysis (Source: Johnson and Scholes 1999).

It is difficult to give guidance on how far such analysis should go. Clearly the less well informed a business is then the less useful such an exercise will be. Predictions using the PEST technique may not always turn out to be totally accurate but they have the potential of allowing businesses to change and adapt with market changes rather than become reactive victims.

15.2.2 Value chain

Considering where a practice sits in the value chain could prove a useful exercise possibly related to house purchase or some other aspect of the business. The example shown in figure 15.2 uses the house purchase process as a context. Here the various organisations involved in the house-buying process are identified. The aim is to identify whether

Sources ⇒	*Practitioners* ⇒	*Customer*
Intermediary – lenders	Valuation surveyors	Insurers
Lenders	Architects	Lenders
Insurers – risk assessment/data	Technicians	Purchasers
Solicitor/conveyancers – sales	Computers (databases)	Vendors
Estate agents – sales	Underwriters	Solicitors
Contractors – specialist requests		
Cross sales – other business clients		
Corporates		
Housing Associations – lettings/maintenance		
Builders – sales		
Funds/investors – residential investment		
Tax requirements – probate		

Channels	
Written reports	Fax
Internet	Verbal reports
Cable	Media

Figure 15.2 Typical 'value chain' diagram for the valuation of residential property.

there are any unexploited opportunities for a business within the chain of events. It is important to bear in mind that not only could direct competitors be doing this exercise but so also could other members of the chain in order to identify where their opportunities might lie.

15.2.3 Five forces

Another system is known as Porter's five forces (Porter 1980). This analyses the state of the industry by looking at the competitive forces within the market place. There are four key areas that combine to produce a fifth force:

- substitute products and services
- potential entrants
- suppliers
- buyers.

These all influence the heart of the model, the impact upon the industry by means of competition and the rivalry, and creates the fifth area. The following is a possible interpretation for a practice in any town or city:

15.2.4 Impact upon a practice

Any practice in any town needs to be aware of the impact that the local economy may have on the local business. In some cases it will be possible to make changes and maintain profitability. In applying the 'five forces model' the competitors should be identified from knowledge of what is happening in the national context and related to local examples. As with any generalised assessment this will raise more questions than answers. These can only be resolved by a detailed analysis of the individual circumstances. The following is an example of how the system can be applied.

New entrants

An example of a threat of new entrants was shown by the increased use of in-house surveyors by many lenders and the corporates acquiring practices. This resulted initially in less work for independents. Some of the practitioners from the acquisitions were released onto the market place to set up their own businesses. This increased the competition and so the rivalry. In order to minimise the impact of this change a number of smaller practices merged, and continue to do so, reinforcing their market position.

Substitutes

Substitutes to producing a valuation by conventional methods include:

- the use of new technologies; and
- a good-quality database that provides a risk assessment for lending organisations.

Even if this is not a practical proposition at the present time in the UK (beware – it is possible to acquire an American version on the Internet), some analysis of this potential threat can still be useful. This may suggest a redirection of a business to suit clients who would not be satisfied by a risk assessment alone. This could result in a product to assess the current condition and the need for remedial works in much more detail. An alternative might be to consider whether the new technology could be turned to the advantage of a practice rather than a threat.

Suppliers

Suppliers relate to those businesses that provide elements of your business. This may be the company providing film for the cameras or garages providing the cars, etc. The key here is that those suppliers really want to keep their customers in business as that obviously helps them. The challenge is to formulate how the two can work together. Bulk purchase is one option but there may be others such as car leasing rather than purchase. Advertising a supplier's products may also allow for discounts. Also looking at the supply chain of goods, can you cut out the 'middle person' or identify a better buying capability than the one previously used?

Buyers

Buyers are those individuals or companies who buy the services offered by a surveying practice. Bear in mind that some of these buyers will in turn be viewing surveying practices as suppliers. Consequently they will be looking for special deals in relation to price and quality. Where corporate buyers are involved these conditions are often linked to volumes of business. Faced with these pressures surveyors have to look for ways of making economies. For example are there technologies that would help provide a more streamlined service for the buyer? Can survey reports be sent in an electronic format to suit the client's own system? Applying the value chain concept can help establish these possibilities.

Competition

When all these contributions are put together a surveying business should be in a better position to assess what the local competition could achieve. Any gaps or niches that could usefully be filled can then be identified. This will then be another step in determining where to position a business and how to target customers.

15.2.5 SWOT

This mnemonic refers to the classic analysis of an individual's or company's Strengths, Weaknesses, Opportunities and Threats. Having done some or all of the other exercises described above it is useful to match the company's strengths and weaknesses. Opportunities are then identified bearing in mind the threats that may also be on the horizon.

The following is an example of some of the factors that may feature in a SWOT analysis. It is by no means comprehensive, nor could it be because every firm will differ in relation to how it finds itself in the market place with the resources it has available. See figure 15.3 for an example of a SWOT analysis.

There are no set answers to any of the situations posed here. The benefit of the exercise is to actually identify what is happening in your area of business. Of course it will only be as good as your knowledge of your business area.

It would have been interesting to compare a SWOT done in the mid to late 1980s with one done in the early 1990s, i.e. the difference between a booming economy and one in recession. The strengths and weaknesses may well have been the same, it is how the opportunities and threats were perceived that could have made the real difference. Some practices saw the threat of over-inflated prices and some went out of business due to an excessive number of negligence cases. There were those that saw an opportunity in handling possession cases, whether by management or auction.

Figure 15.4 includes some questions that you may wish to ask in preparing this type of analysis.

15.2.6 Customers

It would be relatively easy to limit this section to a discussion on what type of report the Surveyor should provide for the customer when undertaking a residential appraisal. This could be done by reference to what is available now and has been, to some extent for the *last* century. However, this would ignore some fundamental issues:

1) Advancement in technology, training techniques, marketing skills, etc. mean that business can offer the customer a better deal on the services or products sold.
2) Customers' expectations are now *far* higher than they have ever been. This will continue with increasing influence from the media and better quality education.
3) The law and its interpretation has also changed to meet changing attitudes. Financial services may have become deregulated in the late 1980s, but other regulatory pressures, such as Ombudsmen and Arbitration schemes, have replaced this. What

Strengths	• A good team of Surveyors, of mixed ages that will give a degree of succession • Skills within the team meet the current business needs and for the foreseeable future • Surveyors have a broad approach to business, are computer literate and able to sell their services • Support staff are of mixed ability and capable of coping with new technology • The client base is wide, but 80% of the work comes from the top 20 suppliers
Weaknesses	• The computer system is 5 years old and has limited capacity • Office locations are in the centre of towns and parking is becoming difficult and expensive • There is currently little need for client access to the office • The business is too focused on one particular area
Opportunities	• A new software is due to become available that will release resources to develop other areas of business • Other practices within the geographical areas are not well positioned to take advantage of the new business areas • The government is to introduce alternatives to the current home buying process and there is scope within this business to take advantage of the new measures
Threats	• The new software package impacts upon our core area of business and means that the company does not need the skilled resources currently available and maintaining them to do that business would make the business uncompetitive • The property market is booming, prices are running ahead of earnings and therefore are unlikely to be sustainable • Increased litigation if evidence cannot be produced to support the need to modify valuations • A significant number of the support staff are approaching retirement age and there is no succession plan in place

Figure 15.3 Example of a SWOT analysis.

this means for any business is that it must maintain a watching brief on how customer expectations are changing and what new technology is available, along with other business techniques, to meet those expectations. The customer has been identified as another key driver of change.

One such technique that can enable organisations to keep up with these changes is known as Total Quality Control. This is a system of management that creates a culture that should be designed to meet business objectives.

15.3 Total quality control (TQC)

Whatever the product, whatever the service, it is important to maintain standards. The current process used by most practices and insisted upon by a large number of corporate clients is some form of Quality Assurance. The following is a brief review of this approach and what it should achieve.

Area	*Question*
Staff	• What range of skills do the staff have, can they be used to diversify the business? • What is their attitude to change? • Are they capable of adapting to new working practices?
Technology	• Are the systems up to date? • Is there potential for growth? • What would be the cost of updating the systems? • Are there new features that would make us more competitive?
Premises	• Are they in the right locations? • Are they cost effective? • Are they freehold or let?
Contracts	• Are they flexible? • Do they define your responsibilities clearly? • Are they sustainable?
Suppliers of business	• What is happening in the corporate area of business? • Are they based overseas, what is the economic situation in their home country? • How is the private customer market changing? • What is happening in respect of household formation? • What type of transactions are taking place? e.g. are they new mortgages or re-mortgages?
Economics	• What is happening to costs? • Are house prices rising? • Is the RPI increasing? • Does the government appear to be in control of the economy e.g. is the situation stable or volatile • How does the national situation compare to the local one?
Litigation	• Is there a trend in recent case law? • What is the current angle of the consumer groups? • How are consumer standards changing?

Figure 15.4 Key questions to ask before carrying out a SWOT analysis.

15.3.1 Reasons for TQC

Customers expect greater durability and reliability from products and services. The price of the service may still be a key criterion but low cost should never be an excuse for poor quality. As with any other industries, when providing surveying services it is essential to continually review all aspects of the business to ensure:

- that it is operating effectively and efficiently; and
- meeting the needs of all clientele.

The development of a quality product or service needs an approach that is not just for the front line surveyor but a philosophy that must be applied to all aspects of the business. This would include all staff, procedures, systems, training, management and marketing.

Various terms have been used to reflect the approach. Total Quality Management (TQM) is one that is currently popular and the Quality Assurance Standard ISO 9000 is a product of that system. However, many practitioners have misunderstood this. The next section attempts to clarify the true concept.

15.3.2 A definition of TQC

One broad definition of this concept is given below:

> Total Quality Control is an effective system for integrating the quality development, quality maintenance and quality improvement efforts of the various groups within an organisation so as to enable marketing, engineering, production and service at the most economical levels that allow for full customer satisfaction.
>
> (Feigenbaum 1998)

Flood (1993) states that quality can be defined in a variety of ways. He summarises the work of other gurus on this subject and the author has interpreted the most appropriate for the surveying business as follows:

- Quality is a predictable degree of uniformity and dependability, at low cost and suited to the market.
- Quality is fitness for use.
- Quality is conformance to requirements.
- Quality is in its essence a way of managing the organisation.
- Quality is correcting and preventing loss, not living with loss.
- Quality is the totality of features and characteristics of a product, service or process, which bear on its ability to satisfy a given need; from the customer's viewpoint (British Standards Definition, HMSO 1987).

In addition, the concept of total quality means that everyone should be involved in quality across all functions of an organisation. The story of the two men painting a space ship at Cape Kennedy is legend and makes this point well.

The President of the USA at that time was visiting the site and asked one of the two what he was doing. He answered that he was painting the space ship. He asked the other who responded that he was helping to put a man on the moon.

The second answer typifies the philosophy that is required to achieve a quality system within an organisation. It not only has to include the management team but all individuals managing their own situations and jobs as well. A key component of the system is the need to review the market place and customer requirements so that their expectations can continue to be met or even exceeded. In successful organisations this will become part of their culture.

15.3.3 Quality systems for surveyors

The RICS provide good guidance on how a quality system should be implemented and accreditation obtained (RICS 1994) for all chartered surveyors. It is not the purpose of this section to detail this process. Instead we provide an outline so that a brief can be prepared for a consultant who might be appointed to help an organisation develop their own quality systems.

In essence, any management systems should suit the business objectives of the organisation. This represents a change from a backward looking management system based on procedures and current services. It is the former approach that is now encapsulated in the European and International Standards of EN 29000 and ISO 9000.

The 1994 standard was aimed to achieve the following:

- compliance with the standard;
- a satisfactory level of performance both internally and externally;
- conformance to the specified or agreed requirements of the client and the firm.

15.3.4 What should a quality system aim to achieve?

A surveying practice should aim to work to a framework that encapsulates the whole business. The main areas for consideration are probably as follows:

- Strategic management and performance setting:
    - Having a system to identify market needs
    - Production of a business vision and mission
    - Production of a business plan
    - Development of an organisational structure
    - Provision for team building
    - Provision of a communication strategy and system.
- Human resources:
    - Recruitment policy and system
    - Induction programme for new staff
    - Provision of a system to set performance targets to an agreed standard
    - Performance appraisals
    - Training and development
    - A method for evaluation of systems and other resources.
- Service proposition:
    - A system to identify the clients' requirements
    - A procedure to develop the service and product proposals
    - Production of service agreements
    - Provision for a system that matches human resources to product needs
    - Management of the client relationship
    - Management of the service delivery
    - Provision for client feedback and resolution of issues.

How all these aspects interrelate and apply to a business will vary depending on the size and nature of the practice. This should provide a basis for discussion and evaluation of needs.

15.3.5 Practical impact

TQM provides among other things an audit trail that is replicated both in the office and in the field. An example of this can be seen with inspection routines. The days of the inspection done on the back of an envelope are hopefully long gone. All inspections should be undertaken with the use of site notes to record what has been seen and what factors have been taken into account to arrive at a valuation. The precise detail of what to record is covered in the Red Book (RICS 1995) and in various sections throughout this book. The overall purpose of maintaining the record in a standard format is to show that a standard procedure is adopted. This means that should there be any question later as to the level of inspection then the standard format can be shown to have been used.

This rather negative way of viewing a quality assured system also has tangible positive effects. Where a large number of surveyors have been trained in 'the process' then an organisation can be reasonably confident that clients across the whole area of operation are getting the same level of service.

Of course all surveyors *are* different and so are the properties so the standardisation should not go to the extent that it removes the thinking capacity of the surveyor. It should, however, provide the routine to the inspection that allows the surveyor to paint a picture of the property and spot the defect and other irregularities.

A further example of this is that the valuer should have recorded sufficient information in a legible form. This should be clear enough for another surveyor to pick up the notes, carry out a subsequent inspection and determine what might have changed since the initial inspection. This would be expected of a medical doctor in assessing the condition of a patient from one visit to the next no matter how long the intervening period. The information on the site notes should be set out in logical sections with the use of bullet points or single word answers wherever possible. The surveyor should take particular care in choice and recording of the comparables as these are the closest match to the subject property.

TQM defines the process that should be adopted in all aspects of the business and the above example is only one part of that. This has resulted in the better recording of information. This should not necessarily mean more paper but an improvement on how it is used. A small practice will need a less sophisticated system than a larger more dispersed organisation that undoubtedly will find one of the most significant difficulties is distance management.

The customer should be assured of a level of organisation that ensures a consistent and quality service. TQM should provide the process to identify the weaknesses and provide remedial measures for making improvements. This is of significant importance in organisations where the human resource is the key to service quality.

15.4 The benefits of new technology

Improvement in technology and corresponding cost reductions means that there are now more forms of media to relay messages to the client. Technology is another key driver of change, in that it is a facilitator allowing things to be done in a different way.

It has taken some time for the technological revolution to take effect but it is now being brought into everyday use. For example in simple form the common photograph and more specifically the Polaroid has become a regular requirement for most lenders. Experimentation with video surveys has found limited appeal although this form of media is having a degree of success in the sales area. Other opportunities can be found with digital photography. These developments are dependent upon the surveyor and client having a reasonable and matching level of IT skill and technology. This will improve over the next decade and it is anticipated that the scope will mushroom.

In the written form, improvements in word processing and electronic communication mean that potentially surveyors are no longer restricted to one specific report format or even to an office location. Systems are now available that can integrate the whole of the administration to support the surveyor and potentially reduce cost. The key question is how to prepare and position an organisation to take advantage of the opportunities.

There are two key points to understanding the benefits of new technology:

- It is only a benefit if the costs of developing and using the system are exceeded by increased revenue generation or reduced costs over the short to medium term.
- Computers are capable of handling extremely complex situations but remember that all the data has to be inputted. This may make it uneconomic for small-scale operations.

However, the IT industry is really still in its infancy, new developments are happening all the time and the potentials, although appreciated, have yet to be fully realised. Therefore, companies developing new systems at the present time should be considered as pioneers – no industry would progress without them. It is also important to remember that any competitive advantage will only be short lived, hence the need for systems to give a return in the short to medium term.

The potential for use of IT is improving and the current areas include:

- Geographical Information Systems, the use of mapping and data to create visual trends within specified areas, e.g. a value map of a town can be created by displaying average property prices by postcode and super-imposing on a map of the town.
- Expert systems, the production of models that use basic information and then replicate those results when the circumstances are repeated, e.g. the criteria for rising damp can be used on a palm-top computer, such that a student or trainee can utilise by answering a series of questions on inspecting a property.
- Statistical evaluation models, such as Regression Analysis, Neural Nets. This is the use of a large volume of data, such as house prices, to interpret market movements and potentially to predict future movements.
- Databases – the use of large volumes of data to record key attributes and use for comparison purposes.
- Electronic communication including the Internet – the transfer of documents in a high-quality format to include photographs and other interactive information.

The combination of these areas will provide the basis of the future. It is important to also consider what other developments may impact the surveying industry. For example, the developments with neural nets have shown to be of particular relevance in the medical area in diagnosing certain illnesses. The principles of diagnosis are similar for property as for the human condition.

The speed of change in technology means that new systems and techniques are always being developed. It is, therefore, important to note that any current positioning statement will always be dated. Consequently a major source of up-to-date information is more likely to be in the technical journals than in a book. It is for this reason that this section has focused on what is possible and how surveyors should position themselves to make the best use of the opportunities.

References

Deming (1986). *Out of the Crisis: Quality, Productivity and Competitive Position*. Cambridge University Press.

Feigenbaum, A.V. (1998). *Total Quality Control*, 3rd edn. McGraw-Hil.

Flood, R.L. (1993). *Beyond TQM*. Wiley.

HMSO (1987). BS5750: Quality Systems. British Standards Institution, London.

Johnson, G., Scholes, K. (1999). *Exploring Corporate Strategy*. Prentice-Hall, London.

Porter, M.E. (1980). *Competitive Strategy*. Freepress, London.

RICS (1994). Quality Assurance: guidelines for interpretation of ISO 9000 for use by Chartered Surveyors and Certification bodies. RICS, London.

Yianni v. *Edwin Evans* (1982). QB 438.

Index